Biomaterials in Regenerative Medicine
and the Immune System

Laura Santambrogio
Editor

Biomaterials in Regenerative Medicine and the Immune System

 Springer

Editor
Laura Santambrogio
Albert Einstein College of Medicine
Bronx, New York, USA

ISBN 978-3-319-18044-1   ISBN 978-3-319-18045-8 (eBook)
DOI 10.1007/978-3-319-18045-8

Library of Congress Control Number: 2015942636

Springer Cham Heidelberg New York Dordrecht London
© Springer International Publishing Switzerland 2015

Springer International Publishing AG Switzerland is part of Springer Science+Business Media (www.springer.com)

Contents

Contributors

Fnu Apoorva Sibley School of Mechanical and Aerospace Engineering, Cornell University, Ithaca, NY, USA

Michael C. Bellavia Wallace H. Coulter Department of Biomedical Engineering, Georgia Institute of Technology and Emory University, Atlanta, GA, USA

Evelyn Bracho-Sanchez J. Crayton Pruitt Family Department of Biomedical Engineering, University of Florida, Gainesville, FL, USA

Yahya E. Choonara Department of Pharmacy and Pharmacology, University of the Witwatersrand, Wits Medical School, Johannesburg, South Africa

Pui Lai Rachel Ee Department of Pharmacy, National University of Singapore, Singapore

Jennifer H. Elisseeff Translational Tissue Engineering Center, Wilmer Eye Institute and Department of Biomedical Engineering, Johns Hopkins University, Baltimore, MD, USA

Changchun Fan Orthopaedic Research Laboratories, Stanford University, Stanford, CA, USA

Donald O. Freytes The New York Stem Cell Foundation Research Institute, New York, NY, USA

Jordan J. Green Departments of Biomedical Engineering, Materials Science and Engineering, Ophthalmology, and Neurosurgery, Translational Tissue Engineering Center, and Institute for Nanobiotechnology, Johns Hopkins School of Medicine, Baltimore, USA

Stuart Goodman Orthopaedic Research Laboratories, Stanford University, Stanford, CA, USA

Franck Housseau Sidney Kimmel Comprehensive Cancer Center, Johns Hopkins University School of Medicine, Baltimore, MD, USA

Benjamin G. Keselowsky J. Crayton Pruitt Family Department of Biomedical Engineering, University of Florida, Gainesville, FL, USA

Jasmeet Singh Khara Department of Pharmacy, National University of Singapore, Singapore

Pradeep Kumar Department of Pharmacy and Pharmacology, University of the Witwatersrand, Wits Medical School, Johannesburg, South Africa

Matthew Levy Department of Biochemistry Albert Einstein College of Medicine, Bronx, NY, USA

Jamal S. Lewis J. Crayton Pruitt Family Department of Biomedical Engineering, University of Florida, Gainesville, FL, USA

Tzu-Hua Lin Orthopaedic Research Laboratories, Stanford University, Stanford, CA, USA

Florence Loi Orthopaedic Research Laboratories, Stanford University, Stanford, CA, USA

Randall A. Meyer Department of Biomedical Engineering, Translational Tissue Engineering Center, and Institute for Nanobiotechnology, Johns Hopkins School of Medicine, Baltimore, MD, USA

Isabella Pallotta The New York Stem Cell Foundation Research Institute, New York, NY, USA

Drew Pardoll Sidney Kimmel Comprehensive Cancer Center, Johns Hopkins University School of Medicine, Baltimore, MD, USA

Jukka Pajarinen Orthopaedic Research Laboratories, Stanford University, Stanford, CA, USA

Deborah Palliser Department of Microbiology and Immunology, Albert Einstein College of Medicine, Bronx, NY, USA

Viness Pillay Department of Pharmacy and Pharmacology, University of the Witwatersrand, Wits Medical School, Johannesburg, South Africa

Kaitlyn Sadtler Translational Tissue Engineering Center, Wilmer Eye Institute and Department of Biomedical Engineering, Johns Hopkins University, Baltimore, MD, USA

Taishi Sato Orthopaedic Research Laboratories, Stanford University, Stanford, CA, USA

Alex Schudel School of Materials Science and Engineering, Parker H. Petit Institute for Bioengineering and Bioscience, Georgia Institute of Technology, Atlanta, GA, USA

Anirban Sen Gupta Department of Biomedical Engineering, Case Western Reserve University, Cleveland, OH, USA

Ankur Singh Sibley School of Mechanical and Aerospace Engineering, Cornell University, Ithaca, NY, USA

Kara Spiller Biomaterials and Regenerative Medicine Laboratory, School of Biomedical Engineering, Science, and Health Systems, Drexel University, Philadelphia, PA, USA

Svetlana A. Sukhishvili Department of Chemistry, Chemical Biology and Biomedical Engineering, Stevens Institute of Technology, Hoboken, NJ, USA

Bruce Sun The New York Stem Cell Foundation Research Institute, New York, NY, USA

Susan N. Thomas George W. Woodruff School of Mechanical Engineering, Parker H. Petit Institute for Bioengineering and Bioscience, Georgia Institute of Technology, Atlanta, GA, USA

Valerie J. Tutwiler Biomaterials and Regenerative Medicine Laboratory, School of Biomedical Engineering, Science, and Health Systems, Drexel University, Philadelphia, PA, USA

Marek W. Urban Department of Materials Science and Engineering, Center for Optical Materials Science and Engineering (COMSET), Clemson University, Clemson, SC, USA

Emily A. Wrona The New York Stem Cell Foundation Research Institute, New York, NY, USA

Zhenyu Yao Orthopaedic Research Laboratories, Stanford University, Stanford, CA, USA

Tony Yu Biomaterials and Regenerative Medicine Laboratory, School of Biomedical Engineering, Science, and Health Systems, Drexel University, Philadelphia, PA, USA

Chapter 1
Role of Mesenchymal Stem Cells, Macrophages, and Biomaterials During Myocardial Repair

Isabella Pallotta, Emily A. Wrona, Bruce Sun and Donald O. Freytes

1.1 Introduction

Myocardial infarction (MI), also more commonly known as a heart attack, is a potentially lethal condition and remains one of the leading causes of death in the USA and other industrialized nations [1]. Since heart tissue has a diminished capacity to regenerate itself, there could be a substantial reduction in heart function if sufficient damage is sustained after the infarct. One approach to help patients who have suffered an MI is to replace the cellular mass lost after the infarction with repair cells that have the capacity to heal and recover heart function to pre-infarct levels. Among the methods being investigated are the use of injected repair cells encapsulated within a hydrogel-like material or the combination of a biocompatible material seeded with repair cells [2]. One attractive repair cell candidate is the human-derived mesenchymal stem cell (MSC). Although these cells have not been shown to effectively produce functional cardiomyocytes, they have shown great potential to improve angiogenesis, reduce the rate of myocardial wall remodeling, reduce cellular death, and, as a result, improve cardiac function [3–9]. In addition to the repair capability of MSCs, their availability from the patient's bone marrow or adipose tissue makes them an ideal candidate as part of an autologous heart repair strategy.

Isabella Pallotta, Emily A. Wrona and Bruce Sun have all equally contributed to this chapter.

D. O. Freytes (✉) · I. Pallotta · E. A. Wrona · B. Sun
The New York Stem Cell Foundation Research Institute, 1995 Broadway Suite 600,
New York, NY 10032, USA
e-mail: dfreytes@nyscf.org

I. Pallotta
e-mail: ipallotta@nyscf.org

E. A. Wrona
e-mail: ewrona@nyscf.org

B. Sun
e-mail: bsun@nyscf.org

© Springer International Publishing Switzerland 2015
L. Santambrogio (ed.), *Biomaterials in Regenerative Medicine and the Immune System*,
DOI 10.1007/978-3-319-18045-8_1

1

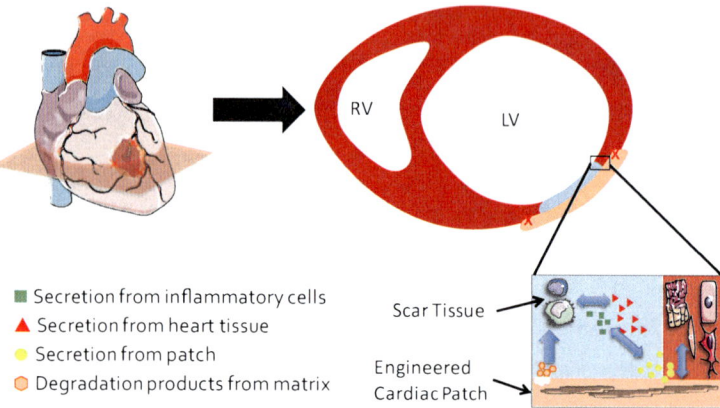

Fig. 1.1 Diagrammatic representation of the dynamic interactions between the infarcted tissue and the surrounding microenvironment composed, in part, of macrophages. Repair cells within the patch will secrete factors that may attract different subpopulations of macrophages. The matrix encapsulating the cells will also play a role in the interactions between the repair cells and the inflammatory cells *via* degradation products or direct contact. *RV* right ventricle, *LV* left ventricle. (Artwork provided by Servier Medical Arts, http://www.servier.com/Powerpoint-image-bank)

Repair cells, used alone or in combination with a biomaterial as part of an engineered cardiac patch, will inevitably be subjected to the inflammatory environment that follows the ischemic insult, as shown in Fig. 1.1 [10, 11]. This cellular environment is very dynamic and is characterized by the presence of multiple pro- and anti-inflammatory cells that appear at different phases of the remodeling process [10]. The inflammatory cells also play a crucial role during the degradation and remodeling of biomaterials [12–14]. This dynamic and critical environment is often overlooked when designing cardiac patches and could potentially dictate the success or failure of a delivered construct by inhibiting survival and/or engraftment of the cells or by modifying the natural degradation and remodeling process of the biomaterial. This chapter describes the events that follow after a myocardial infarct with emphasis on the important inflammatory events that take place during the remodeling process. This is followed by a description of MSCs as repair cells and how MSCs can be combined with biocompatible biomaterials in order to harness or modulate the inflammatory response as part of an engineered cardiac construct.

1.2 Myocardial Infarction and Inflammation

1.2.1 Myocardial Infarction

MI can be defined as damage or death of heart muscle due to deprivation of oxygen and nutrients (ischemia) by the occlusion of one of the arteries supplying blood to the heart tissue. Risk factors for MI include hypertension, cigarette smoking, diabe-

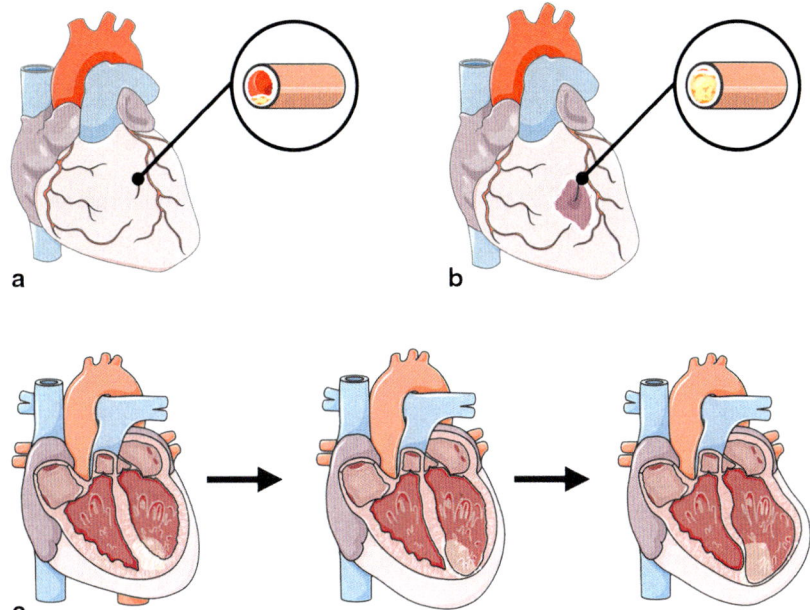

Fig. 1.2 The overall mechanism of myocardial infarction: **a** Fat and plaque build up over time and begin to clog the blood vessel. **b** Occlusion of a major coronary artery causes ischemia and necrosis of the myocardial tissue. **c** Over hours, days, and months, the loss of cardiac muscle leads to ventricular remodeling and the formation of a collagen-rich scar. How this process proceeds greatly determines the patient's prognosis and often results in decreased cardiac output. (Artwork provided by Servier Medical Arts, http://www.servier.com/Powerpoint-image-bank)

tes mellitus, and genetic hypercholesterolemia with an apparent gender bias towards men, showing a higher risk of experiencing an MI than women [15]. Although an MI can occur at any age, the gender bias decreases with age with the chances of an MI increasing equally for both men and women [15].

The overall mechanism of MI can be summarized as shown in Fig. 1.2. The first event is the occlusion of a major coronary artery. Obstruction of the coronary artery is often due to the build up of fatty materials and the rupture of this plaque causes the formation of a blood clot that can grow large enough in size to interrupt blood flow. The essential oxygen and nutrients needed to maintain the continuous muscle contraction of the heart can no longer be delivered, resulting in ischemia [15]. After a prolonged period of time, the occlusion leads to myocardial necrosis, which triggers an inflammatory response that, depending on the extent of the damage, may lead to detrimental ventricular wall remodeling. Additionally, ischemia can produce arrhythmias such as ventricular fibrillation [15]. There is some regenerative potential and necrosis prevention if blood is restored within 20 min of coronary occlusion by saving cells that have not undergone irreparable damage [15]. However, reperfusion itself may impose its own injuries since the introduction of oxygenated blood to ischemic tissue [16, 17], and the enzymes and reactive oxygen species (ROS) released from the initial inflammatory cells [18], may cause separate and secondary inflammatory responses [19].

Current approaches to engineer myocardial tissue replacements fail to address the inflammatory and healing environments after an infarct and the restoration of cardiac function to pre-infarct levels [19]. Since the level of healing achieved by the infarct can affect the overall cardiac function, new therapeutic strategies need to be explored that can restore the cellular mass lost after the infarct and create functional tissue that can restore heart function. Current strategies focus on engineering constructs that take advantage of a cellular component and scaffold material. Regardless of their success under standard culture conditions, any tissue-engineered construct will need to be screened *in vitro* for its survival within an inflammatory environment in order to increase the chances for clinical success.

1.2.2 Inflammation

Following ischemic insult, the cells of the immune system are recruited to the MI to promote an inflammatory response that helps facilitate subsequent tissue remodeling [20]. The inflammatory environment at the site of MI is not completely understood, but it is thought to be composed of sequential recruitment of different blood-derived circulating cells, as well as the activation of resident cells. Necrotic cells release signals that stimulate complement cascades, toll-like receptor (TLR)-mediated pathways, nuclear factor (NF)-κB systems, and the generation of reactive oxygen species (ROS), signaling the initiation of the inflammatory response [20]. These leukocytes are recruited *via* signals released by the necrotic tissue and travel to the infarct site by rolling and migrating across the endothelium with the help of adhesion molecules at the surface of blood vessels [21].

The first cells recruited to the site of MI are neutrophils, which begin to phagocytose cellular debris and release cytokines and proteolytic enzymes, in order to clear dead cells and necrotic tissue [16, 17]. Neutrophils also undergo apoptosis and the signals from these necrotic cells serve as the initial distress call in order to recruit peripheral blood monocytes and tissue-resident macrophages. There is evidence that monocytes are even recruited from the spleen en masse following MI [22], making them one of the most abundant cell types present within the infarcted tissue at early stages. Monocytes differentiate into macrophages *in situ* and perform multiple roles during homeostasis [23], the early stages of inflammation, and throughout the repair of damaged myocardium. Macrophages consume necrotic tissue and foreign materials, while assisting in angiogenesis and extracellular matrix (ECM) formation *via* the release of potent cytokines such as vascular endothelial growth factor-α (VEGF-α) and transforming growth factor-β (TGF-β) [16, 17, 20, 24]. For example, it has been shown that liposome-mediated monocyte depletion reduces neovascularization, myofibroblast formation and recruitment, and collagen accumulation in the area of the infarct [24]. Further understanding of how the classically and alternatively-activated macrophages work has demonstrated that these cells might exist as part of a spectrum of macrophages at multiple polarization stages constantly adjusting to the current inflammatory environment [25]. How these macrophages affect the overall healing process still needs to be investigated.

1.2.3 Macrophages and Myocardial Infarction

After an MI, there are two different surges of macrophages. There is an initial influx of pro-inflammatory (M1) macrophages, also known as classically activated macrophages, which give rise to the macrophage-mediated inflammatory response. Classical activation of M1 macrophages is associated with the release of cytokines such as interleukin (IL)-1β, IL-6, and tumor necrosis factor-α (TNF-α) and is responsible for the initial degradation of biomaterials. M1 macrophages are typically polarized by lipopolysaccharide (LPS) and interferon-γ (IFN-γ) *in vitro*. This initial wave of macrophages is followed by a second wave of anti-inflammatory (M2) macrophages, also known as alternatively-activated macrophages. M2 macrophages are typically associated with tissue repair *via* the stimulation of extracellular matrix deposition and angiogenic effects (although it is still debated if M1 or M2 actually stimulate the initial angiogenic events) [20, 26, 27]. Depending on the pathway that is activated during polarization, some have cataloged M2 macrophages in subsets such as M2a, M2b, and M2c macrophages. A simplified diagram of monocyte-derived macrophages and the related cytokines is shown in Fig. 1.3. M2a refers to the macrophages originally named alternatively-activated macrophages and are typically polarized by IL-4 and IL-13. M2b macrophages are activated by immune complexes, agonists of TLRs, or IL-1R. M2c refers to macrophages activated by IL-10 or glucocorticoid hormones [28]. These subtypes of macrophages all play important but differing roles during the inflammatory response depending on the location and type of injury. The interactions of these subsets of macrophages with repair cells

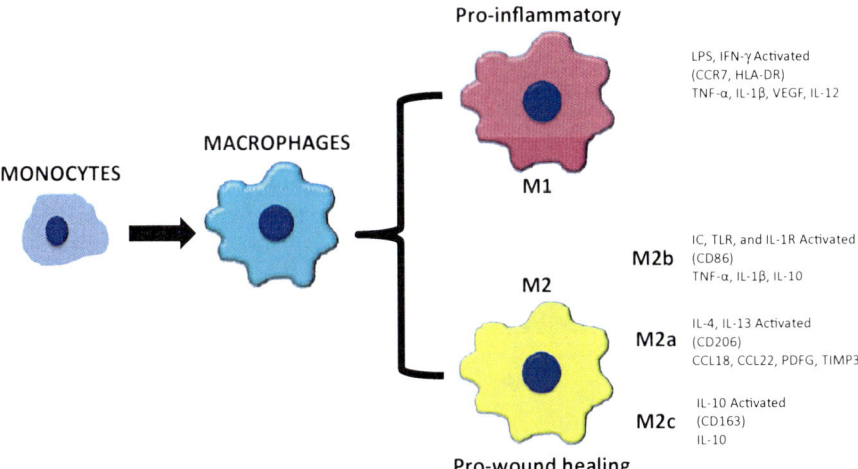

Fig. 1.3 Schematic representation of monocyte-derived macrophage subtypes M1 and M2. The main cytokines released by each subtype are shown. *LPS* lipopolysaccharide, *IFN* interferon-γ *TNF-α* tumor necrosis factor-α, *IL* interleukin, *VEGF* vascular endothelial growth factor alpha, *HLA-DR* human leukocyte antigen-DR, *M1* pro-inflammatory macrophages, *M2* pro-healing macrophages

and biomaterials are still poorly understood [26] and could have profound impact on the survival and integration of repair cells.

1.3   Mesenchymal Stem Cells in Heart Repair

1.3.1   Characteristics of MSCs

MSCs are stromal cells isolated from bone marrow or adipose tissue that have the multipotent potential to differentiate into a variety of cell types *in vitro*. Undifferentiated MSCs can potentially differentiate into different lineages of mesenchymal tissues including bone, cartilage, fat, tendon, muscle, and marrow stroma [29]. Human MSCs cultured *in vitro* appear morphologically heterogeneous and exhibit a spindle-like appearance very similar to that of fibroblast cells [30]. The isolation of MSCs from bone marrow cells *in vitro* is achieved in part thanks to their ability to adhere to tissue culture plastic. However, the use of plastic adherence also yields a heterogeneous population of cells, which includes MSCs and progenitor cells known as colony-forming units (CFUs) [7]. There are no definitive markers for isolating a pure population of MSCs. Rather, cellular purity is determined by their differentiation potential towards bone, adipose, and chondrogenic cells (thus being referred to as "multipotent mesenchymal stromal cells"). Furthermore, MSCs are characterized by the expression of markers such as CD73, SH2, SH3, CD29, CD44, CD71, CD90, CD105, CD106, CD120a, and CD124 [7, 31]. Table 1.1 summarizes the current accepted markers for the identification of MSCs.

In addition to the potential to differentiate into a variety of mesenchymal tissues, MSCs have also been described as immunoprivileged. MSCs lack the expression of histocompatibility complex II (MHC-II), along with co-stimulatory molecule expression for T cell recruitment. This characteristic is particularly beneficial in infarcted cardiac tissue, which is characterized by an intense inflammatory environment [32, 33]. The use of MSCs during preclinical heart repair studies have shown advantages over other cell types due to their potential to modulate the inflammatory

Markers	Marker potential
Stro-1	Enriches colony-forming units-fibroblasts (CFU-F) from whole bone marrow (BM)
SSEA-4	Enriches CFU-Fs from whole BM
CD146	Enriches CFU-Fs from whole BM. Enriches cells with multipotency from BM-MSCs
CD271	Maintains clonogenicity and function of MSCs
GD2	High specificity for isolating BM-MSCs
CD49f	When selected for at first passage, enriches clonogenicity and differentiation
PODXL	Identifies early progenitor MSCs but decreases with high-density culture and passage

Table 1.1 Summary of potential markers for unsorted human bone marrow MSC populations

environment and improve the vascular response following an MI. There have been some initial reports suggesting that human MSCs could differentiate into cardiomyocytes and that the derived cells had potential therapeutic effects after engraftment onto healthy murine hearts [34]. Following this report, various preclinical animal studies were conducted using MSCs as a therapeutic treatment of MI models. These studies resulted in infarct size reduction, improved cardiac contractility and reduced fibrosis [6, 34, 35]. In addition to bone marrow, MSCs from different tissues can serve as potential cell sources. In animal models, MSCs were thought to differentiate into cardiomyocytes by the appearance of spontaneously contracting cells, the expression of cardiac-specific genes, and the expression of cardiac proteins [36]. However, protocols that may lead to differentiated MSCs towards cardiac cells in sufficient numbers have not been adequately described. Researchers have debated the differentiation potential of MSCs towards cardiomyocytes *in vitro* and *in vivo* and their ability to differentiate remains controversial. However, it is well accepted that MSCs play a supporting role during myocardial healing by promoting angiogenesis and serving as a reservoir of growth factors during the host tissue response [37–40].

Although the isolation of MSCs has been well documented with protocols available for clinical-grade applications, some are still cautious about their isolation and urge more characterization and understanding of the cells during culture and implantation. For example, cultures of impure heterogeneous cell population from plastic adherence isolation may need further characterization while maintaining sterility. If expansion is needed, culture medium should not contain xenogeneic ingredients such as fetal bovine serum or animal-based growth factors. Regardless, MSCs remain a very attractive cell population for the treatment of MI.

1.3.2 MSCs in Heart Repair

Even with today's advances in modern medicine and pharmacological administration, the presence of scar tissue renders the heart incapable of performing at optimal levels, resulting in decreased cardiac output. Left ventricular wall remodeling can lead to heart failure with approximately 50% of patients dying within the first 5 years [8]. MSCs have shown promising results in multiple animal studies, and in 2001 it was reported that autologous MSCs transplanted into the heart after MI reduced infarct area and elevated left ventricle (LV) function during a 3 month follow up [3, 4]. Additional studies have shown that MSC transplantation in the heart not only improved LV function, but also was shown to be safe and effective within a small study population [3, 4]. Shortly after, more *in vivo* cases of acute MI, followed by intracoronary transplantation of autologous bone marrow-derived MSCs and bone marrow-derived MSCs combined with endothelial progenitor cells, provided evidence that MSC therapy can increase myocardial viability in a safe and nonfatal manner [41, 42].

One of the first human clinical trials in the USA occurred in 2008 and consisted of 53 patients under Osiris Therapeutics, investigating the use of allogeneic MSC transplantation for acute MI [43]. The phase I clinical trial was encouraging and showed no adverse effects or infusion toxicity in immunocompetent patients. A year later, human clinical trials performed using a dose-ranging, double-blind, and placebo-controlled safety trial concluded that MSCs decreased ventricular arrhythmias, improved pulmonary function, and elevated LV ejection fraction 3 months post therapy. To date, the use of MSC therapy has been shown to be safe with a range of efficacy with many studies still ongoing for an array of cardiomyopathies.

For MSC transplantation to work efficiently and successfully, a variety of factors must be verified and implemented. These factors include the timing of delivery (e.g., post-acute MI or post ischemia), the implantation methodology, MSC characteristic, MSC survival, and also the nature of the patients' pathology. Since the timing of MSC delivery is crucial after an acute ischemic event because of inflammation and cellular necrosis, the best treatment period is prior to fibrosis. However, additional *in vivo* and *in vitro* data are necessary to determine a standardized treatment.

1.4   Interactions Between MSCs and Inflammatory Cells

As mentioned before, the injured myocardium is characterized by the invasion of pro-inflammatory (M1) and pro-healing (M2) macrophages. For this reason, MSCs, once implanted into the injured myocardium, will inevitably interact with the inflammatory cells present at the time of implantation. Because of this close coexistence, cell phenotype and behavior can be reciprocally affected and modulated. The MSC- and macrophage-released cytokines play important roles in mediating their mutual interactions. For example, M2 macrophages and their associated cytokines (e.g., IL-10, TGF-β1, TGF-β3, VEGF) can support the growth of MSCs, while M1 macrophages and their associated pro-inflammatory cytokines (e.g., IL-1β, IL-6, TNF-α and IFN-γ) can inhibit their growth [11].

MSCs have the capacity to modulate the inflammatory cells as well. In particular, MSCs show anti-inflammatory properties by promoting a more pro-healing state. When infused in induced acute myocardial infarcted animal models, MSCs seem to increase the numbers of pro-wound healing monocytes/macrophages *via* the release of IL-10. At the same time, pro-inflammatory monocytes/macrophage numbers are reduced *via* the decrease of pro-inflammatory cytokines such as IL-1β and IL-6 [44]. Additional animal studies demonstrated that MSCs injected within the infarcted myocardium allow for a shift of the infiltrated macrophage phenotype from M1 to M2. The MSCs were surrounded by arginase 1 (Arg1)-expressing (which in the mouse model, is a marker for M2) macrophages when compared to the controls [45]. The shift from a pro-inflammatory to a pro-healing phenotype can also be detected by changes in messenger RNA (mRNA) and protein expression levels. These data suggest that the modulatory action of MSCs can accelerate the

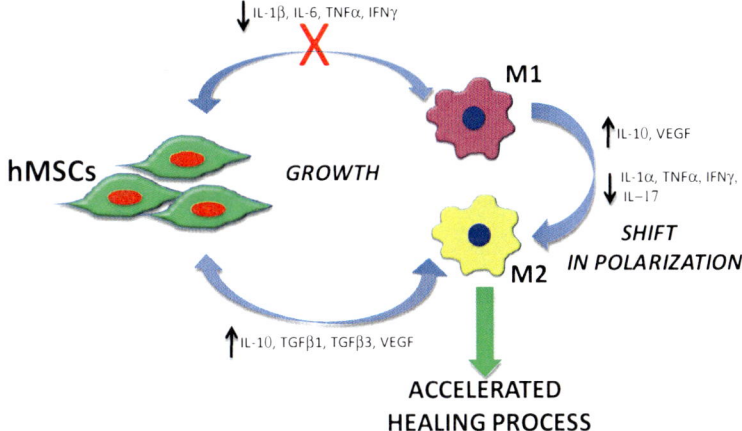

Fig. 1.4 Cross talk between MSCs and inflammatory cells. MSC growth is enhanced by M2 macrophages and inhibited by pro-inflammatory M1-derived cytokines. In turn, MSCs modulate the polarization of macrophages by promoting an M2 phenotype, while reducing an M1 phenotype. These events may result in an accelerated healing process. *hMSCs* human-derived mesenchymal stem cell, *IL* interleukin, *TNF-α* tumor necrosis factor-α, *IFN-γ* interferon-γ, *TGF-β* transforming growth factor-β, *VEGF* vascular endothelial growth factor

healing process in the injured tissue by close modulation of macrophages present at the site of injury.

Interestingly, the interaction between MSCs and macrophages is bidirectional. It has been shown that human adipose-derived mesenchymal stromal cells (hAMSCs) can alter the polarization of macrophages from M1 to M2 by increasing macrophage secretion of anti-inflammatory and angiogenic cytokines, such as IL-10 and VEGF. Moreover, these MSCs can also reduce macrophage secretion of inflammatory cytokines, such as IL-1α, TNF-α, IL-17, and IFN-γ. At the same time, macrophages were shown to upregulate the secretion of IL-4 and IL-13 by MSCs, which are known to polarize macrophages towards an M2 phenotype [46]. Figure 1.4 summarizes the possible bidirectional interactions between MSCs and macrophages.

From a tissue engineering perspective, there is a growing interest in improving the therapeutic potential of injected MSCs. For instance, there is a need to optimize the cell culture conditions before injection into the myocardium and to further improve their survival once delivered into the inflammatory environment of the infarct. Due to the reciprocal interactions between MSCs and macrophages, strategies are needed to exploit these interactions, in order to improve cellular survival and integration. For example, a promising strategy is to culture MSCs as aggregates in three-dimensional (3D) spheroids. This method has been demonstrated to increase the MSC expression of TNF-α-stimulated gene/protein 6 (TSG-6), an anti-inflammatory protein reported to be beneficial in animal models. This technique seems to suppress the inflammatory response both *in vitro* and *in vivo* when compared to MSCs cultured in the classical monolayer. Therefore, this new method of culturing

MSCs can potentially be applied to therapies in which inflammation is present at the site of implantation [47].

Another strategy to optimize the therapeutic outcome of the injected MSCs is to evaluate the source of the MSCs based on their response to the inflammatory environment. The most common source of MSCs is the bone marrow. However, adipose tissue has been recently recognized as a good alternative for its large pool of MSCs. Subcutaneous fat-derived MSCs have exhibited anti-inflammatory properties with low amounts of cytokines secretion and improved cardiac remodeling properties when compared to the right atrium and epicardial fat-derived MSCs. These data suggest that the source and preparation of MSCs may play an important role in modulating the inflammatory environment and should be considered when designing future therapies [48].

1.4.1 MSCs and Biomaterials

Although preclinical and clinical trials of MSC therapy for cardiac diseases have shown highly promising results as previously mentioned [49], some limitations still persist, including low stem cell retention in the infarcted myocardium, poor cell survival [50], and poor cell engraftment [51]. Examples of delivery routes include *via* peripheral intravenous infusion, direct surgical injection during open heart surgery, catheter-based intracoronary infusion, retrograde coronary venous infusion, or transendocardial injection [52, 53]. To overcome these limitations, the use of biomaterials represents a promising strategy to enhance the physical retention and localization of the MSCs at the site of implantation by improving engraftment and differentiation. Biomaterial-based scaffolds can act as mechanical supports that help to localize the cells at the site of injury, thus allowing them to proliferate, improve matrix deposition, and help modulate the host tissue response [53]. By tuning the material properties, the scaffolds can be engineered to control the degradation rate of the material, the diffusion rate of released growth factors, and protect encapsulated cells from the inflammatory environment. This aspect has a particular relevance to MSC transplantation therapy, since the beneficial effects of MSCs are more likely due to paracrine mechanisms *via* the release of growth factors rather than due to their ability to trans-differentiate into cardiomyocytes. Ultimately, engineered biomaterials can potentially confer protection from inflammation, which is known to be crucial during heart damage.

From a clinical point of view, in order to be suitable for regenerative medicine, a biomaterial needs to have peculiar characteristics such as biocompatibility, non-immunogenicity, non-thrombogenicity, non-cytotoxicity, and the ability to support cell differentiation and function. Based on promising results obtained in phase I clinical trials, alginate is an example of a hydrogel with potential cardiac applications, which the US Food and Drug Administration (FDA) has approved for phase II clinical trials for myocardial repair without the use of cells [54]. Engineering of cardiac constructs can take advantage of the growing number of biomaterials that are being shown safe and effective in clinical trials.

Additional characteristics that should be taken into account to improve the therapeutic use of biomaterials are material properties, such as stiffness, elasticity, porosity, and biodegradability. Investigators have characterized the behavior of MSCs cultured on polymer matrices of different stiffness and elasticity, demonstrating that these properties dramatically affect cell fate [54]. In particular, the myogenic differentiation seems to be supported by materials with an intermediate stiffness rather than very soft or hard substrates that support neurogenic and osteogenic differentiation, respectively. In another example, researchers proposed an innovative technique to switch a hydrogel-based material from a self-renewal permissive hydrogel (alginate) to a differentiation-permissive microenvironment (collagen). By fine-tuning the timing of this switch, early lineage specification can be directed, resulting in more control over cellular differentiation at the biomaterial level [56].

To date, repair cells can be delivered to the site of myocardial infarct following two major biomaterial strategies: injectable hydrogels or patches. The different formulation of the biomaterial, hydrogel or patch, will have differing mechanical properties that need to be considered. Researchers have compared hydrogels to two-dimensional (2D) sheets of biomaterials for their ability to support MSC viability and retention after being delivered to the site of MI in a rat model. The biomaterials analyzed differ in formulation and route of administration, as they include an injectable chitosan/β-glycerophosphate (GP) hydrogel, an injectable alginate hydrogel, a collagen patch, and an alginate patch [2]. All the analyzed biomaterials improved MSC viability as compared to monolayer culture, with a greater viability in the long term (6 days) when cells were encapsulated in hydrogels. This might be attributed to the ability of the gel's structure to protect the entrapped cells from oxidative stress. Twenty-four hours after transplantation, an 8-fold and a 14-fold increased MSC retention were reported when delivered *via* alginate and chitosan/β-GP gels, respectively, as compared to cells delivered in saline. Similarly, 47-fold and 59-fold increases were reported for collagen and alginate patches, respectively. Alginate is known to be nonadhesive; however, this limitation is usually overcome through functionalization with arginine–glycine–aspartic acid (RGD)-containing peptides. In addition to alginate and chitosan/β-glycerophosphate hydrogels, alpha-cyclodextrin/methoxy poly(ethylene glycol) (MPEG)–poly(ε-caprolactone) (PCL)–MPEG hydrogels have shown promising results as well. After 4 weeks post implantation, this hydrogel formulation was able to increase cell retention and vessel density in the infarcted myocardium of the injected animals. The LV ejection was improved in combination with an attenuation of the LV dilatation [55]. Similar improvement in myocardial remodeling, such as reduction in LV interior diameter at systole, increased anterior wall thickness, and increase in fractional shortening, has also been observed in experimental MI models treated with a rat-tail collagen patch [56]. More recently, human embryonic stem cell-derived MSCs, which seem to be similar to bone marrow-derived MSCs, have been reported to have beneficial effects when embedded in collagen patches and used for cardiac repair [56]. Taken together, these findings highlight the important role of the biomaterial when designing engineered cardiac constructs.

In addition to collagen and alginate, there are also promising results using de-cellularized sheets of human myocardium as a biomaterial-based scaffold for cell delivery [57]. One interesting example is human mesenchymal progenitor cells (MCPs) entrapped in fibrin hydrogels combined with human myocardial decellu-larized matrix. These hydrogel–ECM constructs were implanted onto an infarct in a nude rat model using MCPs preconditioned with TGF-β. The engineered patch led to enhanced angiogenesis and arteriogenesis through mechanisms that involved the migration of MCPs from the patch into the infarcted myocardium. Interestingly, this platform suggests that cell migration is in part regulated by altering the stromal cell-derived factor 1/C-X-C chemokine receptor type 4 (SDF-1/CXCR4) expression in MCPs. MCPs preconditioned with TGF-β, when encapsulated in the composite scaffold, showed increased CXCR4 expression and suppressed SDF-1 expression. This change in expression results in increased cell responsiveness to a gradient of SDF-1 and consequently migration into the infarcted myocardium. However, once released from the scaffold, MCPs recover the ability to release SDF-1, result-ing in enhanced cell migration, vascularization, and preservation of myocardium functionality, These studies demonstrate, once again, the ability of tuning the cell behavior *via* tissue-engineering approaches that take into account the cell source, the biomaterial, and the inflammatory environment.

1.5   Conclusion

Any implanted engineered cardiac construct will come in contact with the underly-ing inflammatory environment. It is therefore important to have the ability to model and predict the effects of the inflammatory environment on the biomaterial and the repair cells used to create any engineered cardiac construct. It is also important to understand how the inflammatory environment will affect the degradation rate of the biomaterial and what role macrophage polarization will play during the healing process. Since engineered constructs will also have a cellular component, it is also essential that we understand the cellular interactions between the inflammatory en-vironment and the repair cells. MSCs remain a viable candidate for targeted cellular therapies for cardiac applications and, as our understanding of the role of MSCs during myocardial healing improves, new strategies are needed to further test po-tential repair cells in the laboratory. One approach is to mimic important biophysi-cal and cellular events that could have detrimental effects on an engineered cardiac construct using advanced culture systems that recapitulate these events *in vitro*. Better understanding of how engineered cardiac patches behave in an inflammatory environment *in vitro* will provide a better screening tool to predict the potential suc-cess or potential of any cardiac engineered construct.

Acknowledgement This work was supported by NYSTEM C026721A Empire State Stem Cell Scholars: Fellow-to-Faculty Award.

References

1. Roger VL, Go AS, Lloyd-Jones DM, et al. Heart disease and stroke statistics–2012 update: a report from the American Heart Association. Circulation. 2012;125(1):e2–220.
2. Roche ET, Hastings CL, Lewin SA, et al. Comparison of biomaterial delivery vehicles for improving acute retention of stem cells in the infarcted heart. Biomaterials. 2014;35(25): 6850–8.
3. Strauer BE, Steinhoff G. 10 years of intracoronary and intramyocardial bone marrow stem cell therapy of the heart: from the methodological origin to clinical practice. J Am Coll Cardiol. 2011;58(11):1095–104.
4. Strauer BE, Yousef M, Schannwell CM. The acute and long-term effects of intracoronary stem cell transplantation in 191 patients with chronic heart failure: the STAR-heart study. Eur J Heart Fail. 2010;12(7):721–9.
5. Beltrami AP, Barlucchi L, Torella D, et al. Adult cardiac stem cells are multipotent and support myocardial regeneration. Cell. 2003;114(6):763–76.
6. Amado LC, Saliaris AP, Schuleri KH, et al. Cardiac repair with intramyocardial injection of allogeneic mesenchymal stem cells after myocardial infarction. Proc Natl Acad Sci U S A. 2005;102(32):11474–9.
7. Trivedi P, Tray N, Nguyen T, Nigam N, Gallicano GI. Mesenchymal stem cell therapy for treatment of cardiovascular disease: helping people sooner or later. Stem Cells Dev. 2010;19(7):1109–20.
8. Elnakish MT, Hassan F, Dakhlallah D, Marsh CB, Alhaider IA, Khan M. Mesenchymal stem cells for cardiac regeneration: translation to bedside reality. Stem Cells Int. 2012;2012:646038.
9. Shah VK, Shalia KK. Stem cell therapy in acute myocardial infarction: a pot of gold or Pandora's box. Stem Cells Int. 2011;2011:536758.
10. Freytes DO, Santambrogio L, Vunjak-Novakovic G. Optimizing dynamic interactions between a cardiac patch and inflammatory host cells. Cells Tissues Organs. 2012; 195(1–2):171–82.
11. Freytes DO, Kang JW, Marcos-Campos I, Vunjak-Novakovic G. Macrophages modulate the viability and growth of human mesenchymal stem cells. J Cell Biochem. 2013;114(1):220–9.
12. Wolf MT, Dearth CL, Ranallo CA, et al. Macrophage polarization in response to ECM coated polypropylene mesh. Biomaterials. 2014;35(25):6838–49.
13. Brown BN, Londono R, Tottey S, et al. Macrophage phenotype as a predictor of constructive remodeling following the implantation of biologically derived surgical mesh materials. Acta Biomaterialia. 2012;8(3):978–87.
14. Brown BN, Valentin JE, Stewart-Akers AM, McCabe GP, Badylak SF. Macrophage phenotype and remodeling outcomes in response to biologic scaffolds with and without a cellular component. Biomaterials. 2009;30(8):1482–91.
15. Robbins SL, Kumar V, Cotran RS. Robbins and Cotran pathologic basis of disease. 8th ed. Philadelphia: Saunders/Elsevier; 2010.
16. Frangogiannis NG. The immune system and cardiac repair. Pharmacol Res. 2008;58(2): 88–111.
17. Frangogiannis NG. The inflammatory response in myocardial injury, repair, and remodelling. Nat Rev Cardiol. 2014;11(5):255–65.
18. Kempf T, Zarbock A, Vestweber D, Wollert KC. Anti-inflammatory mechanisms and therapeutic opportunities in myocardial infarct healing. J Mol Med. 2012;90(4):361–9.
19. Yan X, Anzai A, Katsumata Y, et al. Temporal dynamics of cardiac immune cell accumulation following acute myocardial infarction. J Mol Cell Cardiol. 2013;62:24–35.
20. Lambert JM, Lopez EF, Lindsey ML. Macrophage roles following myocardial infarction. Intl J Cardiol. 2008;130(2):147–58.
21. Ley K, Laudanna C, Cybulsky MI, Nourshargh S. Getting to the site of inflammation: the leukocyte adhesion cascade updated. Nat Rev Immunol. 2007;7(9):678–89.

22. Swirski FK, Nahrendorf M, Etzrodt M, et al. Identification of splenic reservoir monocytes and their deployment to inflammatory sites. Science. 2009;325(5940):612–6.

23. Wynn TA, Chawla A, Pollard JW. Macrophage biology in development, homeostasis and disease. Nature. 2013;496(7446):445–55.

24. van Amerongen MJ, Harmsen MC, van Rooijen N, Petersen AH, van Luyn MJ. Macrophage depletion impairs wound healing and increases left ventricular remodeling after myocardial injury in mice. Am J Pathol. 2007;170(3):818–29.

25. Lawrence T, Natoli G. Transcriptional regulation of macrophage polarization: enabling diversity with identity. Nat Rev Immunol. 2011;11(11):750–61.

26. Mantovani A, Biswas SK, Galdiero MR, Sica A, Locati M. Macrophage plasticity and polarization in tissue repair and remodelling. J Pathol. 2013;229(2):176–85.

27. Mantovani A, Sozzani S, Locati M, Allavena P, Sica A. Macrophage polarization: tumorassociated macrophages as a paradigm for polarized M2 mononuclear phagocytes. Trends Immunol. 2002;23(11):549–55.

28. Mantovani A, Sica A, Sozzani S, Allavena P, Vecchi A, Locati M. The chemokine system in diverse forms of macrophage activation and polarization. Trends Immunol. 2004;25(12):677–86.

29. Pittenger MF, Mackay AM, Beck SC, et al. Multilineage potential of adult human mesenchymal stem cells. Science. 1999;284(5411):143–7.

30. Chou SH, Lin SZ, Day CH, et al. Mesenchymal stem cell insights: prospects in hematological transplantation. Cell Transplant. 2013;22(4):711–21.

31. Bernardo ME, Avanzini MA, Ciccocioppo R, et al. Phenotypical/functional characterization of in vitro-expanded mesenchymal stromal cells from patients with Crohn's disease. Cytotherapy. 2009;11(7):825–36.

32. Patel SA, Sherman L, Munoz J, Rameshwar P. Immunological properties of mesenchymal stem cells and clinical implications. Arc Immunol Ther Exp. 2008;56(1):1–8.

33. Jiang S, Kh Haider H, Ahmed RP, Idris NM, Salim A, Ashraf M. Transcriptional profiling of young and old mesenchymal stem cells in response to oxygen deprivation and reparability of the infarcted myocardium. J Mol Cell Cardiol. 2008;44(3):582–96.

34. Toma C, Pittenger MF, Cahill KS, Byrne BJ, Kessler PD. Human mesenchymal stem cells differentiate to a cardiomyocyte phenotype in the adult murine heart. Circulation. 2002;105(1):93–8.

35. Shake JG, Gruber PJ, Baumgartner WA, et al. Mesenchymal stem cell implantation in a swine myocardial infarct model: engraftment and functional effects. Ann Thorac Surg. 2002;73(6):1919–25; discussion 26.

36. Xu W, Zhang X, Qian H, et al. Mesenchymal stem cells from adult human bone marrow differentiate into a cardiomyocyte phenotype in vitro. Exp Biol Med. 2004;229(7):623–31.

37. Rose RA, Jiang H, Wang X, et al. Bone marrow-derived mesenchymal stromal cells express cardiac-specific markers, retain the stromal phenotype, and do not become functional cardiomyocytes in vitro. Stem Cells. 2008;26(11):2884–92.

38. Pijnappels DA, Schalij MJ, Ramkisoensing AA, et al. Forced alignment of mesenchymal stem cells undergoing cardiomyogenic differentiation affects functional integration with cardiomyocyte cultures. Circ Res. 2008;103(2):167–76.

39. Grinnemo KH, Mansson-Broberg A, Leblanc K, et al. Human mesenchymal stem cells do not differentiate into cardiomyocytes in a cardiac ischemic xenomodel. Ann Med. 2006;38(2):144–53.

40. Rota M, Kajstura J, Hosoda T, et al. Bone marrow cells adopt the cardiomyogenic fate in vivo. Proc Natl Acad Sci U S A. 2007;104(45):17783–8.

41. Menasche P, Hagege AA, Vilquin JT, et al. Autologous skeletal myoblast transplantation for severe postinfarction left ventricular dysfunction. J Am Coll Cardiol. 2003;41(7):1078–83.

42. Katritsis DG, Sotiropoulou PA, Karvouni E, et al. Transcoronary transplantation of autologous mesenchymal stem cells and endothelial progenitors into infarcted human myocardium. Catheter Cardiovasc Interv. 2005;65(3):321–9.

43. Murry CE, Field LJ, Menasche P. Cell-based cardiac repair: reflections at the 10-year point. Circulation. 2005;112(20):3174–83.
44. Dayan V, Yannarelli G, Billia F, et al. Mesenchymal stromal cells mediate a switch to alternatively activated monocytes/macrophages after acute myocardial infarction. Basic Res Cardiol. 2011;106(6):1299–310.
45. Cho DI, Kim MR, Jeong HY, et al. Mesenchymal stem cells reciprocally regulate the M1/M2 balance in mouse bone marrow-derived macrophages. Exp Mol Med. 2014;46:e70.
46. Adutler-Lieber S, Ben-Mordechai T, Naftali-Shani N, et al. Human macrophage regulation via interaction with cardiac adipose tissue-derived mesenchymal stromal cells. J Cardiovasc Pharmacol Ther. 2013;18(1):78–86.
47. Bartosh TJ, Ylostalo JH, Mohammadipoor A, et al. Aggregation of human mesenchymal stromal cells (MSCs) into 3D spheroids enhances their antiinflammatory properties. Proc Natl Acad Sci U S A. 2010;107(31):13724–9.
48. Naftali-Shani N, Itzhaki-Alfia A, Landa-Rouben N, et al. The origin of human mesenchymal stromal cells dictates their reparative properties. J Am Heart Assoc. 2013;2(5):e000253.
49. Pfister O, Della Verde G, Liao R, Kuster GM. Regenerative therapy for cardiovascular disease. Translational Res. 2014;163(4):307–20.
50. Mangi AA, Noiseux N, Kong D, et al. Mesenchymal stem cells modified with Akt prevent remodeling and restore performance of infarcted hearts. Nat Med. 2003;9(9):1195–201.
51. Murry CE, Soonpaa MH, Reinecke H, et al. Haematopoietic stem cells do not transdifferentiate into cardiac myocytes in myocardial infarcts. Nature. 2004;428(6983):664–8.
52. Williams AR, Hare JM. Mesenchymal stem cells: biology, pathophysiology, translational findings, and therapeutic implications for cardiac disease. Circ Res. 2011;109(8):923–40.
53. Templin C, Krankel N, Luscher TF, Landmesser U. Stem cells in cardiovascular regeneration: from preservation of endogenous repair to future cardiovascular therapies. Curr Pharm Des. 2011;17(30):3280–94.
54. Godier-Furnemont AF, Martens TP, Koeckert MS, et al. Composite scaffold provides a cell delivery platform for cardiovascular repair. Proc Natl Acad Sci U S A. 2011;108(19):7974–9.
55. Wang T, Jiang XJ, Tang QZ, et al. Bone marrow stem cells implantation with alpha-cyclodextrin/MPEG-PCL-MPEG hydrogel improves cardiac function after myocardial infarction. Acta Biomaterialia. 2009;5(8):2939–44.
56. Simpson D, Liu H, Fan TH, Nerem R, Dudley SC, Jr. A tissue engineering approach to progenitor cell delivery results in significant cell engraftment and improved myocardial remodeling. Stem Cells. 2007;25(9):2350–7.
57. Godier-Furnemont AF, Martens TP, Koeckert MS, et al. Composite scaffold provides a cell delivery platform for cardiovascular repair. Proc Natl Acad Sci U S A. 2011;108(19):7974–9.

Dr. Donald O. Freytes heads a tissue engineering group at The New York Stem Cell Foundation Research Institute that focuses on designing new culture systems that mimic the inflammatory environment found at the site of myocardial infarction to better screen engineered cardiac constructs.

Chapter 2
The Role of Macrophages in the Foreign Body Response to Implanted Biomaterials

Tony Yu, Valerie J. Tutwiler and Kara Spiller

2.1 Introduction

Biomaterials are part of the solution to many unmet clinical needs, from implantable sensors to drug delivery devices and engineered tissues. However, biomaterials face an inflammatory environment upon implantation, which represents a potential obstacle to their success [1]. In this chapter, we review the consequences of the foreign body response (FBR) for biomaterial function and strategies that have been used to inhibit the FBR. We focus on the role of the macrophage, the cell at the center of the inflammatory response, as the major regulator of the FBR, and discuss implications of changing macrophage behavior on biomaterial acceptance or rejection. Finally, we discuss recent discoveries in the role of macrophage phenotype, ranging from pro-inflammatory (M1) to anti-inflammatory (M2), and the role it plays in wound healing and biomaterial vascularization and integration. We conclude with a discussion of biomaterial design strategies that have been suggested to positively interact with and potentially control macrophages in order to improve interactions between biomaterials and the inflammatory response.

K. Spiller (✉) · T. Yu · V. J. Tutwiler
Biomaterials and Regenerative Medicine Laboratory, School of Biomedical Engineering,
Science, and Health Systems, Drexel University, Philadelphia, PA, USA
e-mail: spiller@drexel.edu

T. Yu
e-mail: ty63@drexel.edu

V. J. Tutwiler
e-mail: Valerie.j.tutwiler@drexel.edu

© Springer International Publishing Switzerland 2015
L. Santambrogio (ed.), *Biomaterials in Regenerative Medicine and the Immune System,*
DOI 10.1007/978-3-319-18045-8_2

2.2   The Foreign Body Response and Consequences for Implanted Biomaterials

Upon implantation (Fig. 2.1a), the body mounts the FBR against the biomaterial, beginning with protein adsorption to the biomaterial surface (Fig. 2.1b) [2]. This creates a thrombogenic surface and results in the activation and aggregation of a platelet–fibrin meshwork (Fig. 2.1c) [3]. The resulting procoagulant surface results in the infiltration of inflammatory cells. Neutrophils are the first inflammatory cells to arrive at the biomaterial surface [4]. When these cells are unable to phagocytose the foreign body, cytokines are released, which result in the differentiation of macrophages from monocytes [5]. Of the recruited immune cells, macrophages are the main cell type that regulates the FBR. At this point in normal wound healing, the acute inflammation phase (Fig. 2.1d) would ebb, and proliferation of fibroblasts and eventually remodeling of the wound would occur [4]. However, in response to an implanted biomaterial, the macrophages continue to attempt to remove the foreign body via phagocytosis and secrete enzymes and reactive species that aggravate the inflammatory state [6]. As a result, the progression through normal wound healing is disturbed and a chronic inflammation phase ensues (Fig. 2.1e) [7]. If the biomaterial cannot be degraded, the macrophages fuse together to form multinucleated foreign body giant cells (FBGC) that surround the biomaterial [8]. FBGCs and recruited fibroblasts deposit collagen layers around the biomaterial to form granulation tissues [9]. Over time, the granulation tissue becomes a dense collagen capsule, the hallmark of the FBR (Fig. 2.1f) [9]. Isolation of the biomaterial within this

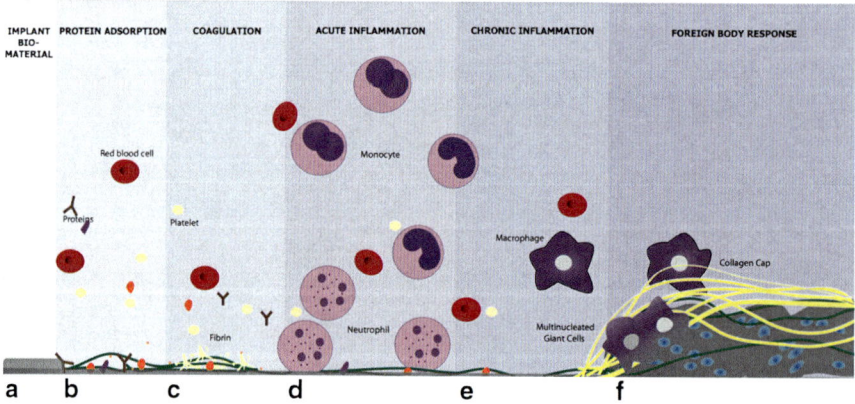

Fig. 2.1 Progression of the foreign body response. Upon implantation of the biomaterial (**a**), proteins from the blood and tissue nonspecifically adsorb to the biomaterial surface (**b**), and the coagulation cascade is initiated (**c**). Neutrophils and monocytes are recruited during the acute inflammatory phase (**d**). Monocytes differentiate into macrophages which attempt to degrade the biomaterial. If the macrophages cannot degrade the material, they fuse into foreign body giant cells, the hallmark of chronic inflammation (**e**). These multinucleated cells stimulate the formation of granulation tissue, which eventually becomes a dense fibrous collagen capsule that isolates the biomaterial from the rest of the body (**f**)

capsule as well as the secretion of damaging enzymes jeopardizes the functioning of the biomaterial. Some biomaterial applications that are particularly sensitive to the FBR include drug delivery devices, sensory devices, electrical devices, and tissue-engineered constructs.

2.2.1 Diffusion-Dependent Biomedical Devices

Biomaterials that depend on diffusion of molecules for their function include drug delivery systems, which are used to locally deliver drugs or growth factors to a particular area of the body, and sensors that measure the level of a molecule in the blood or tissue, including glucose sensors (Fig. 2.2a). Fibrous encapsulation can hinder diffusion, thus adversely affecting the function of the biomaterial [10]. For example, Anderson et al. investigated the drug release of gentamicin, an antibiotic, from silicone rubber rods [11]. Liquid-scintillation counting was performed to determine the release of radiolabeled gentamicin as well as the concentration in tissues adjacent to the implant. The rods were coated with a layer of silicon rubber to reduce the initial burst release and to prolong the drug release over time. The silicon rod drug release system with different gentamicin loading dosages (20, 35, and 40 wt.%) was implanted intramuscularly at the thigh muscles in dogs for 1 day and 1, 2, and 4 weeks. A fibrous capsule of about 5 and 11–15 μm thick was observed at 2 and 4 weeks post-implantation, respectively. Along with the observation of connective tissues forming around the implants, a reduction of the drug release rate was observed over the 4-week period. In addition, the difference in the gentamicin tissue level and serum levels over the first 3 weeks indicated that there was another factor with a different diffusion coefficient that may be responsible for the reduction in the drug release rate. As a result, it was suggested that the formation of the fibrous capsule inhibited the release of gentamicin [11].

One of the most common medical devices to monitor diabetes is the continuous glucose monitor (CGM), which monitors the blood glucose level via diffusion of blood glucose to the glucose sensor [12]. The FBR can hinder this function, as is demonstrated by a study in which macrophage depletion with diphtheria toxin driven by the CD11b promoter improved sensor performance [13]. While the creation of a diffusion barrier by the fibrous capsule likely inhibits sensor performance, it may also be a result of macrophage metabolic activity [14]. Klueh et al. injected macrophages at the implantation site of glucose sensors in mice [12]. The macrophages surrounded the sensors and sensor output quickly diminished, an effect that was not observed with injected lymphocytes. Interestingly, when serum glucose levels were artificially elevated, they were detected by the sensors, suggesting that their function was not permanently impaired by the presence of the macrophages. Moreover, companion in vitro studies showed that the presence of macrophages without the fibrous capsule or other biofouling effects also hindered sensor performance. The authors concluded that metabolism of glucose by macrophages also contribute to decreased sensor performance [12].

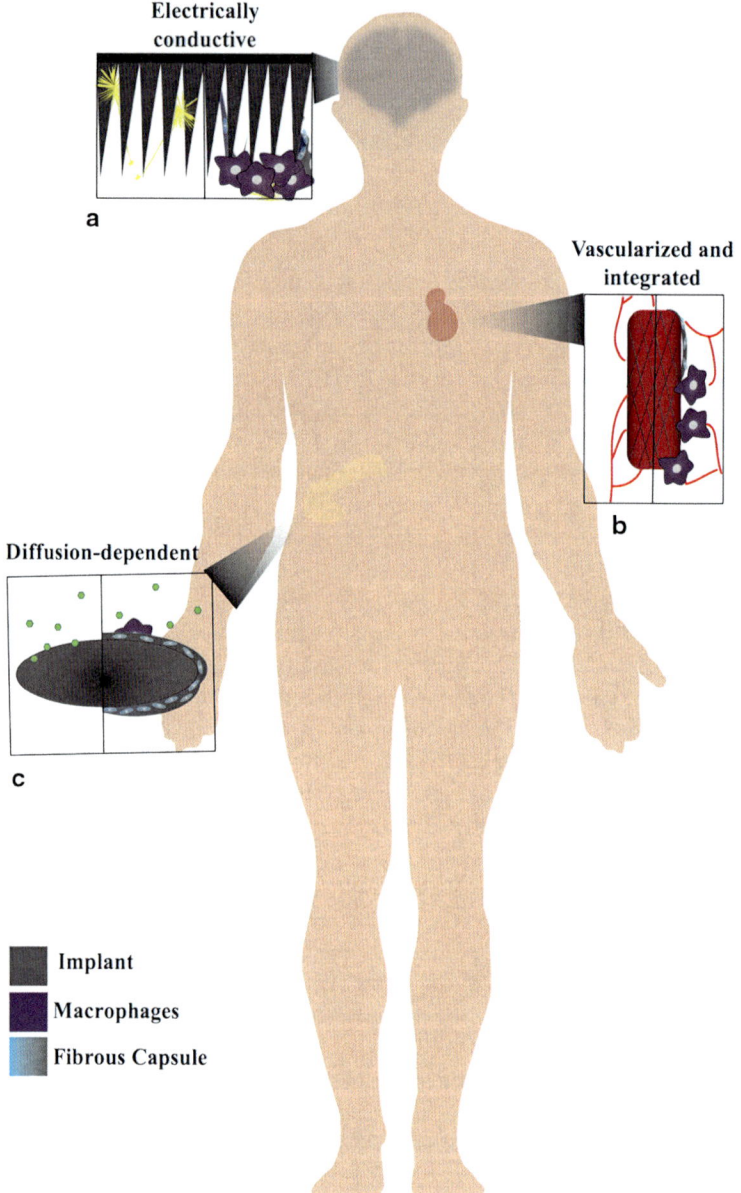

Fig. 2.2 Biomaterials particularly sensitive to the foreign body response. **a** Glial scarring around implanted microelectrodes can block the conductance of neuronal activity. **b** Vascularization, critical for the functionality of some implants, is blocked by fibrous encapsulation. **c** The fibrous capsule acts as a diffusion barrier, decreasing the performance of diffusion-dependent biomaterials such as drug delivery devices and glucose sensors

2.2.2   Transmission of Electrical Signals

Some medical devices require the conductance of electrical signals in order to monitor or pace the electrical activity of the brain, heart, or other muscles in the body, but the transmission of these signals can be inhibited by the presence of a fibrous capsule or a glial scar, as is the case in the central nervous system (Fig. 2.2b) [15]. A common electrical recording device is the silicon microelectrode array, a technology that measures the neuronal activity in the brain that is often used to monitor the activity of neurons and/or to investigate the correlation between the brain activity and behavior. However, one of the main limitations of this technology is inconsistency of performance in long-term applications. In a study by Biran et al. [16], the silicon microelectrode array was implanted into the brains of rats for 2 and 4 weeks to determine the mechanism of failures of the microelectrode arrays. Stab wounds were also created with the same microelectrodes as controls in order to distinguish whether it was the initial penetrating trauma or the FBR to the chronically implanted microelectrodes that caused device failure. After 2 and 4 weeks post-implantation, immunohistochemical analysis of the brain tissue indicated multilayered and dense regions of ED1-positive cells, a pan-macrophage marker in rats, along the implant–brain tissue interface in both the stab wounds and implanted microelectrodes. However, there were more ED1-positive cells surrounding the implanted microelectrodes compared to the stab wound. The intensity of glial fibrillary acidic proteins (GFAP) expression by reactive astrocytes was also significantly higher surrounding implanted microelectrodes compared to the control stab wounds. There was a significant amount of neuronal loss 2 weeks post-implantation in the nearby tissue of the implanted microelectrodes compared to the stab wound. Explants were rinsed with phosphate-buffered saline (PBS) and cultured in media for 24 h to assess cytokine secretion by adherent macrophages, which showed secretions of the inflammatory cytokines tumor necrosis factor-alpha (TNF-α) and monocyte chemotactic protein (MCP-1). Thus, the presence of the foreign body increased inflammation, leading to neuronal loss and device failure [16].

2.2.3   Vascularization and Integration of Biomaterials

Tissue engineering holds tremendous potential to replace damaged tissues and organs. The success of most tissue-engineered constructs requires recruitment of endothelial cells and the formation of new blood vessels to provide nutrients and oxygen transport for implanted cells [17, 18]. However, the FBR and the fibrous capsule prevent direct contact between the biomaterial and the surrounding tissue, so that vascularization and integration are essentially blocked (Fig. 2.2c). Shin et al. showed that fibrous encapsulation of hydrogels based on oligo(poly(ethylene glycol) fumarate) effectively prevented bone formation and vascularization in a rabbit bone defect model [19]. More recently, several studies have confirmed inverse correlation between fibrous capsule thickness and blood vessel ingrowth [5, 20].

2.3   Strategies to Inhibit the FBR

Clearly, the FBR and the formation of the fibrous capsule can drastically inhibit
the function of biomaterials. Thus, researchers have turned to the development of
strategies to inhibit the FBR. Because the FBR begins with protein adsorption and
inflammatory cell interactions at the biomaterial surface, most strategies are based
on modifications of the biomaterial surface [21, 22]. The main strategies include
inhibition of protein adsorption, the use of bioactive coatings, and modifications to
surface topography.

2.3.1   Inhibition of Protein Adsorption

Because the FBR begins with protein adsorption, inhibition of protein adsorption
has been extensively researched as a tool to inhibit the FBR [23]. Polyethylene
glycol (PEG)ylation, hydrogel coatings, plasma treatment, and other methods have
shown substantially decreased protein adsorption in vitro with reduced fibrous capsule formation in vivo [24]. Ultimately, however, the sensors fail because blood
proteins can still adsorb to a certain extent [14]. Recently, ultra-low fouling biomaterials have been prepared from zwitterionic materials [21]. Zwitterionic materials
have both a positive and negative charge that are not dissociated in an aqueous
environment. This property attracts water molecules via charge–dipole interactions
resulting in extremely hydrophilic properties [25]. Thus, adsorption of relatively
hydrophobic proteins is drastically reduced. The zwitterion carboxybetaine was
shown to adsorb <0.3 ng/cm^2 proteins from 100% blood serum, much lower than
the 5 ng/cm^2 of absorbed fibrinogen that is required to initiate platelet adhesion
[22]. When zwitterionic hydrogels based on poly(carboxybetaine methacrylate)
(PCBMA) were implanted subcutaneously in mice for 3 months, the number of pro-
inflammatory macrophages was reduced compared to control hydrogels prepared
from poly(2-hydroxyethyl methacrylate) (PHEMA), and the presence of a fibrous
capsule was not observed [21].

   Thus, the inhibition of protein adsorption is an effective way to mitigate the
FBR. However, without protein adhesion, cells from the body also cannot infiltrate
the material, so these biomaterials may not be appropriate for applications that require integration with the body, such as in tissue engineering. Nonetheless, they
may be extremely useful for applications in which the biomaterials are not intended
to integrate with body, such as catheters.

2.3.2   Surface Modification with Bioactive Coatings

For biomaterials that are intended to integrate with the body, another strategy to
mitigate the FBR and the formation of the fibrous capsule is to make the biomaterial

appear less foreign to immune cells, such as by coating with extracellular matrix (ECM)-derived molecules [26]. The ECM is mainly composed of collagen type I and glycosaminoglycans (GAG) such as hyaluronan (HA), chondroitin sulfate, and dermatan sulfate. Coating titanium rods with collagen chondroitin sulfate has been shown to inhibit fibrous encapsulation and to promote new bone formation in rat tibial defects [27]. Similarly, drug delivery strategies that actively increase integration with the body show decreased fibrous capsule formation. For example, controlled release of vascular endothelial-derived growth factor (VEGF) and nitric oxide (NO) from sensors has been shown to increase vascularization and decrease fibrous capsule formation [28, 29]. Controlled release of anti-inflammatory drugs such as dexamethasone has also been shown to reduce the FBR to sensors [30].

2.3.3   Surface Topography

Modifications to the surface topography of biomaterials have also been shown to affect the FBR [31]. The addition of porous poly(lactic acid) (PLA) coatings to glucose sensors decreased fibrous capsule thickness and increased vascularity following murine implantation [32]. Cao et al. investigated the orientation of the topography of electrospun nanofibrous poly(caprolactone) (PCL) scaffolds and reported the effect on the FBR [31]. The PCL was deposited in three distinct manners: aligned fibers, randomly oriented fibers, and a thin film; the scaffolds were then compared to an arginylglycylaspartic acid (RGD)-coated glass slide as a control. Interleukin (IL)-4 was added to human monocytes in vitro at days 3 and 7 in order to induce the formation of FBGC, mimicking the FBR in vivo. At day 10, the random fiber scaffolds resulted in the highest levels of cell attachment compared to the other scaffolds. In general, the cell density of all surfaces decreased over time as the macrophages fused into FBGC in the presence of IL-4. When the PCL scaffolds were implanted in Sprague-Dawley rats for 1, 2, and 4 weeks, the random fiber scaffold elicited a more severe FBR compared to the other scaffolds, while the aligned fiber scaffold resulted in the thinnest fibrous capsule [31]. Thus, biomaterial topography affects the FBR, and this behavior can be studied using in vitro models of macrophage–biomaterial interactions.

This relationship between in vitro and in vivo results was not supported in another study of the effects of biomaterial topography on the FBR. Expanded polytetrafluoroethylene (ePTFE) membranes with pore sizes of 0.2, 1, and 3 µm were seeded with primary human monocytes in vitro and compared to tissue culture polystyrene as a control [33]. Membranes with 3 µm pore size elicited a significant increase in the secretion of the inflammatory cytokine IL1-beta compared to the other pore sizes. The ePTFE biomaterials were also implanted subcutaneously in mice for 4 weeks to evaluate the formation of the fibrous capsule. Interestingly, despite showing more inflammatory activity in vitro, ePTFE membranes with 3 µm pores resulted in a significantly thinner fibrous capsule than the nonporous ePTFE [33]. Although these findings did show that biomaterial topography affects the FBR, they also highlight the complexity of the relationship between inflammatory cell–biomaterial interactions and the FBR.

While the studies of the effects of biomaterial surface topography on the FBR have been largely empirical, they do suggest that modulation of topography may be a potential tool for mitigating the FBR. More systematic analyses are required to determine the mechanism of topographical effects on macrophage behavior and the FBR.

2.4 Macrophage Biology

It has been shown through many studies that macrophages play a crucial role in regulating the FBR [5]. A better understanding of macrophage dynamics may be the key to overcoming their ability to impair biomaterial performance. To understand the behavior of macrophages in response to biomaterials, it is helpful to consider biomaterial implantation as a chronic wound. Then, the behavior of macrophages can be assessed in comparison to normal wound healing in order to discover the mechanisms of impaired healing.

2.4.1 Macrophage Phenotypes in Normal Wound Healing

Normal wound healing in response to an injury generally consists of four distinct stages: hemostasis, the inflammatory stage, the proliferation stage, and the remodeling stage [4, 34, 35]. Macrophages can be polarized into a spectrum of phenotypes ranging from pro-inflammatory to anti-inflammatory and pro-healing depending on the environmental stimulus [36]. Pro-inflammatory macrophages are often referred to as "classically activated," or M1, while the anti-inflammatory macrophages are referred to as "alternatively activated," or M2 phenotype, following the T helper cell nomenclature of Th1 and Th2 [36]. At early stages of normal wound healing, M1 macrophages infiltrate the wound to promote inflammation and to stimulate the wound healing process (Fig. 2.3). M2 macrophages begin to accumulate around day 3 or 4 post-injury, while the level of M1 macrophages decreases [36]. M2 macrophages may accumulate via the direct transition of M1 to M2, the polarization of newly arriving macrophages to M2, and proliferation of other M2 macrophages [37]. The accumulated macrophages eventually emigrate to the draining lymph nodes returning back to the pre-injury state of resident macrophages after the wound is completely remodeled and healed [38].

Macrophages have been widely recognized as major regulators of wound healing and tissue regeneration over the past few decades. However, much of macrophage biology is still not well understood. Although the classification of the different macrophage phenotypes is widely accepted, a consensus has not yet been reached as to the overall effects and consequences of the diverse macrophages phenotypes on wound healing.

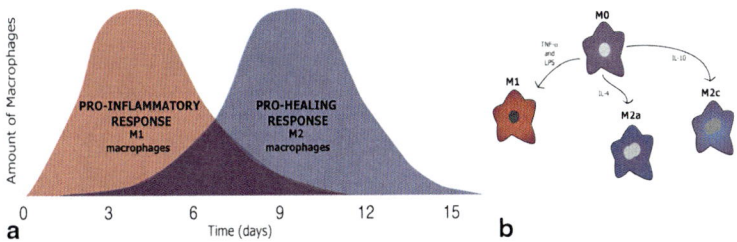

Fig. 2.3 Macrophages in normal wound healing. **a** In normal wound healing, macrophages initially express a pro-inflammatory response, but as time progress they transition to a pro-healing response. **b** Macrophages can be polarized into a spectrum of phenotypes that range from a pro-inflammatory to a pro-healing response. Those discussed here include M1 macrophages which are induced through by TNF-α and LPS, M2a macrophages which are stimulated by IL-4, and M2c macrophages which are activated by IL-10. *LPS* lipopolysaccharide

2.4.2   The Role of M1 Macrophages in Healing

Macrophages are polarized to the M1 phenotype by pro-inflammatory stimuli and cytokines such as bacterial lipopolysaccharide (LPS), TNF-α, and interferon-gamma (IFN-γ, Fig. 2.3b) [5]. M1 macrophages attempt to phagocytose any bacteria, cellular debris, and foreign invaders. However, controversies surround the role of M1 macrophages in wound healing. On one hand, chronic inflammation, characterized by persistent numbers of M1 macrophages is known to impair wound healing. For example, Kigerl et al. studied the effect of macrophage activation on central nervous system injury of C57BL/6 mice [39]. Moderate midthoracic spinal cord injury (SCI) was inflicted on the mice, while the sham mice receive a laminectomy without SCI. The tissue samples were collected at day 1, 3, 7, 14, and 28 post-SCI for immunohistochemical analysis. M1 macrophages were predominant at the sites of SCI as indicated by CD86 staining. The RNA from each of the wound sites at each time point was extracted for gene expression and showed that the genes associated with the M2 macrophages returned to pre-injury level at day 7 post-SCI, while genes associated with the M1 macrophages were maintained for 1 month post-SCI. These results suggest that the M1 macrophages were responsible for the defective wound healing over time. Furthermore, macrophage-conditioned media (MCM) was also collected from the supernatant of polarized macrophages to determine the effect of M1 and M2 MCM on cortical neurons in vitro. The M1—but not the M2—MCM was neurotoxic to cortical neurons [39]. Consequently, this study suggests that the M1 macrophages are detrimental to healing.

On the other hand, M1 macrophages have also been shown to be beneficial for wound healing [40]. M1 macrophages are highly angiogenic, stimulating endothelial cell sprout formation in vitro and in vivo in part by secretion of VEGF [5, 41]. When M1 macrophages were depleted in a mouse model of skeletal muscle injury via CD11b-diptheria toxin, muscle regeneration was completely prevented [42]. In contrast, when M2 macrophages were depleted, muscle regeneration was still possible, but was significantly impaired. However, persistent numbers of M1 macrophages mark chronic inflammation and impaired healing, highlighting the importance of the correct M1-to-M2 sequence in tissue repair.

2.4.3   The Role of M2 Macrophages in Wound Healing

M2 macrophages are generally associated with healing and tissue remodeling of the wound and are the dominant phenotype in the proliferation and remodeling stages of wound healing [43]. The M2 macrophages are usually responsible for the formation of connective tissue [44]. However, the granulation tissue may eventually lead to the formation of scar tissue or a fibrous capsule. For this reason, it is believed that the M2 macrophages contribute to the fibrous capsule formation and the FBR [4].

The M2 macrophage phenotype can be further classified into three subpopulations: M2a, M2b, and M2c [5, 45]. Macrophages are polarized to the M2a and M2c phenotype by environmental stimulation of IL-4 and IL-10, respectively (Fig. 2.3b) [5, 45]. The M2b phenotype is polarized by toll-like receptors (TLC) or other immune complexes [45]. Although the M2b is categorized within the M2 phenotype, the M2b macrophages are activated by an inflammatory environmental stimulus more similar to that of the M1 macrophages [46]. Their role in wound healing is not known. The traditional alternatively activated M2 macrophages are now referred to as the M2a phenotype, which promotes the production of ECM and collagen, a necessary part of healing [5, 46].

Preliminary studies have attempted to explain the functioning of M1, M2a, and M2c subpopulations in wound healing and vascularization [5]; however, there is still a great need for further research in this area. M1 macrophages secrete VEGF to initiate angiogenesis. M2a macrophages secrete platelet-derived growth factor (PDGF), a chemoattractant that stabilizes growing blood vessels and promotes anastomosis of new blood vessels into networks [5]. M2c macrophages secrete high levels of matrix metalloprotease 9 (MMP-9), a protease that is involved in the breakdown and remodeling of the ECM and vasculature [5]. M2c macrophages also express high levels of CD163, which has been shown to be associated with tissue repair and remodeling of the wound and promoting cell proliferation in mice [45]. Thus, it appears that M1, M2a, and M2c macrophages appear sequentially in normal wound healing, but more studies are required that distinguish between M2a and M2c macrophages in order to confirm this hypothesis.

2.4.4   Role of M1 and M2 Macrophages in the FBR
to Biomaterials

Surprisingly, it is still not clear which macrophage phenotype is responsible for the formation of the fibrous capsule. M1 macrophages are widely believed to be the cause of the fibrous capsule as the inflammatory response upregulates the FBR, thereby increasing the thickness of the fibrous capsule [47]. However, IL-4, a cytokine that induces the activation to the M2a phenotype, has been shown to promote the formation of FBGC, which ultimately causes the formation of the fibrous capsule [5, 48].

To investigate the association of macrophage phenotype and tissue remodeling of biological scaffolds in the FBR, Badylak et al. implanted porcine small intestine submucosa (SIS), carbodiimide-crosslinked porcine-derived SIS (CDI-SIS), and autologous tissue graft as a control subcutaneously in Sprague-Dawley rats for 1, 2, 4, and 16 weeks [49]. Immunostaining of CCR7 and CD163, representing M1 and M2 markers, respectively, showed an infiltration of M2 macrophages in the SIS scaffolds at all time points, along with an organized layer of connective tissue at the 16-week time point, which suggests constructive remodeling. The CDI-SIS scaffolds were initially infiltrated with equal amounts of M1 and M2 macrophages at 2 weeks post-implantation, but were eventually dominated by the M1 macrophages at 4 weeks post-implantation leading to the formation of FBGC and fibrosis at 16 weeks post-implantation. The autologous graft scaffold showed a high level of M2 macrophages at the 1- and 2-week time points, but over time, an even distribution of M1 and M2 macrophages was observed. At the 16-week post-implantation, less organized collagenous connective tissue was also observed. The authors concluded that biomaterials that promote the M2 phenotype are associated with constructive tissue remodeling compared to those that promote the M1 phenotype [49]. This study was one of the first to suggest macrophage phenotype as a predictor of the success or failure of biomaterials.

Brown et al. evaluated the effects of 14 FDA (the Food and Drug Administration)-approved biological scaffolds on macrophage phenotype and the FBR in rats [50]. All scaffolds were composed of naturally occurring material but differed in the source and species of the tissue. The 14 scaffolds included AlloMax, AlloDerm, Avaulta Plus, CollaMend, Flex HD, InteXen LP, MatriStem, PelviSoft, Strattice Firm, Strattice Pliable, Sugisis, SurgiMend, Veritas, and Xenform. Immunohistochemical analysis of CD68, CCR7, and CD206 was performed on the samples after 14 and 35 days of implantation to quantify the relative proportions of M1 and M2 macrophages at each time point. In general, a higher M2 to M1 ratio was associated with more constructive remodeling [50].

Goreish et al. evaluated the effects of phosphorylcholine (PC)-coated biomaterials on the FBR and fibrous capsule formation via intramuscular implantation in rabbits for 4 and 13 weeks and found that there were less M1 macrophages and a reduced fibrous capsule on the PC-coated biomaterial compared to the control of high-density polyethylene, but were not significantly different at 4 days post-implantation [51]. At 13 weeks post-implantation, little to no M1 macrophages were found at the implantation site for both control and PC-coated biomaterials. The fibrous capsule thickness of the PC-coated biomaterial was reduced to no capsule or partial encapsulation at the 13-week time point, while the fibrous capsule thickness of the control was thinner but still present [51]. This study indicated a direct correlation of inflammatory macrophages to the thickness of the fibrous capsule.

On the other hand, M2 macrophages have also been implicated in the formation of the fibrous capsule. M2 macrophages secrete PDGF and transforming growth factor beta (TGF-β), known mediators of the fibrous capsule [52]. Several stud-ies have shown that IL-4, which causes M2 polarization, induces the formation of FBGC and subsequently the fibrous capsule [52]. A study by Kao et al. implanted a cage system subcutaneously in rats, providing a prolonged inflammatory environ-ment [52]. Anti-murine IL-4 (IL-4Ab), murine IL-4 (muIL-4), normal goal nonspe-cific control IgG (gtIgG), and PBS were injected into the cages and released into the implant site to determine the effects of IL-4 on the FBR. Implants releasing muIL-4 showed a significant increase in the formation of FBGC and fibrous capsule com-pared to the IL-4Ab and controls. The IL-4Ab resulted in a significantly decreased formation of FBGC and fibrous capsule compared to the controls [52].

A recent study was designed to directly assess the role of macrophage pheno-type in vascularization and fibrous encapsulation of tissue engineering scaffolds [5]. Model biomaterials chosen to elicit a range of macrophage responses were im-planted subcutaneously into mice for 10 days. Porous collagen scaffolds were ex-pected to elicit an M2 response, LPS-coated scaffolds were expected to elicit an M1 response, and glutaraldehyde-crosslinked collagen scaffolds were expected to elicit a mixed M1/M2 response. Immunohistochemical analysis demonstrated that the unmodified scaffolds were surrounded by the highest numbers of M2 macrophages, indicated by three markers of the M2 phenotype, coincident with a dense fibrous capsule. The LPS-coated scaffolds were infiltrated by high numbers of M1 macro-phages, as indicated by three markers of the M1 phenotype, but without evidence of fibrous encapsulation. Interestingly, the glutaraldehyde-crosslinked scaffolds stained strongly for both M1 and M2 markers, and were infiltrated by high numbers of blood vessels, in keeping with the idea that both M1 and M2 macrophages are required for vascularization [5]. Clearly, more studies are required to elucidate the relative contributions of M1 and M2 macrophages to the FBR and fibrous capsule formation.

2.5   Macrophage Polarization in Response to Biomaterials

Understanding of the role of macrophage phenotype on the spectrum from healing to fibrous encapsulation will allow us to design biomaterials with macrophages in mind. As an initial step towards controlling macrophage behavior, several studies have systematically evaluated the response of macrophages to changing biomaterial properties.

2.5.1   Decellularized Biological Scaffolds

Decellularized ECM scaffolds have a strong clinical potential to mitigate the FBR and to promote wound healing. Decellularized bladder ECM has been used to re-pair damaged skeletal muscle of humans, resulting in the formation of new muscle tissue, neovascularization, recruitment of endogenous myogenic progenitor cells, and functional recovery [53]. These scaffolds are prepared by removing cells and cellular debris from tissues using detergents, often followed by chemical cross-linking. Several studies have shown their potential to modulate inflammation in comparison to cellular scaffolds. Xu et al. used a primate model to show that decellularized scaffolds does not induce significant chronic inflammation or promote a severe immune response compared to the cellular scaffolds [54]. These findings are supported by the findings of Brown et al., in which the inflammatory response associated with decellularized allografts (rat body wall ECM) compared to cellular autograft (autologous body wall tissue) and porcine bladder xenografts for 28 days post-implantation was examined in Sprague-Dawley rats [55]. Immunohistochemical analysis revealed that the cellular scaffolds had a macrophage population that favored the M1 phenotype at the majority of time points, associated with a dense, poorly organized collagenous response, while the decellularized scaffolds had a macrophage population that favored the M2 phenotype, which was associated with neomatrix formation by the end of the observation period. Interestingly, only the decellularized scaffolds resulted in angiogenesis as early as 3 days post-implantation [55]. Thus, the use of decellularized scaffolds to regenerate tissue is an active area of research [56].

2.5.2 Porosity of Biomaterials

In addition to allowing for ingrowth of tissue and blood vessels, the porosity and pore diameter of a biomaterial has been shown to be critical in the FBR [20]. For example, Madden et al. showed that porous hydrogel scaffolds promoted a pro-healing response compared to nonporous biomaterials in vivo in two different species [20]. Microtemplated hydrogel scaffolds were fabricated from poly(2-hydroxyethyl methacrylate-co-methacrylic acid) (PHEMA-co-MAA) scaffolds with pore diameters of 20–80 μm. Cardiac implantation of the acellular scaffolds of varying pore sizes was evaluated for 4 weeks to determine the effects of pore diameter on the formation of the fibrous capsule, neovascularization, and the macrophage phenotype profile. The nonporous scaffolds promoted a more severe inflammatory response indicated by a significantly higher expression of the M1 marker inducible NO synthase (iNOS) compared to the porous scaffolds. Although there was not a significant difference, there was a general increase in the expression of MMR, an M2 marker, in the porous scaffolds. In addition, the porous scaffolds with larger pores (30–60 μm in mice and 40–80 μm in rats) resulted in more neovascularization as indicated by significantly higher expression of rat endothelial cell antibody-1 (RECA-1), a marker for endothelial cells, compared to nonporous scaffolds and scaffolds with 20 μm pores. At the same time, nonporous scaffolds and those with 20 μm pores resulted in a significant increase in the fibrous capsule thickness. There seemed to be an optimal pore diameter of the scaffold, approximately 30–40 μm, that reduced the fibrous capsule thickness and the number of M1 macrophages and enhanced neovascularization. Interestingly, the authors noted a much higher number of macrophages expressing markers of both the M1 and M2 phenotype compared to the numbers of more homogenous populations, indicating significant contribution of hybrid phenotypes to neovascularization. This study proposed that porous biomaterials increase the activation of M2 macrophages, which coincides with enhanced vascularization of the biomaterial [20].

2.6   Future Directions: Active Control over Macrophage Phenotype

With the recent increase in the number of studies examining the relationships between biomaterial properties, macrophage phenotype, and healing outcome, the possibility of controlling macrophage behavior for a beneficial response is becoming a reality. Strategies are moving away from attempts at inhibition of the inflammatory response towards those that actively work with macrophages for therapeutic effects.

Preliminary studies of drug delivery strategies that release immunomodulatory factors have shown proof of concept that macrophages can be harnessed. For example, the release of chemotactic factors for macrophages, including MCP1 and agonists of the S1P receptor, cause increased scaffold integration, vascularization, and tissue regeneration in cardiac and bone defects [57–59]. Mokarram et al. evaluated the effects of macrophage phenotype manipulation via local delivery of immunomodulatory cytokines IFN- γ and IL-4 from agarose hydrogels on peripheral nerve regeneration in critical-sized 15-mm sciatic nerve gap defects in rats [60]. IFN-γ, which promotes the M1 phenotype, or IL-4, which promotes the M2a phenotype, was mixed with agarose hydrogel and inserted into a hollow polysulfone tube as a nerve guidance channel and were implanted in critical nerve gap of rats for 3 weeks. The release of IL-4 resulted in a significant increase in the M2 macrophages and enhanced Schwann cell migration to the middle of the scaffold compared to the blank control and the release of IFN-γ. Furthermore, axon regeneration was observed in the IL-4 samples, and the amount of axons in the IL-4 samples was approximately 20 times greater than the control and IFN-γ samples. The ratio of M2 to M1 markers was linearly correlated with axon regeneration [60]. Controlled release strategies like these show that it is possible to control macrophage behavior for beneficial effects on biomaterials.

2.7   Conclusions

Macrophages are at the center of the FBR and consequently hold the key to the success or failure of a biomaterial. We now know that macrophage phenotype plays a critical role in determining the therapeutic outcome. More studies are required to delineate the roles of macrophage phenotype in the FBR, particularly in fibrous capsule formation versus vascularization and integration, and the roles that biomaterials properties like surface structure and pore size play in dictating phenotypic changes. Future generations of biomaterials will consider macrophage phenotypic response as an important design consideration and will even control macrophage behavior for beneficial outcomes.

References

1. Higgins DM, Basaraba RJ, Hohnbaum AC, Lee EJ, Grainger DW, Gonzalez-Juarrero M. Localized immunosuppressive environment in the foreign body response to implanted biomaterials. Am J Pathol. 2009;175(1):161–70.
2. Schmidt D, Waldeck H, Kao W. Protein adsorption to biomaterials. In: Puleo DA, Bizios R, editors. Biological interactions on materials surfaces. New York: Springer; 2009. pp. 1–18.
3. Verheye S, Markou CP, Salame MY, Wan B, King SB, 3rd, Robinson KA, et al. Reduced thrombus formation by hyaluronic acid coating of endovascular devices. Arterioscler Thromb Vasc Biol. 2000;20(4):1168–72.
4. Chamberlain CS, Leiferman EM, Frisch KE, Duenwald-Kuehl SE, Brickson SL, Murphy WL, et al. Interleukin-1 receptor antagonist modulates inflammation and scarring after ligament injury. Connect Tissue Res. 2014;55(3):177–86.
5. Spiller KL, Anfang RR, Spiller KJ, Ng J, Nakazawa KR, Daulton JW, et al. The role of macrophage phenotype in vascularization of tissue engineering scaffolds. Biomaterials. 2014;35(15):4477–88.
6. Baker DW, Zhou J, Tsai YT, Patty KM, Weng H, Tang EN, et al. Development of optical probes for in vivo imaging of polarized macrophages during foreign body reactions. Acta biomaterialia. 2014;10(7):2945–55.
7. Lickorish D, Chan J, Song J, Davies JE. An in-vivo model to interrogate the transition from acute to chronic inflammation. Eur Cells Mater. 2004;8:12–9; discussion 20.
8. Yang J, Jao B, McNally AK, Anderson JM. In vivo quantitative and qualitative assessment of foreign body giant cell formation on biomaterials in mice deficient in natural killer lymphocyte subsets, mast cells, or the interleukin-4 receptoralpha and in severe combined immunodeficient mice. J Biomed Mater Res Part A. 2014;102(6):2017–23.
9. Bakker D, van Blitterswijk CA, Hesseling SC, Grote JJ. Effect of implantation site on phagocyte/polymer interaction and fibrous capsule formation. Biomaterials. 1988;9(1):14–23.
10. Ratner BD. Reducing capsular thickness and enhancing angiogenesis around implant drug release systems. J Control Release. 2002;78(1–3):211–8.
11. Anderson JM, Niven H, Pelagalli J, Olanoff LS, Jones RD. The role of the fibrous capsule in the function of implanted drug-polymer sustained release systems. J Biomed Mater Res. 1981;15(6):889–902.
12. Klueh U, Frailey JT, Qiao Y, Antar O, Kreutzer DL. Cell based metabolic barriers to glucose diffusion: macrophages and continuous glucose monitoring. Biomaterials. 2014;35(10):3145–53.
13. Klueh U, Qiao Y, Frailey JT, Kreutzer DL. Impact of macrophage deficiency and depletion on continuous glucose monitoring in vivo. Biomaterials. 2014;35(6):1789–96.
14. Frost M, Meyerhoff ME. In vivo chemical sensors: tackling biocompatibility. Anal Chem. 2006;78(21):7370–7.
15. Moshayedi P, Ng G, Kwok JC, Yeo GS, Bryant CE, Fawcett JW, et al. The relationship between glial cell mechanosensitivity and foreign body reactions in the central nervous system. Biomaterials. 2014;35(13):3919–25.
16. Biran R, Martin DC, Tresco PA. Neuronal cell loss accompanies the brain tissue response to chronically implanted silicon microelectrode arrays. Exp Neurol. 2005;195(1):115–26.

17. Padera RF, Colton CK. Time course of membrane microarchitecture-driven neovascularization. Biomaterials. 1996;17(3):277–84.
18. Maki T, Otsu I, O'Neil JJ, Dunleavy K, Mullon CJ, Solomon BA, et al. Treatment of diabetes by xenogeneic islets without immunosuppression. Use of a vascularized bioartificial pancreas. Diabetes. 1996;45(3):342–7.
19. Shin H, Quinten Ruhe P, Mikos AG, Jansen JA. In vivo bone and soft tissue response to injectable, biodegradable oligo(poly(ethylene glycol) fumarate) hydrogels. Biomaterials. 2003;24(19):3201–11.
20. Madden LR, Mortisen DJ, Sussman EM, Dupras SK, Fugate JA, Cuy JL, et al. Proangiogenic scaffolds as functional templates for cardiac tissue engineering. Proc Natl Acad Sci U S A. 2010;107(34):15211–6.
21. Zhang L, Cao Z, Bai T, Carr L, Ella-Menye JR, Irvin C, et al. Zwitterionic hydrogels implanted in mice resist the foreign-body reaction. Nat Biotechnol. 2013;31(6):553–6.
22. Tsai WB, Grunkemeier JM, Horbett TA. Human plasma fibrinogen adsorption and platelet adhesion to polystyrene. J Biomed Mater Res. 1999;44(2):130–9.
23. Wisniewski N, Reichert M. Methods for reducing biosensor membrane biofouling. Colloids Surf B Biointerfaces. 2000;18(3–4):197–219.
24. Quinn CA, Connor RE, Heller A. Biocompatible, glucose-permeable hydrogel for in situ coating of implantable biosensors. Biomaterials. 1997;18(24):1665–70.
25. Wang H, Yue G, Dong C, Wu F, Wei J, Yang Y, et al. Carboxybetaine methacrylate-modified nylon surface for circulating tumor cell capture. ACS Appl Mater Interfaces. 2014;6(6):4550–9.
26. Schulz MC, Korn P, Stadlinger B, Range U, Moller S, Becher J, et al. Coating with artificial matrices from collagen and sulfated hyaluronan influences the osseointegration of dental implants. J Mater Sci Mater Med. 2014;25(1):247–58.
27. Rammelt S, Illert T, Bierbaum S, Scharnweber D, Zwipp H, Schneiders W. Coating of titanium implants with collagen, RGD peptide and chondroitin sulfate. Biomaterials. 2006;27(32):5561–71.
28. Ward WK, Quinn MJ, Wood MD, Tiekotter KL, Pidikiti S, Gallagher JA. Vascularizing the tissue surrounding a model biosensor: how localized is the effect of a subcutaneous infusion of vascular endothelial growth factor (VEGF)? Biosens Bioelectron. 2003;19(3):155–63.
29. Gifford R, Batchelor MM, Lee Y, Gokulrangan G, Meyerhoff ME, Wilson GS. Mediation of in vivo glucose sensor inflammatory response via nitric oxide release. J Biomed Mater Res Part A. 2005;75(4):755–66.
30. Patil SD, Papadimitrakopoulos F, Burgess DJ. Dexamethasone-loaded poly(lactic-co-glycolic) acid microspheres/poly(vinyl alcohol) hydrogel composite coatings for inflammation control. Diabetes Technol Ther. 2004;6(6):887–97.
31. Cao H, McHugh K, Chew SY, Anderson JM. The topographical effect of electrospun nanofibrous scaffolds on the in vivo and in vitro foreign body reaction. J Biomed Mater Res Part A. 2010;93(3):1151–9.
32. Koschwanez HE, Yap FY, Klitzman B, Reichert WM. In vitro and in vivo characterization of porous poly-L-lactic acid coatings for subcutaneously implanted glucose sensors. J Biomed Mater Res Part A. 2008;87(3):792–807.
33. Bota PC, Collie AM, Puolakkainen P, Vernon RB, Sage EH, Ratner BD, et al. Biomaterial topography alters healing in vivo and monocyte/macrophage activation in vitro. J Biomed Mater Res Part A. 2010;95(2):649–57.
34. Verhamme P, Hoylaerts MF. Hemostasis and inflammation: two of a kind? Thromb J. 2009;7:15.
35. Hubner G, Brauchle M, Smola H, Madlener M, Fassler R, Werner S. Differential regulation of pro-inflammatory cytokines during wound healing in normal and glucocorticoid-treated mice. Cytokine. 1996;8(7):548–56.
36. Song E, Ouyang N, Horbett M, Antus B, Wang M, Exton MS. Influence of alternatively and classically activated macrophages on fibrogenic activities of human fibroblasts. Cell Immunol. 2000;204(1):19–28.

37. Jenkins SJ, Ruckerl D, Cook PC, Jones LH, Finkelman FD, van Rooijen N, et al. Local macrophage proliferation, rather than recruitment from the blood, is a signature of TH2 inflammation. Science. 2011;332(6035):1284–8.
38. Bellingan GJ, Caldwell H, Howie SE, Dransfield I, Haslett C. In vivo fate of the inflammatory macrophage during the resolution of inflammation: inflammatory macrophages do not die locally, but emigrate to the draining lymph nodes. J Immunol. 1996;157(6):2577–85.
39. Kigerl KA, Gensel JC, Ankeny DP, Alexander JK, Donnelly DJ, Popovich PG. Identification of two distinct macrophage subsets with divergent effects causing either neurotoxicity or regeneration in the injured mouse spinal cord. J Neurosci. 2009;29(43):13435–44.
40. Novak ML, Weinheimer-Haus EM, Koh TJ. Macrophage activation and skeletal muscle healing following traumatic injury. J Pathol. 2014;232(3):344–55.
41. Willenborg S, Lucas T, van Loo G, Knipper JA, Krieg T, Haase I, et al. CCR2 recruits an inflammatory macrophage subpopulation critical for angiogenesis in tissue repair. Blood. 2012;120(3):613–25.
42. Arnold L, Henry A, Poron F, Baba-Amer Y, van Rooijen N, Plonquet A, et al. Inflammatory monocytes recruited after skeletal muscle injury switch into antiinflammatory macrophages to support myogenesis. J Exp Med. 2007;204(5):1057–69.
43. Bullers SJ, Baker SC, Ingham E, Southgate J. The human tissue-biomaterial interface: a role for PPARgamma-dependent glucocorticoid receptor activation in regulating the CD163 M2 macrophage phenotype. Tissue Eng Part A. 2014;20(17–18):2390–2401.
44. Ito T, Kaneko T, Yamanaka Y, Shigetani Y, Yoshiba K, Okiji T. M2 macrophages participate in the biological tissue healing reaction to mineral trioxide aggregate. J Endod. 2014;40(3):379–83.
45. Villalta SA, Nguyen HX, Deng B, Gotoh T, Tidball JG. Shifts in macrophage phenotypes and macrophage competition for arginine metabolism affect the severity of muscle pathology in muscular dystrophy. Hum Mol Genet. 2009;18(3):482–96.
46. Gea-Sorli S, Guillamat R, Serrano-Mollar A, Closa D. Activation of lung macrophage subpopulations in experimental acute pancreatitis. J Pathol. 2011;223(3):417–24.
47. Vistnes LM, Ksander GA, Kosek J. Study of encapsulation of silicone rubber implants in animals. A foreign-body reaction. Plast Reconstr Surg. 1978;62(4):580–8.
48. McNally AK, Anderson JM. Interleukin-4 induces foreign body giant cells from human monocytes/macrophages. Differential lymphokine regulation of macrophage fusion leads to morphological variants of multinucleated giant cells. Am J Pathol. 1995;147(5):1487–99.
49. Badylak SF, Valentin JE, Ravindra AK, McCabe GP, Stewart-Akers AM. Macrophage phenotype as a determinant of biologic scaffold remodeling. Tissue Eng Part A. 2008;14(11):1835–42.
50. Brown BN, Londono R, Tottey S, Zhang L, Kukla KA, Wolf MT, et al. Macrophage phenotype as a predictor of constructive remodeling following the implantation of biologically derived surgical mesh materials. Acta biomaterialia. 2012;8(3):978–87.
51. Goreish HH, Lewis AL, Rose S, Lloyd AW. The effect of phosphorylcholine-coated materials on the inflammatory response and fibrous capsule formation: in vitro and in vivo observations. J Biomed Mater Res Part A. 2004;68(1):1–9.
52. Kao WJ, McNally AK, Hiltner A, Anderson JM. Role for interleukin-4 in foreign-body giant cell formation on a poly(etherurethane urea) in vivo. J Biomed Mater Res. 1995;29(10):1267–75.
53. Sicari BM, Rubin JP, Dearth CL, Wolf MT, Ambrosio F, Boninger M, et al. An acellular biologic scaffold promotes skeletal muscle formation in mice and humans with volumetric muscle loss. Sci Transl Med. 2014;6(234):234–58.
54. Xu H, Wan H, Sandor M, Qi S, Ervin F, Harper JR, et al. Host response to human acellular dermal matrix transplantation in a primate model of abdominal wall repair. Tissue Eng Part A. 2008;14(12):2009–19.
55. Brown BN, Valentin JE, Stewart-Akers AM, McCabe GP, Badylak SF. Macrophage phenotype and remodeling outcomes in response to biologic scaffolds with and without a cellular component. Biomaterials. 2009;30(8):1482–91.
56. Badylak SF, Taylor D, Uygun K. Whole-organ tissue engineering: decellularization and re-cellularization of three-dimensional matrix scaffolds. Ann Rev Biomed Eng. 2011;13:27–53.

57. Roh JD, Sawh-Martinez R, Brennan MP, Jay SM, Devine L, Rao DA, et al. Tissue-engineered vascular grafts transform into mature blood vessels via an inflammation-mediated process of vascular remodeling. Proc Natl Acad Sci U S A. 2010;107(10):4669–74.
58. Petrie Aronin CE, Shin SJ, Naden KB, Rios Jr PD, Sefcik LS, Zawodny SR, et al. The enhancement of bone allograft incorporation by the local delivery of the sphingosine 1-phosphate receptor targeted drug FTY720. Biomaterials. 2010;31(25):6417–24.
59. Awojoodu AO, Ogle ME, Sefcik LS, Bowers DT, Martin K, Brayman KL, et al. Sphingosine 1-phosphate receptor 3 regulates recruitment of anti-inflammatory monocytes to microvessels during implant arteriogenesis. Proc Natl Acad Sci U S A. 2013;110(34):13785–90.
60. Mokarram N, Merchant A, Mukhatyar V, Patel G, Bellamkonda RV. Effect of modulating macrophage phenotype on peripheral nerve repair. Biomaterials. 2012;33(34):8793–801.

Dr. Kara Spiller is an assistant professor in Drexel University's School of Biomedical Engineering, Science, and Health Systems. Her current research interests include the design of immunomodulatory biomaterials, drug delivery in tissue engineering, and international engineering education.

Chapter 3
Integrating Tissue Microenvironment with Scaffold Design to Promote Immune-Mediated Regeneration

Kaitlyn Sadtler, Franck Housseau, Drew Pardoll and Jennifer H. Elisseeff

3.1 Introduction

As the field of biomedical engineering further develops the ability to engineer diverse organs, the need to understand the role of the tissue microenvironment in host immune responses to the engineered tissues or scaffolds has surfaced. It has been appreciated that there are varying responses to pathogens dependent upon the location in which that pathogen is encountered. This principle also applies to biomaterials such as synthetics and tissue-derived scaffolds made from extracellular matrix (ECM) when used in regenerative medicine. We begin our discussion on immune responses to these materials by first considering the ability of the host immune system to cater its response appropriately to the tissue microenvironment and how that environment in turn alters the response.

J. H. Elisseeff (✉)
Translational Tissue Engineering Center, Wilmer Eye Institute and Department of Biomedical Engineering, Johns Hopkins University, Smith Building Rm. 5035, 400 N Broadway, Baltimore, MD 21231, USA
e-mail: jhe@jhu.edu

K. Sadtler
Translational Tissue Engineering Center, Wilmer Eye Institute and Department of Biomedical Engineering, Johns Hopkins University, Baltimore, MD 21287, USA
e-mail: ksadtle3@jhmi.edu

F. Housseau · D. Pardoll
Sidney Kimmel Comprehensive Cancer Center, Johns Hopkins University School of Medicine, Baltimore, MD 21231, USA

F. Housseau
e-mail: fhousse1@jhmi.edu

D. Pardoll
e-mail: dpardol1@jhmi.edu

© Springer International Publishing Switzerland 2015   35
L. Santambrogio (ed.), *Biomaterials in Regenerative Medicine and the Immune System*,
DOI 10.1007/978-3-319-18045-8_3

3.1.1   Tissue-Based Class Control

Depending upon the location in which the host receives a danger signal, the immune system can alter the magnitude and polarization of that response. This signal can arise from specific sources, such as invading pathogens, or from damage-associated molecular patterns (DAMPs) created upon tissue damage [1]. After receiving this signal, cells present in the vicinity will integrate the signals from their location as well as from the type of danger that is present and respond appropriately. This is the basis of a theory known as tissue-based class control [2].

Three commonly defined immune polarizations are T_H1, T_H2, and T_H17, referring to the T cells involved in the class of response. Broadly, T_H1 and T_H2 immune polarizations are also known as type-1 and type-2 responses. There is a fourth type, T_{reg}, that is also discussed in this chapter. Major cytokines and cells that are associated with these phenotypes are shown in Table 3.1. These type-1 and type-2 immune polarizations also involve differential activation of myeloid cells such as macrophages, neutrophils, and eosinophils. It is important to note that there is a large continuum in immune polarization, where individual cells rarely fit directly into these categorizations, and intermediate phenotypes are also observed [3]. One example is T_H3 responses induced in situations of oral tolerance [4]. Oral tolerance is when an antigen is given orally and in the gut induces a response that has several hallmarks of both the T_H2 response, interleukin-4 (IL-4) and interleukin-5 (IL-5) production, as well as the T_{reg} response, interleukin-10 (IL-10), and transforming growth factor-beta (TGF-β), inducing a tolerance to that antigen. Engineers and immunologists exploited this phenomenon of oral tolerance to treat individuals that suffer from allergic reactions, such as peanut allergies [5]. In this example, patients

Table 3.1 Immune polarizations and their effector function

	Cytokines and chemokines	Major cells	Antibody class	Main tissues/ targets	Pathogenesis
T_H1	IFN-γ, TNF-α, IL-1β, IL-6, IL-12	CTLs, NK cells, macrophages expressing iNOS, dendritic cells	IgG	Skin, intracellular pathogens	Local: damaging inflammation, cell death Systemic: cachexia
T_H2	IL-4, IL-5, IL-13, IL-10	Eosinophils, basophils, macrophages expressing ARG1, B cells	IgE	Mucosal surfaces, helminths, allergy	Local: fibrosis, scar deposition Systemic: asthma/ allergy, diarrhea, promotes cancer survival
T_H17	IL-6, TGF-β, IL-17, IL-22, IL-23	Neutrophils	IgG	Skin, extracellular pathogens	Autoimmunity
T_{reg}	IL-10, TGF-β	T cells, macrophages	IgA	All organs, immunosuppression	Local: antigenic ignorance Systemic: can promote cancer survival

are orally exposed to low doses of antigen that slowly tolerize their immune reaction, reducing or preventing the strong eosinophilic and IgE-induced allergen hypersensitivity response.

There are other factors that help direct the antigen-experienced cells to the correct location for their response. Such factors include expression of adhesion proteins on the immune cell surface that are specific for proteins or carbohydrates present in the target location. Cells that have received an antigen signal from the intestine upregulate $\alpha_4\beta_7$ integrin that binds to specific sialyl-Lewis^x acids present on intestinal epithelia [6]. When these immune phenotypes are presented in the wrong location, they could result in severe inflammation and damage of delicate tissues or in an incomplete clearance of an infection. In addition to cell adhesion molecules, chemokines (CCL or CXCL) and chemokine receptors (CCR or CXCR) are another set of ligands and receptors important for tissue-specific homing of lymphocytes [7]. Chemokines, such as CCL25, are expressed by small intestinal epithelial cells, endothelial cells, and other associated cells, which bind receptors on CCR9 + T cells, lymphocytes that are primed for small intestinal homing. In another example, $\alpha_4\beta_1$ + CLA + cutaneous lymphocyte antigen, a glycoprotein (CCR4 + CCR10 + T cells) bind carbohydrates, known as selectins, CCL17 and CCL27, which are expressed by cutaneous cells. These ligands are differentially expressed in homeostatic and inflamed tissue, demonstrating how an alteration in the immune status of tissues changes the recruitment of different effector cells. Chemokines can also serve to recruit specific cell types to an area such as neutrophils and eosinophils [8].

The various cells of the immune system can have both positive and negative effects on tissue regeneration (Table 3.2). For tissue engineers, these intricacies of the host immune response complicate the goal of immune acceptance and remodeling by introducing variables dependent upon the host tissue in which the engineered tissue is designed to integrate. If we can leverage the polarization tendencies of varying tissues, and the optimal immune response for regeneration, we will be able to create a more rational scaffold design that incorporates immunological instructions to stimulate or guide tissue repair.

3.1.2 Immune-Privileged Sites and Clonal Ignorance

Integration of signals received from both the challenge encountered, as well as the location in which this challenge occurs, ultimately determines the phenotype of the immune response. In certain tissues, there is another method to avoid excessive immune-mediated tissue damage, a phenotype known as clonal ignorance. Here, antigens from organs that remain sequestered from the rest of the body, these "immune-privileged" organs, such as brain, eye, and testes, are ignored by the host immune system [9]. At immune-privileged sites, these antigens are not "tolerized" but the host does not mount a response to them. Misregulation of this system can result in various autoimmune pathologies, such as multiple sclerosis.

Table 3.2 Role of immune population in inflammation and tissue regeneration

	Immune function	Constructive phenotypes	Destructive phenotypes
Macrophages	Phagocytic cells of the innate immune system that are capable of clearing debris and secreting inflammatory messengers as well as antigen presentation	Inducing helpful inflammation to clear pathogens and recruit adaptive immune cells. Deposition of collagenous scars during wound healing and required for limb regeneration in axolotls	M1 macrophages express iNOS which can lead to faster scaffold degradation, overactive inflammation, M2 macrophages can induce overscarring and fibrosis
PMN's	Includes neutrophils, basophils, eosinophils. Phagocytic and chemotactic cells of the innate immune system that are involved in clearing debris, activation and polarization of adaptive immune system	Recruitment of other cells to the site of injury or infection. Polarizing to secrete cytokines such as IL-4 to help induce wound healing responses	Increased PMN infiltration and cell death can lead to exudate accumulation
T cells	Major mediators of cellular adaptive immunity. Help polarize immune response and subsets are responsible for killing of target cells	Secretion of cytokines that help polarize immune response, act as regulators to prevent an overactive response	Overactive immune response can induce inflammation, heightened regulatory response

If we are better able to understand the intricacies of immune responses to biomaterials as a function of their tissue microenvironment, we will be able to design enhanced scaffolds and better predict the host response to these regenerative grafts. Through manipulation of the biomaterials-response system, we could also create an appropriate immune response to help guide proper regeneration and formation of a functional replacement tissue. A delicate balance between promoting immune recruitment while avoiding excessive inflammation will have to be reached to optimize the remodeling of these tissues. This immune modulation could be obtained through modification of scaffolds, changing the biomaterial as a function of the tissue to be regenerated, while driving the proper inflammatory response and immune environment.

3.2 Development and the Immune System

As tissue engineers drive the growth of new tissue development to replace a nonfunctional tissue or organ, we can use insights from naturally occurring regeneration and development to guide scaffold design. We will first look at immunology as it pertains to developmental biology, including several signaling molecules that have different roles in both systems, and then look at direct evidence of immune system involvement in natural and engineered tissue regeneration.

3.2.1 Shared Signaling Molecules: Immune Response and Developmental Biology

One of the most interesting evolutionary adaptations is the use of the "Toll" and "Toll-like receptors (TLR)" signaling pathways for dorsal–ventral patterning in *Drosophila melanogaster* and innate immune receptor signaling in vertebrates. In *Drosophila,* the Toll receptor binds a protein called Spätzle, which activates a downstream cascade of proteins, resulting in degradation of cactus, and nuclear translocation of Dorsal inducing a ventral cell fate. This pathway is also important in antifungal response in adult *Drosophila* [10]. TLR's also induce the nuclear factor-kappaB (NFκB) signaling pathway, which results in degradation of nuclear factor of kappa light polypeptide gene enhancer in B cells inhibitor, alpha (IκBα), and nuclear translocation of p65/RelA, activating the expression of inflammatory genes [11]. In this situation, Toll is orthologous to TLR, and Dorsal is orthologous to p65/RelA. At the evolutionary level, the immune system and tissue development are tightly integrated, shown in this adaptation of a signaling pathway for both embryonic development, and immune response.

Looking closer at the dynamics within a single organism, we can see functions for various proteins in both development and the immune response. The TGF-β family of proteins has many roles in immunology, including the induction of regulatory responses along with IL-10, and has been associated with the alternative polarization of macrophages that do not produce an inflammatory phenotype [12]. The TGF-β family of proteins is also important in differentiation between the endoderm, mesoderm, and ectoderm germ layers in various organisms [13]. In amphibians, TGF-β-like signals in the organizer (a cellular structure in embryos) specify the dorsal mesoderm. In birds, signaling by proteins in this family is important in primitive streak formation, a structure in which there is epithelial-to-mesenchymal transition of mesodermal precursors. Another important cytokine, IL-4, promotes type-2 immunity and alternative activation of macrophages, and is also associated with allergy and hypersensitivity reactions. IL-4 has also been implicated as a driving force in myotube formation during muscle development [14]. Developing myotubes secrete IL-4, recruiting endogenous progenitor cells and inducing fusion with existing myoblasts to promote mature myotube formation. Interestingly, one of the largest inhibitory signals of muscle growth comes from myostatin, which is also in the TGF-β family of proteins. The overlap of signaling molecules that are important in both embryonic development and immune response suggests that a varying immune microenvironment during development will alter the ultimate phenotype of that tissue. This association between development and the immune system also has implications for regenerative medicine. Just as researchers have attempted to harness elements of normal developmental biology pathways to enhance new tissue development and repair, the immune system may also be critical to consider when engineering strategies for regeneration.

Scientists studying natural limb reconstruction in amphibians have directly implicated the role of the immune system in regeneration [15]. After severing the limb of an axolotl, the limb bud was injected with liposomes, containing clodronate, a cytotoxic compound that is released into the cytoplasm of phagocytic cells, such as macrophages, after uptake of liposomes. Upon local depletion of macrophages, limb bud formation and proper limb regeneration was inhibited. This study demonstrated that local macrophages were necessary for proper limb bud formation and limb regeneration in axolotls. Without immune interference, axolotls are capable of regenerating a normal limb about 2.5 months post amputation.

Through the use of techniques implemented by developmental biologists and integration of their findings on immune-mediated tissue development, engineers could better mimic the local environment in natural tissue regeneration. Alterations of the immune microenvironment of engineered tissues could promote or inhibit differentiation tendencies of various progenitor cells dependent upon the signals they receive from resident immune cells.

3.2.2   Functional Healing: Tissue Engineering's Approach to Development

One major goal of regenerative medicine is to create a functional replacement for missing or damaged tissue. Since the development of new tissue has many corollaries with embryonic development, we can assume that the immune system and associated signaling molecules play a pivotal role in repair and regeneration. Several studies on the role of the immune system in tissue engineering, and a new field of "immunoengineering" have appeared in recent years [16, 17]. The goal of immunoengineering is to create an immune response through biological or pharmacological intervention that will yield a desired outcome, such as decreased inflammation or targeted destruction of cancer cells. Working in the field of macrophage biology, several groups have characterized various polarizations of macrophage response that can alter the outcome of regeneration; more specifically, studying the M1–M2 axis of macrophage polarization [18, 19]. Others have discussed the possibility of manipulating the macrophage response to biomaterials to induce a better outcome of host immune response [20]. In their view, macrophages are the driving force behind both positive and negative aspects of the host response to biomaterials. Since macrophages are responsible for both inflammation and scarring, they could potentially be a good target for manipulation and improving integration of new tissues with the surrounding environment. A plethora of heterogeneity exists in these polarizations; varying scaffolds, both natural and synthetic, will elicit varying responses.

Macrophage biology has been implicated in several aspects of development and regeneration [21, 22]. Eosinophils, another immune cell type associated with type-2 polarization, are important mediators of both muscle and liver regeneration. As mentioned previously, IL-4 signaling is required for myotube fusion and formation of a mature muscle fiber. Through various mouse models that eliminated IL-4 and IL-13 signaling, Heredia et al. showed that IL-4 secreted by eosinophils was required for differentiation of satellite cells (muscle stem cells) through signals received from fibro/adipogenic progenitors (FAPs) resident in muscle tissue

[23]. Furthermore, IL-4 prevented adipogenic differentiation of these precursors and therefore prevented accumulation of ectopic adipocytes in muscle. These type-2 signals also promote clearance of necrotic debris resulting from muscle trauma or degeneration. In this study, a cascade of signals was outlined starting with a type-2 innate immune response signaling to FAPs through IL-4 and IL-13, which then promoted satellite cell differentiation into regenerated muscle. Type-2 signals were generated by eosinophils in cardiotoxin-mediated muscle damage. IL-4 and IL-13 signaling was not required in satellite cells for induction of regeneration.

Eosinophils and type-2 immunity have also been associated with liver regeneration in partial hepatectomy and toxin-mediated liver damage models [24]. Through the use of an eosinophil knockout (ΔdblGATA) mouse model and an IL-4 signaling knockout *(IL-4Rα-/-)* model, Goh et al. concluded that both eosinophil and IL-4 signaling were required for liver regeneration. In this case, selective inhibition of IL-4 signaling in hepatocytes decreases proliferation measured by Ki67 (a cellular proliferation marker) and 5-bromo-2'-deoxyuridine (BrdU) incorporation.

3.3 Polymer Scaffolds and the Surrounding Tissue Environment

Many synthetic polymers are used as scaffolds and drug delivery devices in tissue engineering (Table 3.3). Even relatively inert polymeric scaffolds are capable of eliciting varying types of immune responses. Evaluation of these responses is important for the selection of biomaterials for use in drug delivery or as a scaffold, and screening methods are being developed to provide insight into proper material design [25]. Outside of the typical foreign body response (FBR), host immune polarization is dependent upon the type of biomaterial and location of the synthetic scaffold implant. This includes a varying degree of immune cell activation, inflammation and subsequent scaffold degradation rates dependent upon the tissue environment [26]. With one compound capable of eliciting different responses, it is important that we understand these responses and better cater our implants to the immune microenvironment that it will be subjected to.

3.3.1 Polyethylene Glycol Hydrogels

Synthetic polymers are used in many locations in the body due to the flexibility of their design and low immunogenicity. One polymer that is well established in tissue engineering is poly(ethylene glycol) (PEG). Studies conducted by Hillel et al. (2011) on a PEG-based soft tissue replacement showed a variance in immune response dependent upon location of injection. Implants placed in the dorsal subcutaneous area of rats showed a decreased inflammatory response compared to those adjacent to adipose tissue in preliminary clinical trials [27]. This observation leads to

Table 3.3 Common materials used in tissue engineering

		Material	Applications	Commercial products
	PLGA	Poly(lactic-*co*-glycolic acid)	Drug delivery, nanofiber scaffolds, nanoparticles	Vicryl
Synthetics	PCL	Polycaprolactone	Nanofiber scaffolds	Capronor
	PLLA	Poly-L-lactide	Dermal filler, nanofiber scaffolds	Dacron, Bio-Anchor, DEXON
	ACD	Acellular dermis	Hernial repair, skin lesion repair, abdominal cavity defect repair, volumetric muscle loss, burns, multipurpose degradable scaffolds	SurgiMend®, Alloderm, FlexHD®
ECM	SIS	Small intestinal submucosa		Oasis® Wound Matrix, CorMatrix®, DynaMatrix®
	UBM	Urinary bladder matrix		Matristem®

further investigation of the location-regulated immune response through implantation of PEG hydrogels in varying locations of Sprague-Dawley rats [28]. These rats received implants subcutaneously, intraperitoneally, and postlaterally in the thigh near a large adipose depot. After 1 week of incubation the implants were excised to evaluate the early immune response. Subcutaneous implants showed the least inflammatory cell recruitments, followed by intraperitoneal implants, with adipose inducing the largest response as reported previously. Cells of the immune system such as macrophages can produce reactive oxygen species (ROS) that can lead to hydrolysis of PEG hydrogels. M1 macrophages express iNOS (inducible nitric oxide synthase, NOS2), which will catalyze the production of nitric oxide (an ROS) whose primary function is to defend against bacteria. Neutrophils and other immune cells are also capable of producing ROS, leading to altered implant degradation kinetics. Not only will the immune response directly affect the degradation and lifetime of biodegradable synthetics, it will also produce different signals that cells residing in that scaffold will receive, depending on the biomaterial's tissue location.

3.3.2   FBR Microenvironment

Within the context of a FBR [29] to materials such as PEG, there are both systemic and local effects of the immune response. Locally, the biomaterial or implant is surrounded by a dense fibrous capsule and often induces the fusion of macrophages to form foreign body giant cells (FBGCs), signaling poor graft integration. Proximal to the implant the FBR is dominated by type-2 cytokines that help induce collagen deposition and wall off the foreign body. Bacterial infection at the site of implantation can yield an inflammatory type-1 immune response, causing damaging release of free radicals that increase degradation of the biomaterial. Immune cells that interact with foreign bodies often have a mixed phenotype, showing signs of both collagen deposition and scar formation (arginase-1, IL-4) characteristic of a type-2 response, as well as inflammatory mediators (interferon-γ (IFN-γ), tumor necrosis factor α (TNF-α), iNOS characteristic of a type-1 response.

3.3.3 Synthetics Coated with ECM Scaffolds

To promote the acceptance of polymer-based scaffolds, researchers have coated synthetic mesh with a decellularized urinary bladder matrix (UBM, 4.2) to delay the inflammatory reaction associated with a FBR to polymer scaffolds [30, 31]. In these studies, polypropylene surgical mesh was coated with UBM or dermal ECM that had been digested with pepsin. An abdominal wall injury study showed that proximity of macrophages to the synthetic mesh was correlated with a variation in their polarization. Macrophages that were prevented from interacting with the synthetic material and instead remained on the ECM coating showed a more M2-like polarization, whereas those in direct contact with the mesh were more inflammatory, showing an M1-like phenotype. This coating thereby moderated the host-implant response and allowed for a decrease in inflammation at the implant location.

3.4 Decellularized ECM: A Model and Scaffold

One promising material for use in regenerative medicine is decellularized ECM. ECM-based scaffolds are derived from native tissue that has been thoroughly cleansed of cellular material, leaving behind large structural proteins and associated growth factors (GFs) and small molecules [32, 33]. Decellularization is achieved by treating the tissue with a variety of solvents meant to disrupt cell membranes and remove cells and nucleic acids. These solvents include, but are not limited to, acids, detergents, and enzyme solutions. Current ECM scaffolds that are commercially available and approved for human clinical use include acellular dermis (Collamend™), UBM (Matristem®), and small intestinal submucosa (SIS, OASIS® Wound Matrix, Table 3.3). Concerns have been raised about the efficiency of decellularization [34] and possible xenogeneic effects, but clinically tissue-derived ECM appears to be a promising scaffold with specific immunomodulatory properties that promote regeneration.

3.4.1 Small Intestinal Submucosa

The Badylak group at University of Pittsburgh is a pioneer in the field of decellularized ECM and has studied the properties of porcine SIS. SIS has been evaluated in many animal models such as murine and canine, and also used in human tendon repair as well as abdominal cavity defects and other applications. Currently they are investigating the use of various pepsin-digested ECM scaffolds as hydrogels for regeneration [35]. ECM hydrogels have been tested with success in skeletal muscle regeneration after acute large-scale trauma [36]. Their research has also investigated the polarization of macrophages along the M1–M2 axis and its implications in graft acceptance and remodeling versus inflammation and rejection. They have associated the alternative (M2) state of macrophage activation with the use of SIS as opposed to the inflammatory (M1) activation state. SIS contains several GFs that are known to be important in tissue regeneration, including TGF-β and vascular endothelial growth factor (VEGF), which are key regulators in wound repair and vascularization. The matrix has also been shown to include fibroblast growth factor-2 and ECM proteins such as glycosaminoglycans and proteoglycans, though its structure is predominated by collagens. SIS is also available commercially through various companies.

3.4.2   Urinary Bladder Matrix

Decellularized UBM has been made commercially available through the company ACell marketed under the product name Matristem. UBM is used in both sheet and particulate forms as a scaffold for regeneration and wound healing. Much like SIS and dermal ECM products, UBM in its sheet form is used in abdominal wall repair and skin lesion repair. The particulate form was tested in topical applications to aide in complex tissue regeneration. UBM hydrogels that are formed by pepsin digestion of ECM are under evaluation for the treatment of traumatic brain injury (TBI) [37] and tested as an undigested gel as well as a lyophilized powder. UBM has also been combined with synthetics through incorporation via electrospinning [38]. These scaffolds represent a hybrid between typical synthetic scaffolds and biologically derived ECM, offering some of the signals and ECM macromolecules to promote healing and regeneration.

3.4.3   Acellular Adipose Tissue

Though largely composed of lipid triglycerides for energy storage, white adipose tissue also has an associated connective tissue that can be isolated. This acellular tissue is obtained by extrusion of lipids, followed by standard decellularization procedures such as acid and detergent treatments. Acellular adipose tissue (AAT) is currently being investigated in its promise as a soft-tissue replacement [39] as it forms a thick gel when hydrated.

In a porcine model, both human and porcine AAT were injected subcutaneously and then excised to gauge the integration with host tissue as well as the inflammatory response. In most cases, there were a large number of cell-dense aggregates that appeared on the periphery of the xenogeneic human-derived ECM, while fewer aggregates formed around the allogeneic implant. These cellular aggregates stained densely for the macrophage marker CD11b in both xenogeneic and allogeneic implants. After a 2-week incubation of porcine AAT in rats, de novo adipose tissue was observed in the soft tissue implants. Vascularization of subcutaneous grafts was observed in an athymic mouse model as early as 1 week post implantation.

AAT has also been tested as a model for breast cancer progression and metastasis in response to various drugs used to treat these malignancies [40]. Breast cancer cells were cultured in human adipose tissue-derived ECM (hDAM) in the presence of doxorubicin and lapatinib to measure their behavior by migration, proliferation, and cell shape. This model is stated to be more tissue-mimetic than current ex vivo cancer models such as three-dimensional Matrigel culture and conventional two-di-mensional (2D) cell culture. Inhibition of epidermal growth factor receptor (EGFR) signaling (a target effect of lapatinib) was increased in hDAM cultures compared to a 2D system, whereas the effect of doxorubicin was decreased in hDAM compared to 2D assays. The increased ability to study the tumor microenvironment will allow for an easier and more effective technique for engineering cancer therapies.

3.4.4   As an Immunomodulatory Reagent and Tissue Model

The role of the tissue microenvironment in host responses to pathogens or injury has been an exciting area of research in the field of immunology. It is well accepted that the surrounding tissues in which that response is occurring largely dictate the polarization of that response [41]. This is especially relevant in immune-privileged organs such as the brain and eye, where an inflammatory response would be cata-strophic to the function of a tissue. These effects are caused not only by the cellular component of an organ but also the ECM and associated small proteins, sugars, and other molecules. Decellularized ECM is a promising tool for bioengineers as a scaffold and has recently demonstrated utility as an immunomodulatory reagent. A more thorough understanding of the ability of ECM to induce these polarizations, as well as an in-depth characterization of the responses of cells to varying tissue ECM, have yet to be demonstrated. With further understanding of these interac-tions, we could leverage the polarization tendencies of various ECM scaffolds with a synthetic ability to promote a desired immune phenotype in regenerative medicine and disease models.

Macrophage polarization in response to a scaffold directly affects the outcome of that scaffold being either rejected or remodeled. In general, the field has concluded that an abundance of M1 macrophages, expressing iNOS as well as a host of inflam-matory cytokines, TNFα, interleukin 1 beta (IL-1β), and interleukin 6 (IL-6) among others, signals poor scaffold integration with surrounding tissue. Conversely, M2 macrophages expressing arginase 1 (ARG1), Fizz1, and cytokines such as IL-10 are important in suppressing inflammation and healing damaged tissue. Previous research has verified the ability of decellularized SIS to promote an M2 macrophage phenotype in vivo. Using SIS as both a scaffold and a powder for wound healing has shown the promise of ECM as a powerful immunomodulatory reagent.

Further studies would need to involve the characterization of this response as a whole, with a deeper understanding of the intricacies of immune polarization. There are some pitfalls using a purely M2-promoting scaffold, including the essence of the type-2 phenotype: collagen deposition. In tissue engineering, our goal is to create a functional replacement tissue, which is hindered by the deposition of large collagenous scarring. Pathogenic arginase expression, which leads to collagen deposition, is present in fibrotic diseases such as pulmonary fibrosis [42]. As previously mentioned, type-2 cytokines are present in the FBR to synthetic materials, which yields fibrotic encapsulation of engineered devices. In order for a scaffold to create an optimal immune microenvironment, we would need it to promote a balance between the M1 and M2 phenotypes. This balance would prevent any damaging inflammation caused by a strong M1 reaction, while avoiding a large M2-derived scarring response rendering the tissue nonfunctional. Through the use of decellularized ECM scaffolds, researchers have demonstrated the ability of varying tissues to exert polarization influence on the M1/M2 axis of macrophage activation. These polarizations are dependent upon both the ECM tissue source, as well as whether this source is xenogeneic or allogeneic in nature.

3.5   Biochemical and Physical Alteration of Scaffolds

Through the integration of the physical and chemical properties of scaffolds, it is possible to modify biomaterials to better promote the desired cellular outcome [43, 44]. Physical modification of scaffolds includes the use of particle and fiber-based polymers, as well as particle and sheet-based ECM scaffolds. These materials can be further modified biochemically with proteins and small molecules that promote cell polarization to a desired lineage or allow for a more conducive environment for progenitor cells to grow and differentiate.

3.5.1   Physical Alteration of ECM and Synthetics

Decellularized ECM has been used in sheet and particulate forms, depending upon the desired application. ECM sheets are typically created out of a thin section of tissue that has been decellularized, maintaining its overall architecture. This sheet can be fenestrated, with openings introduced to its structure, promoting vascularization and cell migration. Particulate ECM is typically tissue that has been decellularized and lyophilized prior to milling, either through a liquid nitrogen "cryo" mill or through a knife mill. The latter produces larger particle sizes and does not require the ECM to be lyophilized. Cryo-milling involves cooling lyophilized ECM with liquid nitrogen to make it very brittle, then milling it with a magnetic rod. This yields a material with a size range in the micrometer scale. It is possible to further decrease particle size by hydraulic shearing to create nanometer-sized ECM.

Recently, researchers have shown that different physical properties of synthetic scaffolds can alter the behavior of cells that come in contact with that material. Density and torsion of polycaprolactone (PCL)-based scaffolds have been changed to modify the polarization of immune cells such as macrophages [45]. The authors claim that differing cyclic strain on human peripheral blood mononuclear cells (PBMCs) induces the presence of M2-polarization, promoting the deposition of new ECM. It is important to note that in all cases, strained and unstrained, CCR7 + M1 macrophages were present in addition to the presence or absence of CD206 + M2 macrophages. The majority of cells were M1 phenotype at day 1, however, under intermediate (7 %) torsional strain the population of M2 cells increased over time.

3.5.2 Cyclodextrin and Arginine–Glycine–Aspartic Acid Modification of Polymeric Scaffolds

Through the use of polyethylene glycol diacrylate (PEG-DA)- and polyethylene glycol monoacrylate (PEG-MA)-modified Matrigel hydrogels, Beck et al. demonstrated a varying cellular response to modified physical and adhesive properties of these hydrogels [46]. PEG-DA is able to crosslink and will increase hydrogel stiffness, while PEG-MA is not able to crosslink and therefore limits hydrogel stiffness. With increasing concentrations of PEG-DA, Beck et al. observed altered growth and differentiation of mammary epithelial organoids in vitro. With greater scaffold stiffness, they observed less differentiation and formation of primitive branching. They also tested the effect of epithelial cell dissemination from PEG-Matrigel constructs by varying the concentration of adhesion motifs. These motifs were added through the use of a cyclodextrin ring molecule with covalently attached arginine–glycine–aspartic acid (RGD), which is a peptide sequence that is recognized adhesion proteins known as integrins. With the addition of these extracellular adhesion signals, the authors showed that cell dissemination was inhibited by stiffness of the hydrogel, and induced by chemically added adhesion signals.

3.5.3 GF Modification and Modulation of Microenvironment

Through manipulation of GFs, researchers have been able to administer exogenous signaling proteins to an area of tissue damage to promote proper repair [47]. In this example, an engineered fibronectin (FN) fragment was modified with factor XIIIa. The segment of factor XIIIa that was introduced to the recombinant FN fragment polymerizes fibrin and is cross-linked to the scaffold. The modified FN fragment displays both integrin and GF binding domains. When applied exogenously, therapeutic GFs will associate with the GF domain of the FN fragment, which has been linked to the host ECM through factor XIIIa cross-linking as described previously. This localized the GF near integrin-binding domains of the recombinant protein, bringing cells into close contact with the GF and allowing signaling to induce proliferation and differentiation. GFs tested with the binding peptide included VEGF-A, platelet-derived growth factor BB (PDGF-BB) and bone morphogenetic protein-2 (BMP-2). Combinations of these GFs were combined to promote skin and bone repair.

Application of these peptide-manipulations has also been tested in osteoarthritis (OA) treatment. Modification of a hyaluronic acid (HA)-binding peptide allowed researchers to extend the lifetime of HA injected into the synovium of the knee joint in an OA model. Using a biochemical approach, in several platforms, it is possible to localize signaling and structural proteins to a location of interest, which could be exploited for use in immunomodulation by synthetically creating an optimal immune microenvironment [48].

3.6   Summary

The tissue microenvironment plays a large role in the host response to biomaterials. This phenomenon has been observed with other immunological challenges such as pathogens that elicit different responses dependent upon the location in which the host encounters that pathogen. Current research has shown a variance in the extent of the inflammatory response to a scaffold as well as its degradation kinetics when applied in different tissue locations. These altered immunologic reactions also occur with biologically derived scaffolds, such as ECM. SIS, UBM, and AAT are all examples of ECM-derived scaffold that have been used successfully in the clinic and are under evaluation for their precise role in wound healing and regeneration of new tissue. ECM can act both as a regenerative scaffold and a bioengineering tool to mimic the tissue environment, promoting certain immune phenotypes that will provide a conducive setting for regeneration. These tendencies for certain immune polarizations will vary depending upon the tissue source of the ECM, much like how the immune response to a scaffold changes depending upon the location within a cellular tissue. Through physical and chemical manipulation of the biomaterial with reference to its location of implantation, researchers will be able to create an optimal setting for functional healing and generate a more effective replacement tissue.

References

1. Bianchi ME. DAMPs, PAMPs and alarmins: all we need to know about danger. J Leuk Bio. 2007;81:1–5.
2. Matzinger P, Kamala T. Tissue-based class control: the other side of tolerance. Nat Rev Immunol. 2011;11:221–30.
3. Murray PJ, Allen JE, Biswas SK, Fisher EA, Gilroy DW, Goerdt S, Gordon S, Hamilton JA, Ivashkiv LB, Lawrence T, Locati M, Mantovani A, Martinez FO, Mege JL, Mosser DM, Natoli G, Saeij JP, Schultze JL, Shirey KA, Sica A, Suttles J, Udalova I, van Ginderachter JA, Vogel SN, Wynn TA. Macrophage activation and polarization: nomenclature and experimental guidelines. Immunity. 2014;41:14–20.
4. Weiner HL. Oral tolerance: immune mechanisms and the generation of Th3-type TGF-beta-secreting regulatory cells. Microbes Infect. 2001;3:947–54.
5. Clark AT, Islam S, King Y, Deighton J, Anagnostou K, Ewan PW. Successful oral tolerance induction in severe peanut allergy. Allergy. 2009;64:1218–20.

6. Berlin C, Bargatze RF, Campbell JJ, von Andrian UH, Szabo MC, Hasslen SR, Nelson RD, Ber EL, Erlandsen SL, Butcher EC. Alpha 4 integrins mediate lymphocyte attachment and rolling under physiologic flow. Cell. 1995;80:413–22.
7. Rosa Bono M, Elgueta R, Sauma D, Pino K, Osorio F, Michea P, Fierro A, Rosemblatt M. The essential role of chemokines in the selective regulation of lymphocyte homing. Cytokine Growth Factor Rev. 2007;18:33–43.
8. Ono SJ, Nakamura T, Miyazaki D, Ohbayashi M, Dawson M, Toda M. Chemokines: roles in leukocyte development, trafficking, and effector function. J Allergy Clin Immunol. 2003;111:1185–99.
9. Niederkorn JY. See no evil, hear no evil, do no evil: the lessons of immune privilege. Nat Immunol. 2006;7:354–9.
10. Lemaitre B, Nicolas E, Michaut L, Reichhart JM, Hoffman JA. The dorsoventral regulatory gene cassette spätzle/Toll/cactus controls the potent antifungal response in Drosophila adults. Cell. 1996;86:973–83.
11. Medzhitov R, Preston-Hurlbert P, Janeway CA Jr. A human homologue of the Drosophila Toll protein signals activation of adaptive immunity. Nature. 1997;388:394–97.
12. Ma J, Chen T, Mandelin J, Ceponis A, Miller NE, Hukkanen M, Ma GF, Konttinen YT. Regulation of macrophage activation. Cell Mol Life Sci. 2003;60:2334–46.
13. Roberts AB, Flanders KC, Heine UI, Jakowlew S, Kondaiah P, Kim SJ, Sporn, MB. Transforming growth factor-β multifunctional regulator of differentiation and development. Philos Trans R Soc Lond B, Biol Sci. 1990;327:145–54.
14. Horsley V, Jansen KM, Mills ST, Pavlath GK. IL-4 acts as a myoblast recruitment factor during mammalian muscle growth. Cell. 2003;113:483–94.
15. Godwin JW, Pinto AR, Rosenthal NA. Macrophages are required for adult salamander limb regeneration. Proc Natl Acad Sci. 2013;110:9415–20.
16. Hubbell JA, Thomas SN, Swartz MA. Materials engineering for immunomodulation. Nature. 2009;462:449–60.
17. Leleux J, Roy K. Micro and nanoparticle-based delivery systems for vaccine immunotherapy: an immunological and materials perspective. Adv Healthc Mater. 2013;2:72–94.
18. Badylak SF, Valentin JE, Ravindra AK, McCabe GP, Stewart-Akers AM. Macrophage phenotype as a determinant of biologic scaffold remodeling. Tissue Eng Part A. 2008;14:1835–42.
19. Brown BN, Londono R, Tottey S, Zhang L, Kukla KA, Wolf MT, Daly KA, Reing JE, Badylak SF. Macrophage phenotype as a predictor of constructive remodeling following the implantation of biologically derived surgical mesh materials. Acta Biomater. 2012;8:978–87.
20. Brown BN, Ratner BD, Goodman SB, Amar S, Badylak SF. Macrophage polarization: an opportunity for improved outcomes in biomaterials and regenerative medicine. Biomaterials. 2012;33:3792–802.
21. Wynn T, Chawla A, Pollard JW. Macrophage biology in development, homeostasis and disease. Nature. 2013;496:445–55.
22. Stefater III JA, Ren S, Lang RA, Duffield JS. Metchnikoff's policemen: macrophages in development, homeostasis and regeneration. Cell Press: Trends Mol Med. 2011;17:743–52.
23. Heredia JE, Mukundan L, Chen FM, Mueller AA, Deo RC, Locksley RM, Rando TA, Chawla A. Type 2 innate signals stimulate fibro/adipogenic progenitors to facilitate muscle regeneration. Cell. 2013;153:376–88.
24. Goh SYP, Henderson NC, Heredia JE, Red Eagle A, Odegaard JI, Lehwald N, Sheppard D, Mukundan L, Locksley RM, Chawla A. Eosinophils secrete IL-4 to facilitate liver regeneration. Proc Natl Acad Sci. 2013;110:9914–9.
25. Kou PM, Babensee JE. Validation of a high throughput methodology to assess the biomaterials on dendritic cell phenotype. Acta Biomater. 2010;6:2621–30.
26. Artzi N, Olivia N, Puron C, Shitreet S, Artzi S, bon Ramos A, Groothuis A, Sahagian G, Edelman ER. In vivo and in vitro tracking of erosion in biodegradable materials using noninvasive fluorescence imaging. Nat Mater. 2011;10:704–9.
27. Hillel AT, Unterman S, Nahas Z, Reid B, Coburn JM, Axelman J, Chae JJ, Guo Q, Trow R, Thomas A, Hou Z, Lichtsteiner S, Sutton D, Matheson C, Walker P, David N, Mori S, Taube

JM, Elisseeff JH. Photoactivated composite biomaterial for soft tissue restoration in rodents and in humans. Sci Transl Med. 2011;3:93ra67

28. Reid B, Gibson M, Singh A, Taube J, Furlong C, Murcia M, Elisseeff JH. PEG hydrogel degradation and the role of the surrounding tissue environment. J Tissue Eng Regen Med. 2013. http://dx.doi.org/10.1002/term.1688.

29. Anderson JM, Rodriguez A, Chang DT. Foreign body reaction to biomaterials. Semin Immunol. 2008;20:86–100.

30. Wolf MT, Carruthers CA, Dearth CL, Crapo PM, Huber A, Burnsed OA, Londono R, Johnson SA, Daly KA, Stahl EC, Freund JM, Medberry CJ, Carey LE, Nieponice A, Amoroso NJ, Badylak SF. Polypropylene surgical mesh coated with extracellular matrix mitigates the host foreign body response. J Biomed Mater Res Part A. 2014;102:234–46.

31. Faulk DM, Londono R, Wolf MT, Ranallo CA, Carruthers CA, Wildemann JD, Dearth CL, Badylak SF. ECM hydrogel coating mitigates the chronic inflammatory response to polypropylene mesh. Biomaterials. 2014;35:8585–95.

32. Gilbert, TW, Sellaro TL, Badylak SF. Decellularization of tissues and organs. Biomaterials. 2006;27:3675–83.

33. Badylak, SF, Freytes DO, Gilbert, TW. Extracellular matrix as a biological scaffold material: structure and function. Acta Biomater. 2009;5:1–13.

34. Zheng MH, Chen J, Kirilak Y, Willers C, Xu J, Wood D. Porcine small intestine submucosa (SIS) is not an acellular collagenous matrix and contains porcine DNA: possible implications in human implantation. J Biomed Mater Res. 2005;73B:61–7.

35. Wolf MT, Daly KA, Brennan-Pierce EP, Johnson SA, Carruthers CA, D'Amore A, Nagarkar SP, Velankar SS, Badylak SF. A hydrogel derived from decellularized dermal extracellular matrix. Biomaterials. 2012;33:7028–38.

36. Sicari BM, Rubin JP, Dearth CL, Wolf MT, Ambrosio F, Boninger M, Turner NJ, Weber DJ, Simpson TW, Wyse A, Brown EH, Dziki JL, Fisher LE, Brown S, Badylak SF. An acellular biologic scaffold promotes skeletal muscle formation in mice and humans with volumetric muscle loss. Sci Transl Med. 2014;6:234ra58.

37. Zhang L, Zhang F, Weng Z, Brown BN, Yan H, Ma X, Vosler PS, Badylak SF, Dixon CE, Cui XT, Chen J. Effect of an inductive hydrogel composed of urinary bladder matrix upon functional recovery following traumatic brain injury. Tissue Eng Part A. 2013;19:1909–18.

38. Stankus JJ, Freytes DO, Badylak SF, Wagner WR. Hybrid nanofibrous scaffolds from electrospinning of a synthetic biodegradable elastomer and urinary bladder matrix. J Biomater Sci, Polymer Edition. 2008;19:635 52.

39. Wu I, Nahas Z, Kimmerling KA, Rosson GD, Elisseeff JH. An injectable adipose matrix for soft tissue reconstruction. Plast Reconstr Surg. 2012;129:1247–57.

40. Dunne LW, Huang Z, Meng W, Fan X, Zhange Q, An Z. Human decellularized adipose tissue scaffold as a model for breast cancer cell growth and drug treatments. Biomaterials. 2014;35:4940–9.

41. Matzinger, P. Friendly and dangerous signals: is the tissue in control? Nat Immunol. 2007;8:11–13.

42. Pechkovsky DV, Prasse A, Kollert F, Engel KMY, Dentler J, Luttmann W, Friedrich K, Müller-Quernheim J, Zissel G. Clin Immunol. 2010;137:89–101.

43. Bryers JD, Giachelli CM, Ratner BD. Engineering biomaterials to integrate and heal: the biocompatibility paradigm shifts. Biotechnol Bioeng. 2012;109:1898–911.

44. Zakrzewski JL, van den Brink MRM, Hubbell JA. Overcoming immunological barriers in regenerative medicine. Nat Biotechnol. 2014;32:786–94.

45. Ballotta V, Driessen-Mol A, Bouten CV, Baaijens FP. Strain-dependent modulation of macrophage polarization within scaffolds. Biomaterials. 2014;35:4919–28.

46. Beck JN, Singh A, Rothenberg AR, Elisseeff JH, Ewald AJ. The independent roles of mechanical, structural and adhesion characteristics of 3D hydrogels on the regulation of cancer invasion and dissemination. Biomaterials. 2013;34:9486–95.

47. Martino MM, Torelli F, Mochizuki M, Traub S, Ben-David D, Kuhn G, Müller R, Livne E, Eming SA, Hubbell JA. Engineering the growth factor microenvironment with fibronectin domains to promote wound and bone tissue healing. Sci Transl Med. 2011;3:100ra89.
48. Singh A, Corvelli M, Unterman SA, Wepasnick KA, McDonnell P, Elisseeff JH. Enhanced lubrication on tissue and biomaterial surfaces through peptide-mediated binding of hyaluronic acid. Nat Mater. 2014;13:988–95.

Jennifer H. Elisseeff, Ph.D   is the founder and director, Translational Tissue Engineering Center, Wilmer Eye Institute, and the Department of Biomedical Engineering, Johns Hopkins.

Chapter 4
Advances in Molecular Design of Polymer Surfaces with Antimicrobial, Anticoagulant, and Antifouling Properties

Marek W. Urban

4.1 Introduction

Polymers that exhibit tunable physical and chemical properties offer a number of unique applications when utilized in biological systems. Notable applications range from medical implants [1–3], microarrays [4], biosensors, and bio-actuators [5] to tissue engineering [6–8], or gene and drug delivery systems [7, 9–11], to name just a few. Inertness, mechanical strength, or biocompatibility are those attributes that allow many polymeric materials to be successfully integrated into biological systems, but not without adverse effects. Polymers that are often used in biomedical applications include poly(tetrafluoroethylene) (PTFE), poly(ethylene terephthalate) (PET), ultra-high molecular weight poly(ethylene) (UHMWPE), poly(polypropylene) (PP), polyether ether ketone (PEEK), poly(methylmethacrylate) (PMMA), polypyrrole (PPy), polythiophene (PT), poly(lactide-co-glycotide) (PLGA), poly(N-isopropylacrylamide) (PNIPAAm), poly-L-lactic acid (PLLA), poly(N,N'-(dimethylamino)ethyl methacrylate) (pDMAEMA), poly(ethylene glycol) (PEG), poly(ε-caprolactone) (PCL), poly(acrylic acid) (PAA), poly(allylamine) (PAH), and polyaniline (PANI). There are others, and each polymer offers specific and unique attributes that are pertinent to their functions.

Keeping in mind that a thermoplastic polymer will serve specific medical functions, desirable mechanical strength, low toxicity, and inertness are essential attributes in biological environments. For that reason, nontoxic, chemically and thermally resistant polymers, such as PTFE and PET, are utilized in implant vascular grafts [1, 4]. On the other hand, UHMWPE is a self-lubricating polymer with low moisture adsorption, which found applications in hip and knee replacements as well

M. W. Urban (✉)
Department of Materials Science and Engineering, Center for Optical Materials Science and Engineering (COMSET), Clemson University, Clemson, SC 29634, USA
e-mail: mareku@clemson.edu

© Springer International Publishing Switzerland 2015
L. Santambrogio (ed.), *Biomaterials in Regenerative Medicine and the Immune System*,
DOI 10.1007/978-3-319-18045-8_4

as artificial joint implants [1, 3]. The primary attributes are high impact strength, low friction coefficient, and resistance to abrasion. Due to its good physical strength and excellent impact resistance, PP [4] is often the polymer of choice for heart valves [1], hemodialysis membranes, and mesh for hernia repairs. Mechanical strength and biocompatibility make PEEK useful in long-term medical implant devices [1]. In contrast, PMMA is desired for its transparency and UV resistance and has been used in intraocular implants, bone cement, and dentures [1, 4]. PMMA also found applications in oligonucleotide microarrays [4], which facilitate the immobilization of DNA. In contrast, biosensors require PPy and PT that are integrated with electrical transducers capable of sensing biomolecules such as DNA or glucose [5]. On the other end of the spectrum of bio-applicable polymers are biodegradable polymers represented by PLGA, PLLA, and PCL [2, 6, 7, 9]. These materials are often used in sutures, stents, and scaffolding materials for tissue engineering, where biodegradation kinetics, good mechanical properties, low immunogenicity, and nontoxicity are essential properties [12, 13]. On the other hand, stimuli-responsive polymers, such as PNIPAAm, PAA, and PAH, play a crucial role in drug delivery systems [7, 10] for their ability to undergo shrinkage/expansion at specific pHs or temperatures, thus allowing controllable releases of their drug cargo. Other stimuli-responsive polymers such as cationic pDMAEMA are reportedly an effective transfection agent utilized in gene delivery [9], whereas PEG is a widely exploited polymer for enhancing biocompatibility attributed to its antifouling properties and the ability to greatly increase circulation times within the body [7]. The growing field of artificial muscle actuators [5] requires the use of conducting polymers, such as PPy and PANI, where mechanical forces are produced by the application of electrical current.

It is thus apparent that polymeric materials are highly attractive, but when in contact with biological organisms they may and will become susceptible to microbial growth and/or undesirable protein adsorption. Although the primary driving force for attracting other species is the low surface energy of polymer surfaces [14], surface topography and morphological features are also significant. We focus on the control of morphology by chemical modifications, which ultimately impacts interactions with biological systems. Thus, in order to circumvent interactions with bio-organisms while maintaining useful bulk characteristics, polymer surfaces can be physically or chemically altered to exhibit antimicrobial, anticoagulant, as well as antifouling properties. Taking this one step further, and creating stimuli-responsive [15, 16] and self-healing [17] surfaces is particularly appealing.

Modification of polymeric surfaces to create biocompatible or bioactive materials has been explored utilizing an array of approaches. Numerous studies involving attachments of biologically active species which exhibit antimicrobial [14, 18, 19] and anticoagulant [14] properties have been conducted in various biomedical application areas [20–23]. Reactions leading to polymer surface modifications may range from the simple addition of bioactive molecules to a polymer matrix during processing [21, 22] or non-covalent physisorption using layer-by-layer deposition [11, 24, 25] and self-assembled monolayers (SAMs) [23, 26, 27] as well as covalent bonding using grafting-to [21, 28] and grafting-from [21, 22, 29] approaches. The latter may offer greater control over surface chemistry and morphologies [14] as well as the use of chemical spacers to impart mobility and accessibility to the bioactive species of interest [14]. In view of the recent advances, this chapter focuses on stimuli-responsive polymer surfaces of common thermoplastics utilized in biomedical applications and is organized into the following sections: selected polymer surface modifications, bioactivity of modified polymer surfaces, stimuli-responsive polymer surfaces, and surface attachments of bioactive species.

4.2   Selected Polymer Surface Modifications

Since the main characteristics of biocompatible polymeric systems within a biological environment are surface properties, approaches to immobilize bioactive molecules on polymeric surfaces by either non-covalent or covalent bonding have been of significant interest. Non-covalent bonding methods include physisorption [1, 11], electrostatic interactions [30], and ligand–receptor recognition [27], while covalent bonding [1] is achieved by chemical reactions of functional groups to form stable surface entities. Figure 4.1 depicts physical adsorption and covalent attachment methods to immobilize molecules onto surfaces which may or may not be beneficial in long-range surface stability.

Physisorption of bioactive molecules onto polymer surfaces is the simplest method of temporarily altering biopolymer surface properties. For example, gentamicin [21] buffer solutions are being used to soak PP mesh implants before surgery to prevent infections. Similarly, by combining mammalian cells with polymers such as PLA, it is possible to create new skin for burn patients [7].

Another short-term stability surfaces are achieved using layer-by-layer (LBL) approach which takes advantage of electrostatic interactions by dipping of a charged substrate into dilute aqueous solutions of polyelectrolyte with opposite charges, thus allowing adsorption of alternating charges on the substrate [11, 31, 32]. Multilayer films can be produced by sequential adsorptions of anionic and cationic polyelectrolytes, which result in heterogeneous surfaces with limited stability [32]. An example of LBL surface modification is multilayer thrombresistant thin films con-

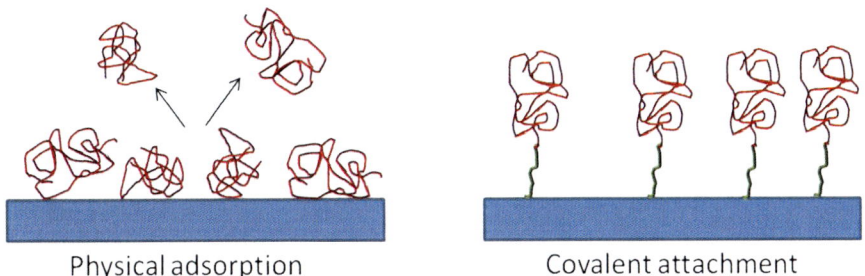

Physical adsorption   Covalent attachment

Fig. 4.1 Schematic illustration of physical adsorption (non-covalent) and covalent attachment

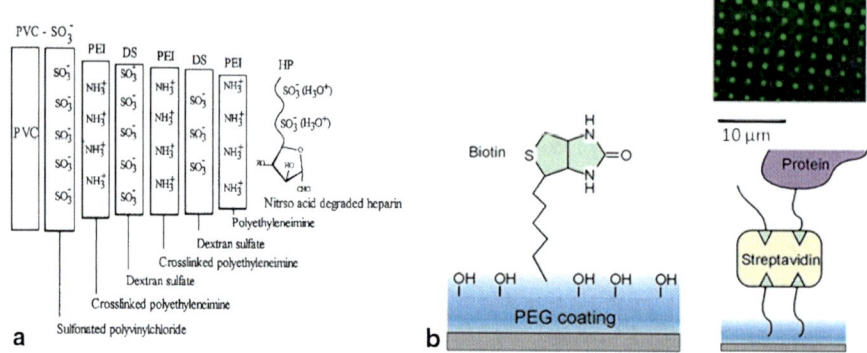

Fig. 4.2 **a** Schematic diagram of LBL structure on PVC surface [25]. **b** Schematic diagram of biotinylated green fluorescent proteins deposited on SAMs through biotin–avidin recognition and fluorescent nanopatterns generated [27]

taining poly(ethylenimine) (PEI), dextran sulfate (DS), and heparin (HP), deposited on poly(vinyl chloride) (PVC) surfaces which is depicted in Fig. 4.2a [25].

In addition to electrostatic interactions, one of the strongest non-covalent bonding with the disassociation constant K_d of 10^{-15} M [33] is the ligand–receptor recognition through biotin–streptavidin interactions. Streptavidin exhibits a specific affinity for biotin molecules with four receptor sites leading to specific orientation of immobilized species onto surfaces [33]. Figure 4.2b illustrates the deposition process of biotinylated green fluorescent proteins on SAMs through streptavidin-biotinylated ligands to generate surface nanopatterns [27]. Although non-covalent bonding can effectively bind biomolecules to polymer surfaces, covalent attachments can be utilized to achieve significantly more stable surfaces.

To achieve covalently modified polymer surfaces, the first step is to generate a surface reactive species from which bio-relevant molecules can be anchored. One approach is to react maleic anhydride to form acid groups, which can be followed by further reactions of a spacer [19] to provide mobility of the species of interest attached to the other end of the spacer. Well-studied chemical reactions for surface attachment via linkages between functional groups include esterification [19], amidation [34], and etherification [35].

Photografting via UV radiation has been used to affix bioactive molecules to polymer surfaces using photoinitiators and photosensitizers [36]. This approach has been useful for the fabrication of DNA oligonucleotide arrays on PMMA utilizing UV treatment to generate carboxylic acid functionality [4, 37]. In addition to photografting, surface-initiated polymerization, such as controlled living polymerizations including reversible addition–fragmentation chain transfer (RAFT) and atom transfer radical polymerization (ATRP), has gained popularity due to their ability to provide controllable chain lengths as well as higher grafting density. Examples of these processes include RAFT surface grafting polymerization of pDMAEMA on cellulose fiber surfaces to produce antimicrobial quaternary ammonium groups

[28]. RAFT polymerization was also utilized to synthesize cell mimicking polymer brushes on the surface of polysulfone [38]. Additionally, surface-initiated polymerization of ATRP-grafted poly(poly(ethylene glycol) methyl ether monomethacrylate) *(PPEGMA)* to poly(dimethyl siloxane) (PDMS) was utilized to impart non-fouling properties [39].

Solvent-based chemical functionalization can be utilized to modify polymer surfaces with the use of liquid reagents that treat polymer surfaces to create reactive groups via aminolysis, alkaline and acidic hydrolysis, as well as hydrogen peroxide reactions [4, 36]. Selected examples of chemical modifications include submerging polyethylene (PE) into an aqueous solution of chromium trioxide and sulfuric acid to generate COOH groups [40] or the introduction of oxygen-containing moieties on PE and PP by reacting with a mixture of chromic acid, potassium permanganate, and sulfuric acid [4]. These methods offer an easy approach to polymer surface modification; however, aside from nonspecific and rather uncontrollable functionalization, they produce toxic chemical waste and often cause etching of the surface resulting in significant morphological alterations [36].

Other solvent-based functionalization methods include click chemistry reactions [41] which are relatively fast, high-yielding reactions achieved under mild conditions [42]. Examples of click reactions include Diels–Alder reactions, nucleophilic substitution with exopy and aziridine compounds, thiol-yne reactions, and Cu(I)-catalyzed Huisgen 1,3-dipolar cycloaddition (azide-alkyne) [43]. These reactions have been utilized in the functionalization of polymeric micelles and vesicles [44]; however, their use on polymeric substrates has not been explored. For these reactions to be useful in polymer surface reactions, the surfaces must contain moieties necessary for click reactions to occur.

Utilizing high-energy radiation is an effective method for modifying the surfaces of biopolymers [36]. They include electron beam radiation [45] and radio or microwave plasma reactions [46, 47]. In particular, microwave plasma offers a clean, fast, and solventless route to surface functionalization without affecting bulk properties. Plasma reactions require vacuum conditions in gaseous environments, typically N_2, O_2, CO_2, He, or Ar. Here, excited ionized gas is utilized to create reactive groups in the presence of a desirable monomer, thus providing controllable chemical conditions for covalently attaching desirable chemical entities which generally occur within 0.5–10 s [46, 48]. Glow discharge plasma is created in which the gases are ionized to generate highly reactive species within the reactor [36]. These reactive species interact with the polymer surface in the reaction chamber to create both chemical and physical alterations of the surface. Plasma-induced changes at the substrate surface can reach depths of several hundred angstroms [36], but it may be in the micron range if high energy input per area and extended exposure times are applied. The first step is the attachment of an anchor molecule through the reaction of maleic anhydride to form acid groups, followed by reactions of a spacer to provide mobility to the active antibiotic molecule attached to the other end of the spacer. Previous studies have employed plasma reactions for the modification of biopolymer surfaces including patterning of PDMS followed by ammonolysis with amoxicillin to create antimicrobial surfaces [18, 49–51]. Patterning can be achieved

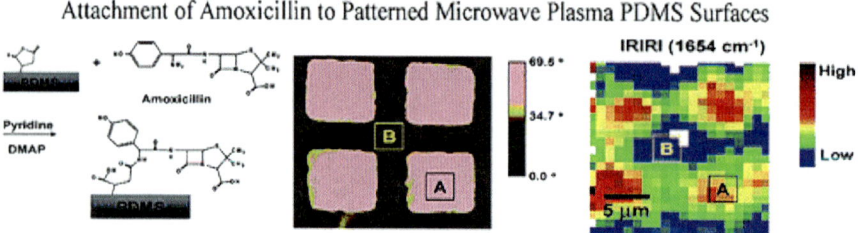

Fig. 4.3 PDMS surface patterning by microwave plasma reactions using a masking technique to create distinct areas of functional surface groups to which the antibiotic amoxicillin is attached. Surfaces patterns were analyzed by AFM and IRIRI spectroscopy [18]

by masking PDMS substrates during microwave plasma reaction as illustrated in Fig. 4.3 [18] and it clearly shows that masked areas of the substrate are not chemically altered. Prevention of undesirable protein adsorption, blood coagulation, or bacterial biofilm formation is critical in many applications, and incorporation of these attributes is necessary for polymer substrates in contact with biological environments.

4.3   Bioactivity of Modified Polymer Surfaces

Owing to low surface energies, polymers, when in contact with biological systems, become susceptible to microbial growth and/or undesirable protein adsorption [14]. Microbial growth, thrombosis, and protein absorption are biological events that must be minimized or eliminated in order for a polymer surface to be functional in biological environments. Several specific strategies have been taken to prevent these problems, and modified polymer surfaces play an essential role in enhancing antimicrobial, anticoagulant, and antifouling properties. The significance of surface modifications is manifested by the fact that each year, the number of deaths resulting from infection continues to increase, with approximately 64 % of infections acquired at hospitals being attributed to the attachment of viable bacteria to medical devices and implants [31]. Many of these infections could be avoided by taking the appropriate preventative measures, such as the introduction of antimicrobial agents onto the surfaces of medical devices and implants which would minimize or prevent bacterial attachment and biofilm formation. The summary of chemicals that exhibit anticoagulant, antifouling, and antimicrobial properties is shown in Table 4.1. It should be noted that the majority of these chemicals are utilized in a liquid phase. The challenge is to anchor these materials on polymeric surfaces in such a fashion that they will retain their anticoagulant, antifouling, and antimicrobial properties.

Table 4.1 Examples of chemicals for preventing coagulant, fouling, and microbial processes [14]

Anticoagulant		Antifouling	
Name	Chemical Structure	Name	Chemical Structure
Warfarin		Poly(oligo (ethylene glycol) methyl ether methacrylate)[78]	
Heparin			
		Zwitterionic polymer	
Hirudin		• Phosphoryl-choline (PC)	
		• Sulfobetaine (SB)	
Argatroban		•Carboxybetaine (CB)	

Antifouling		Antimicrobial	
Name	Chemical Structure	Name	Chemical Structure
Chlorothalonil		Metals: Silver ion, Tin, Mercury	Ag⁺, Sn, Hg
Diuron		Quaternary ammonium	
Dichlofluanid		Beta-lactams • Penicillin	
		• Ampicillin	
Phosphoryl-choline-PDMS[83]		• Amoxicillin	
		Aminoglycosides • Gentamicin	
PEG-fluoropolymer[74]		• Streptomycin	
		Chitosan	

4.3.1 Antimicrobial Surfaces

Several methods have been employed in an attempt to introduce antimicrobial agents into biomaterials in order to prevent biofilm formation. A simple approach is adding antimicrobial agents into the polymer during processing [52]. Their release is controlled by diffusion from a polymer matrix [21]. Many antimicrobial agents have been used for this purpose. They include quaternary ammonium (or phosphonium) salts (QAS) [30, 52], chitosan [53], antimicrobial peptides [52, 54], silver ions (Ag^+) [55], bacteriophages [56], and antibiotics [14]. As expected, they exhibit different and for the most part unknown mechanisms of inhibition of bacterial growth. While metals cause the displacement of essential Ca^{2+} and Zn^+ ions in bacteria [57], QAS [30] and chitosan [53] bind to negatively charged bacterial surfaces causing the leakage of intracellular components. The attachment of poly(quaternary ammonium) to PP surfaces using photochemical synthesis and ATRP was successful against *Staphylococcus aureus* and *Escherichia coli*, and molecular weight and the density of the QA can be controlled [58]. An example of antimicrobial QAS on surfaces is illustrated in Fig. 4.4 and involves an LBL assembly of alternating anionic PAA and cationic cetyltrimethylammonium bromide (CTAB) layers that demonstrate antimicrobial activity as the CTAB moiety diffuses to the surface of the assembly [30]. In contrast to the above approaches, Ag^+ ions bind to electron donor groups containing sulfur, oxygen, or nitrogen, which are present in biological molecules as thio, amino, imidazole, carboxylate, and phosphate groups [57]. Presumably, by displacing other ions, such as Ca^{2+} and Zn^+, Ag^+ effectively interrupts a number of cellular transport and oxidation processes [57]. For that reason, Ag^+ ions have been incorporated into PET films by LBL surface modification with

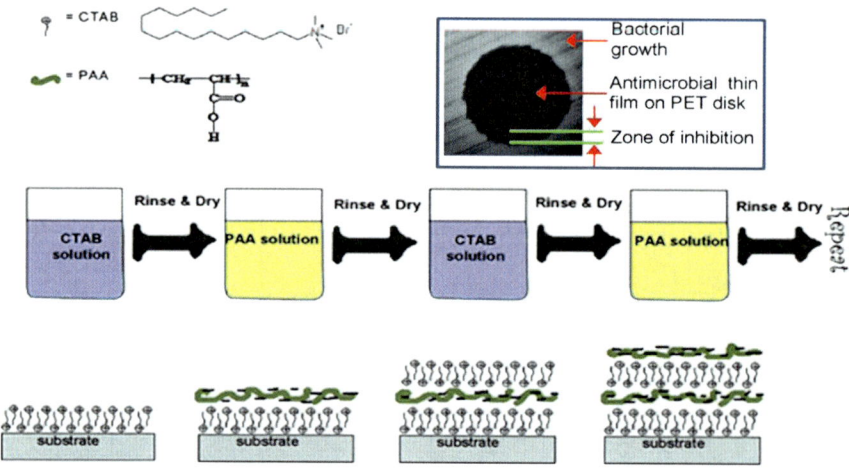

Fig. 4.4 Schematic diagram of LBL of PET film consisting of alternating layers of PAA and antimicrobial CTAB. Antimicrobial activity is shown by the zone of inhibition of bacterial growth [30]

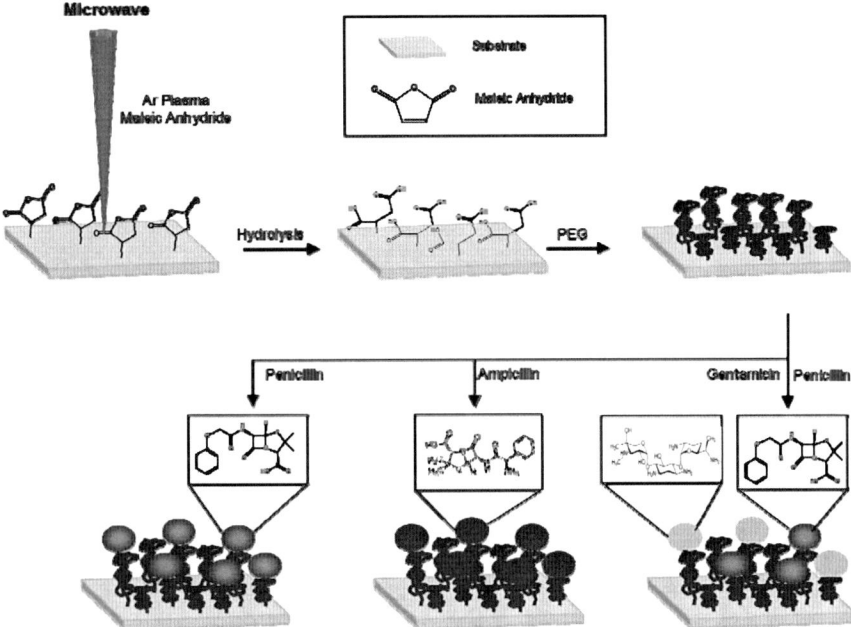

Fig. 4.5 Schematic diagram of the covalent attachment of antibiotics to polymer surfaces via microwave plasma reactions [14]

PEI and PAA which exhibits biocidal activity against *S. aureus* and *E. coli* [55], but there are a number of poorly understood potential side effects that may prohibit the use of silver.

On the contrary, antibiotics have different modes of action for antimicrobial activity, with the common classes being aminoglycoside and beta-lactam-based antibiotics. Aminoglycosides, such as gentamicin, kanamycin, and streptomycin, prevent bacterial protein synthesis by entering the bacteria and binding to ribosomes within the cell, whereas beta-lactams, such as penicillin (PEN), amoxicillin, and ampicillin (AMP), inhibit bacterial cell wall formation. PEN [19, 59] and AMP [60] were successfully attached to the ePTFE surfaces via prior microwave plasma reactions leading to grafted carboxylic acid groups. These studies showed high effectiveness against both Gram-positive and Gram-negative bacteria. Microwave plasma reactions and subsequent attachment of antibiotics are illustrated in Fig. 4.5. The first step is the attachment of an anchor molecule through the reaction of maleic anhydride to form acid groups. This solventless and sterilized process takes less than 10 s. The second step involves reactions of a spacer to provide mobility to the active antibiotic molecule attached to the other end of the spacer. The covalent attachment of poly(vinyl-N-hexylpyridinium) onto HDPE and PET surfaces also indicated bactericidal activity against *S. aureus* and *E. coli* [61].

4.3.2 Anticoagulant Surfaces

Due to low polymer surface energies, most polymeric materials exhibit highly thrombogenic surfaces, and consequently may adsorb fibrin, thrombus, or other proteins resulting in clot formation [22]. Thrombosis, or blood clotting within minor wounds, is necessary for hemostasis to allow healing. After the clot has served its purpose, it is dissolved through a process called fibrinolysis [62, 63]. Naturally occurring hirudin and synthetically attained bivalirudin peptides prevent thrombosis [22]. Covalent immobilization of these peptides onto polymer surfaces such as PET [64] and PLGA [65] has been widely studied in order to inhibit thrombus formation. Another anticoagulant species, heparin (HEP), is a linear polysaccharide consisting of uronic acid-(1,4)-D-glucosamine repeating disaccharide subunits [66], which acts by binding to a thrombin inhibitor known as antithrombin III (AT-III) [66, 67]. As illustrated in Fig. 4.6, HEP molecule dissociates from the complex and can be reutilized [67]. Several studies have sought to develop hemocompatible devices such as dialysis membranes, catheters, coronary stents, and vascular grafts by immobilizing HEP onto polymer surfaces including PET [68], PTFE [68, 69], and PU [70] through LBL surface modification to prevent blood coagulation.

   Acetyl salicylic acid (aspirin) and dipyridamole are also important anti-thrombosis agents. Their anticoagulant activities involve inactivation of clotting enzymes which facilitate thrombus formation. Aspirin has been incorporated into polymer matrix such as PLGA [71] and poly(vinyl alcohol) (PVA) [72] to enhance blood compatibility. Similarly, dipyridamole has been covalently attached onto polyurethane (PUR) surfaces via photomodification [73]. Minimization of biofouling and blood coagulation at the surface through various surface modifications greatly enhances biocompatibility of polymeric materials. Selected chemicals offering anticoagulant properties are listed in Table 4.1.

Fig. 4.6 Inactivation of clotting enzymes by HEP [67]. a No binding of clotting enzyme to AT-III. b Binding of clotting enzymes to AT-III due to the presence of HEP. c Dissociation of heparin from the AT-III/clotting enzyme complex

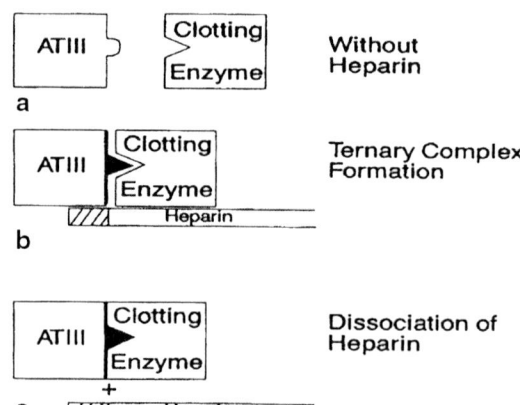

4.3.3 Antifouling Surfaces

The adsorption of proteins to a polymer substrate in contact with biological environments may be detrimental to biocompatibility since the adsorbed proteins may initiate platelet adhesion and activation [74, 75]. The complexity of protein adsorption is due to electrostatic interactions between the surface and protein, protein concentration levels, and surface energy [36]. There are typically two recognized modes of interaction involved in protein adsorption on the surface, which are depicted in Fig. 4.7 [76]. Primary adsorption (a) occurs at the polymer surface–brush interface and occurs when small proteins are able to penetrate through the polymer brush and adsorb to the substrate surface, while secondary adsorption (b) is due to larger proteins being attracted by van der Waals interactions at the solvent–brush interface [4, 76].

Since adhesion strength of biomolecules to polymer surfaces is a function of surface energy and polymer matrix modulus, attachment of hydrophilic PEG with a low surface energy will result in resistance to biofouling, providing a high activation barrier for protein adsorption as well as steric repulsions which are also important for protein resistance [14, 74]. PEG-functionalized polymer brushes synthesized by surface-initiated ATRP have been intensively examined which have resulted in surfaces with significant resistance to protein adsorption as well as cell adhesion [39, 77–80]. On the other hand, elastomeric polymers such as PDMS exhibit favorable low modulus, but high surface energy which results in an increase in protein adhesion strength. In order to circumvent these problems, several polymer architectures and surface modification techniques have been explored involving PDMS [81–83]. For example, zwitterionic phosphorylcholine polymer that resembles the structure of natural membrane lipids [84] has been grafted onto PDMS via photo-induced polymerization, resulting in greatly improved surface hydrophilicity and antifoul-

Fig. 4.7 Two modes of adsorption of a particle on a polymeric layer grafted to a substrate: **a** primary adsorption in which proteins penetrate the polymer brush and **b** secondary adsorption at the brush–solvent interface [76]

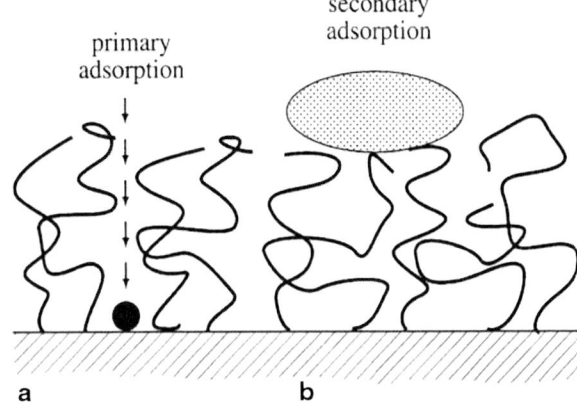

ing properties [83]. Several zwitterionic containing polymers, including phosphor-ylcholine, polybetaine, carboxybetaine, and sulfobetaine, are utilized as antifouling polymers due to their hydrophilicity as well as electroneutrality [84] and are shown in Table 4.1 [14].

Significant advances have been made in surface modifications of commodity polymers to achieve antimicrobial, anticoagulant, and antifouling properties. In spite of the efforts, there are major unresolved scientific and technological problems associated with the safe use of polymers in biological environments. Understanding surface modifications of polymers and the development of reproducible technolo-gies is timely due to devastating effects of microbial infections and thrombosis.

4.4 Stimuli-Responsive Polymer Surfaces

The majority of physicochemical approaches, such as layer-by-layer deposition [85], chemical etching [86], or radiation grafting [45], appear to have short life span effectiveness. Although in the layer-by-layer approach, multiple layers are depos-ited onto surfaces by dipping, and multilayers are held together by opposite electro-static charges, the resulting films are mechanically unstable and often may exhibit chemically heterogeneous surfaces [87]. Chemical etching alters hydrophobicity via surface oxidation and morphology changes, whereas radiation grafting uses ra-diation sources such as infrared, visible light, ultraviolet, and γ radiation as well as high-energy electrons to generate reactive groups on surfaces in the presence of monomer to graft species to/from the surface [45, 86]. In contrast, microwave plasma surface reactions offer a fast, solventless, and sterile method for covalently attaching carboxylic acid groups to almost any polymer substrate.

If aliphatic polymer substrates, such as PE and PP, exhibit –COOH surface groups, their presence facilitates an opportunity for further reactions. Although sev-eral studies demonstrated that indeed, various surface entities can be covalently at-tached, it was apparent that lower reaction yields were achieved as more layers were reacted. Since click chemistry [41, 42] represents one of the highly efficient (>95%) synthetic routes of achieving high reaction yields in relatively short time under mild conditions, we will take advantage of combining both approaches to create antimi-crobial PE and PP surfaces. Reactions that meet "click" chemistry criteria, but are not limited to, include Huisgen 1,3-dipolar cycloaddition [42], Diels–Alder reactions [41], nucleophilic substitution with epoxy and aziridine compounds [41], sharpless dihydroxylation [41], and thiol-yne reactions [88]. Notably, the Huisgen 1,3-dipo-lar cycloaddition involves a copper(I)-catalyzed [43] 1,2,3-triazole formation from azides and alkyne functionalities with complete specificity of reactants [89]. In terms of surface reactions, cycloaddition click reactions yielding high efficiencies with Cu [42], Au [90, 91], Si [88, 92], glass [93], and even carbon nanotube [94] surfaces have shown promising results, such as the biofunctionalization of Si surfaces with "clicked" biotin and glucose was successful [42, 44, 92]. These unique attributes of click chemistry also broke new grounds for selective reactions with complex den-drimers [95, 96], hydrogels [97], vesicles [98], and nanoparticles [91, 99].

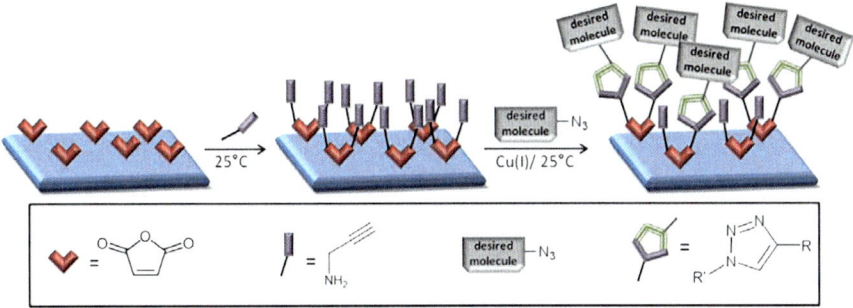

Fig. 4.8 Schematic diagram of "clickable" polymeric surfaces exhibiting alkyne functionalities for "clicking" any azide-containing molecule [100]

Although click chemistry has been utilized in a variety of model studies, their practicality has been limited by one major roadblock: the ability to react to typically inert polymeric surfaces. In view of the previous studies utilizing microwave plasma reactions leading to –COOH functionalization of aliphatic polymer surfaces, simple and clean surface reactions were developed and this formulated a platform for almost any polymer surface modifications without adverse effects on polymer bulk properties [100]. Figure 4.8 schematically depicts a sequence of surface reactions using two simple steps: microwave plasma reactions that lead to –COOH formation, followed by covalent attachment of alkyne moieties resulting in polymeric surfaces to which any azide-containing molecule may be "clicked." Selected examples of "desired molecules" include peptides utilized in stem cell adhesion [101], fluorophores for labeling hydrogels [102, 103], polyhedral oligomeric silsequioxane (POSS) [88], biotin [93, 104], bacteriophages [105], or DNA [106], to name just a few. In these studies, we focused on attachment of antibiotics, such as AMP which exhibits Gram (+) and (−) antimicrobial functions.

Figure 4.9 illustrates the two-step process in which maleic anhydride (MA) was reacted to PE and PP (a) [47, 107] to obtain –COOH groups, followed by their conversion to acid chloride. The second step relies on reactions of propargylamine (b) to obtain alkyne-functionalized polymeric surfaces to which any azide-containing molecule may be reacted.

Figure 4.10a illustrates structural features that develop from reactions depicted in steps A and B of Fig. 4.8. To examine the effectiveness of AMP covalently attached to polymer surfaces against *S. aureus* bacteria, PE-MA-polymer processing additive (PPA)-AMP and PP-MA-PPA-AMP surfaces were tested for antimicrobial activity against *S. aureus*. The results are shown in Fig. 4.10b in which the colony-forming units (CFU) per square centimeter were enumerated using a drop plate method for PE/PP (a), PE-MA/PP-MA (b), PE-MA-PPA/PP-MA-PPA (c), and PE-MA-PPA-AMP/PP-MA-PPA-AMP (d) surfaces. Analysis of the data indicates that clicked AMP on both PE and PP facilitates a major enhancement of the antimicrobial activity manifested by a drop of CFUs from 4000 to 5000 for PE-MA-PPA and PP-MA-PPA specimens to 10–100 for PE-MA-PPA-AMP and PP-MA-PPA-AMP

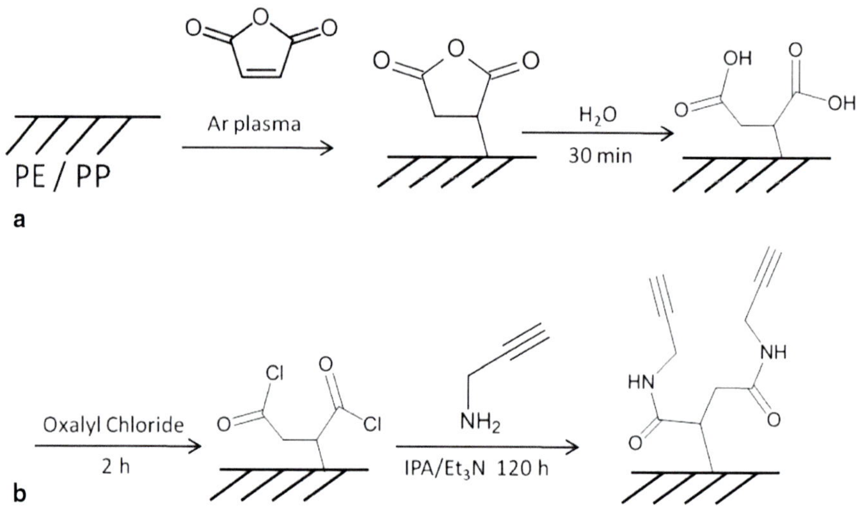

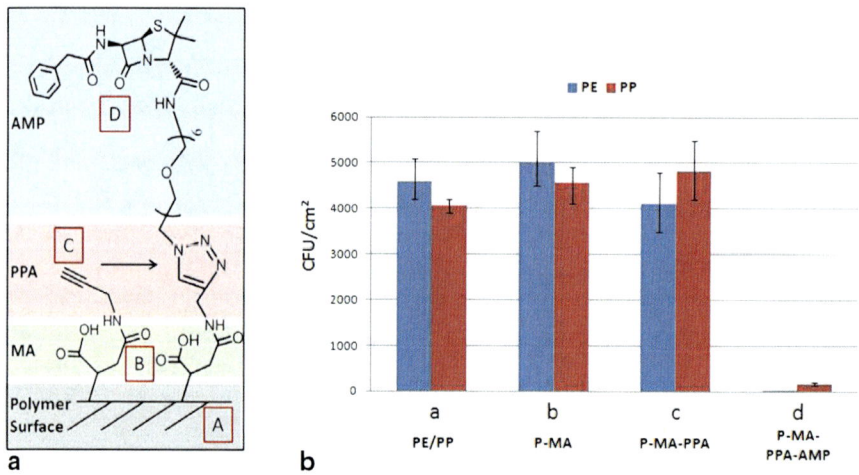

Fig. 4.9 Surface reactions leading to the formation of alkyne surface groups; microwave plasma reactions in the presence of MA (**a**) and alkyne functionalization using propargylamine (**b**)

Fig. 4.10 a Reaction sequences on PE and PP substrates leading to AMP-clicked surfaces; **b** Antimicrobial activity against *S. aureus* of (**a**) PE *(blue)* and PP *(red)* surfaces; (**b**) PE-MA and PP-MA surfaces; (**c**) PE-MA-PPA and PP-MA-PPA surfaces; (**d**) PE-MA-PPA-AMP and PP-MA-PPA-AMP surfaces [108]

specimens. These AMP-clicked surfaces exhibit highly efficient antimicrobial activity against *S. aureus* with a 97–99.8 % decrease of bacterial growth.

Although numerous studies have been conducted on polymer surface modifications, recent advances [108] in surface and interfacial reactions have resulted in the development of a new class of materials that exhibit stimuli-responsive characteristics. For example, by creating a switchable surfaces through the use of polyelectrolytes [109], followed by functionalizing the end groups of these polyelectrolytes, a given surface may adapt itself to environmental pH changes. Taking this concept, a step further allows creating not only functional surfaces but also engineering surface responsiveness to an array of external stimuli. The majority of surface reactions have focused on silicon and gold surfaces [110], but reactions on polymer substrates are challenging due to the inert nature of polymer surfaces and their morphological heterogeneities. This is particularly important if responses to an external stimulus, such as temperature [15, 111–113], ionic strength [114], UV radiation [113], or pH [15, 109, 111], are required. In the recent studies, stimuli-responsive polymeric surfaces "decorated" with cationic and anionic polyelectrolytes terminated by –COOH or –NH$_2$ groups were developed [100]. The advantages of oppositely charged polyelectrolytes not only facilitate the formation of hydrophilic surfaces on hydrophobic polymer substrates but also introduce surface dynamic properties as a function of pH. Specifically when cationic segments extend, anionic tethers collapse and vice versa. Figure 4.11a illustrates an overall scheme of the coupling reactions which facilitate the attachment of poly(2-vinyl pyridine) (P2VP) and PAA. Since P2VP and PAA chain ends are terminated with –NH$_2$ and –COOH functionalities, respectively, the scope of these studies is further expanded to the attachment of HEP (on P2VP) AMP (on PAA; Fig. 4.11b) to achieve antimicrobial and anticoagulant pH-responsive polymeric surfaces with antimicrobial and anticoagulant moieties.

While Fig. 4.11 depicted the general theme to achieve pH-responsive anticoagulant and antimicrobial surfaces. Figure 4.12 illustrates reaction sequences leading to the formation of covalently attached multilayers (CAM) tethered to Si, PE, and PTFE surfaces.

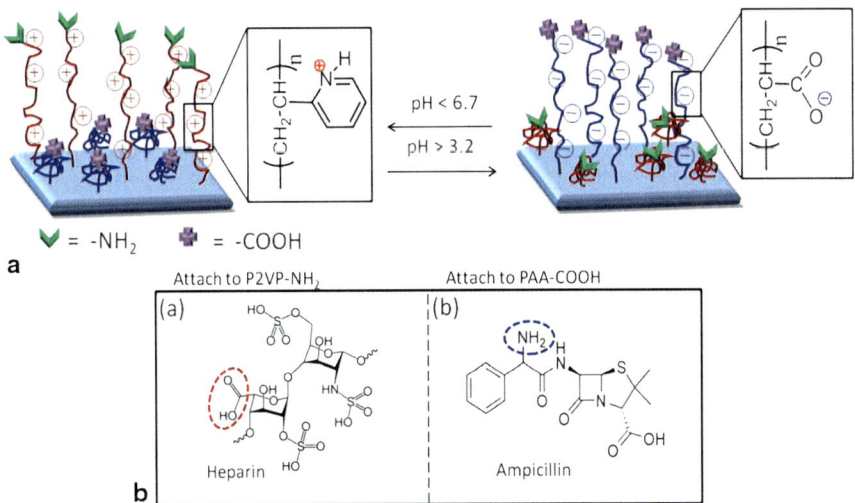

Fig. 4.11 a Schematic diagram of dual stimuli-responsive polyelectrolyte surfaces terminated with –NH$_2$ and –COOH moieties. **b** Structures of **a** HEP and **b** AMP; *circled* moieties are available for attachment [108]

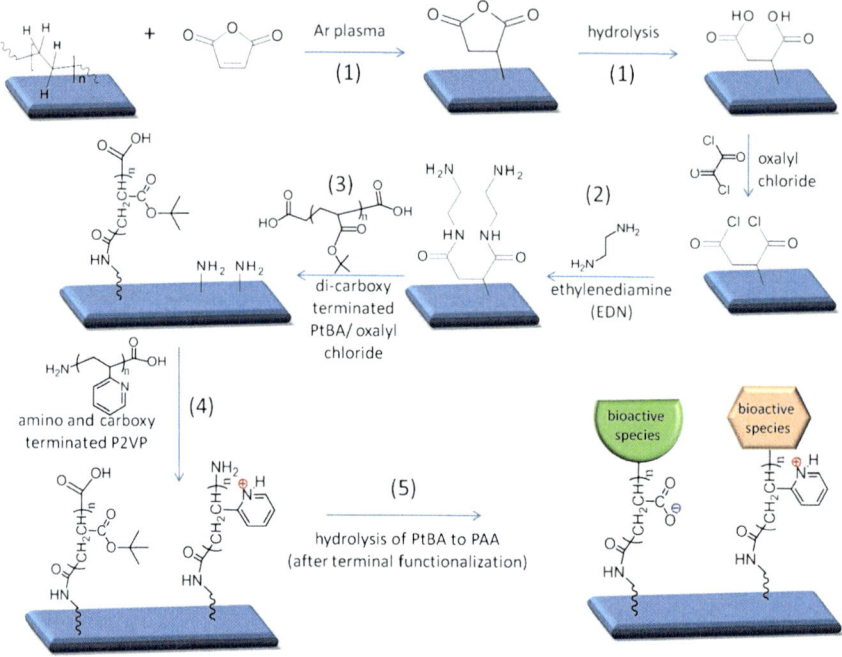

Fig. 4.12 Reaction sequences on polymeric substrates leading to the formation of stimuli-respon-sive polyelectrolyte surfaces terminated with –COOH and –NH₂ functionalities and ultimately exhibiting bioactive molecules

Similar to the previous studies (step 1), MA was reacted to a substrate [47, 107], followed by hydrolysis to obtain –COOH groups, which were then converted to acid chloride. In the step 2, ethylenediamine (EDN) was reacted with –COCl groups resulting in –NH₂ terminated surfaces. The next step (3) involved the attachment of di-carboxy-terminated poly(tert-butyl acrylate) (PtBA) to the –NH₂ terminated surfaces via amide linkages, followed by the attachment of P2VP (4), carried out by reacting amide linkages between –COOH groups on P2VP and the –NH₂ groups. Upon terminal functionalization with desired bioactive species, PtBA can be hydrolyzed under mild conditions to PAA polyelectrolyte. This choice of sequences with polyelectrolyte chain ends containing –NH₂ and –COOH terminal groups facilitated to achieve two objectives: the ability to attach bioactive species as well as facilitate pH-sensitive expansion/collapse responses. Figure 4.13a illustrates a sequence of CAM reactions leading to –COOH terminal entities. The first step involved MA reactions to a substrate, followed by EDN and PtBA (PAA) functionalization. Reactions of P2VP to polymer surfaces facilitate pH responsiveness in the pH range of 1 < pH < 6.7 as well as the opportunity for terminating each tether with –NH₂ functionalities for further reactions. Figure 4.13b illustrates a sequence of CAM reactions leading to –NH₂ terminal groups.

The covalent attachment of AMP and HEP to modified polyethylene (PE-MA-EDN-PAA or PE-MA-EDN-P2VP) or polytetrafluoroethylene (PTFE-MA-EDN-

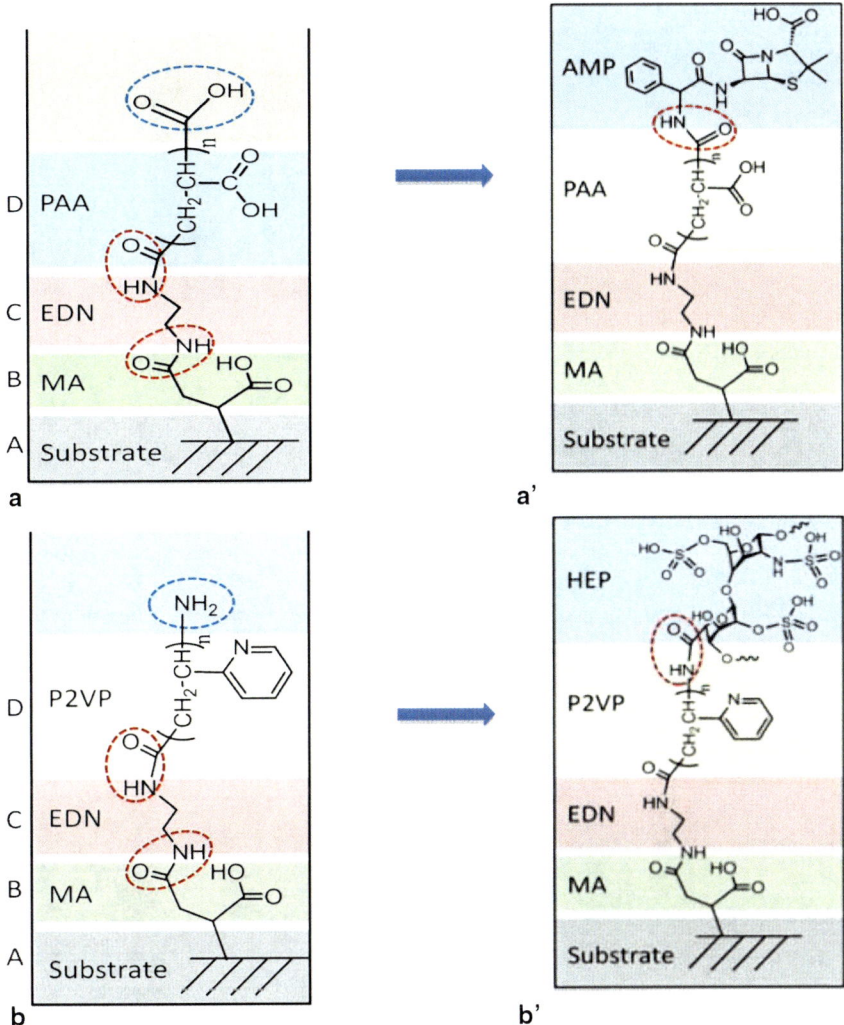

Fig. 4.13 Chemical structures resulting from reactions with AMP to PTFE-MA-EDN-PAA (**a/a'**) and HEP to PTFE-MA-EDN-P2VP (**b/b'**) (*circles* represent reaction sites)

PAA/PTFE-MA-EDN-P2VP) surfaces is shown in Fig. 4.13. Figure 4.14 illustrates the CFU per square centimeter enumerated using a drop plate method for PE/PTFE (a), PE-MA/PTFE-MA (b), PE-MA-EDN-PAA/PTFE-MA-EDN-PAA (c), and PE-MA-EDN-PAA-AMP/PTFE-MA-EDN-PAA-AMP (d) surfaces. Although the analysis of the data indicates that in general, PE exhibits better antimicrobial resistance than PTFE, it is apparent that covalent attachment of AMP facilitates a major enhancement of antimicrobial activities. This is illustrated in Fig. 4.14d which shows significant drop of CFUs from 4000–5000 to 10–40.

Fig. 4.14 Antimicrobial
activity against *S. aureus* of
(**a**) PE *(blue)* and PTFE *(red)*
surfaces; (**b**) PE-MA and
PTFE-MA surfaces; (**c**) PE-
MA-EDN-PAA and PTFE-
MA-EDN-PAA surfaces; (**d**)
PE-MA-EDN-PAA-AMP and
PTFE-MA-EDN-PAA-AMP
surfaces

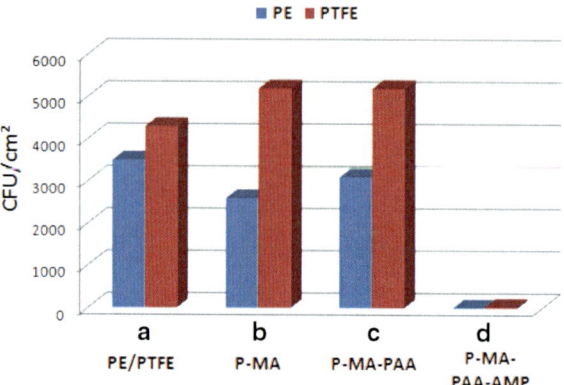

Anticoagulant activity of PE-MA-P2VP and PTFE-MA-P2VP surfaces functionalized with HEP was determined using Raman imaging in the presence of whole
lagomorph blood. This can be achieved by tuning to the 1620 cm^{-1} band of the
specimens obtained from incubating each surface in lagomorph blood. This is illustrated in Fig. 4.15, in which PE (a), PE-MA (b), PE-MA-EDN-P2VP (c), and
PE-MA-EDN-P2VP-HEP (d) are Raman images of the substrates before/after blood
exposure. As shown by color scale bars, red color represents high intensity corresponding to 100% coagulation. As coagulation decreases, color changes which

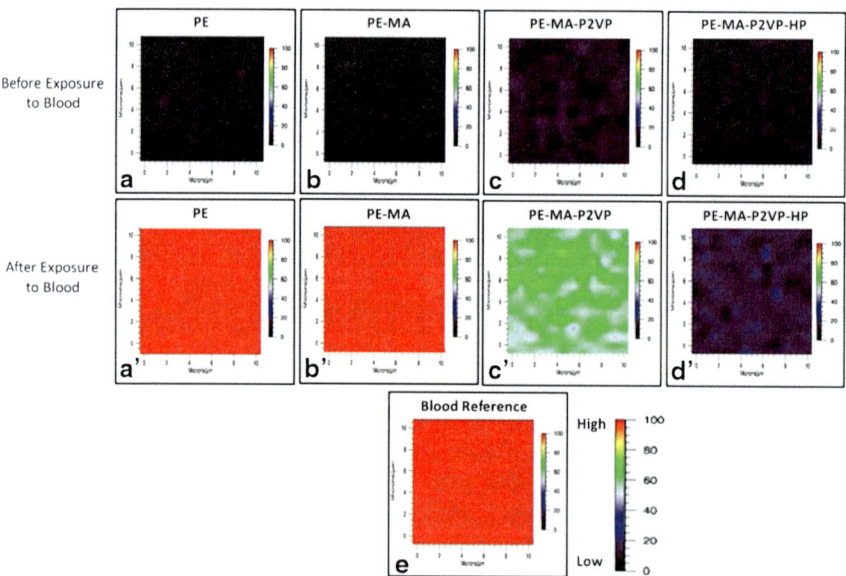

Fig. 4.15 Raman images of the 1620 cm^{-1} band before/after exposure to whole blood to (**a/a′**)
PTFE; (**b/b′**) PTFE-MA; (**c/c′**) PTFE-P2VP; (**d/d′**) PTFE-P2VP-HEP; (**e**) blood reference. Each
image represents a 10 × 10 μm area

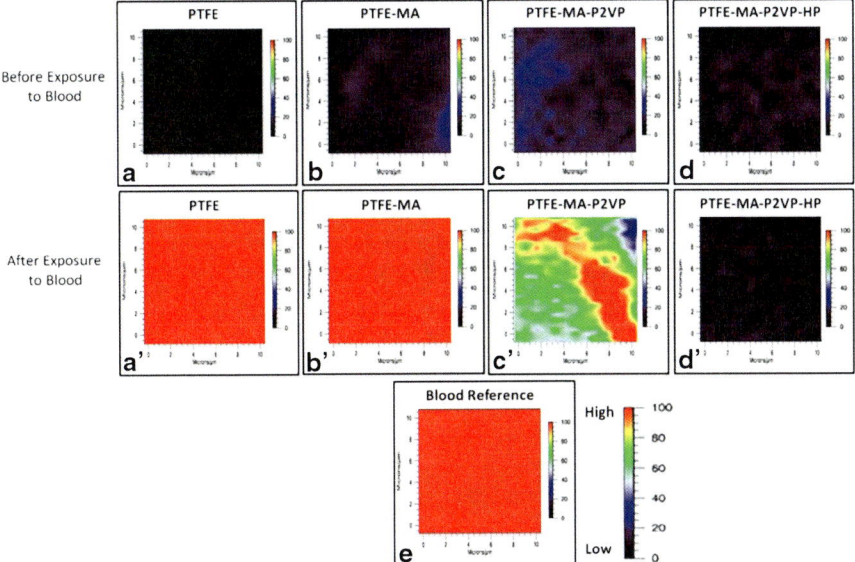

Fig. 4.16 Raman images of the 1620 cm^{-1} band before/after exposure to whole blood to (**a/a'**) PE; (**b/b'**) PE-MA; (**c/c'**) PE-P2VP; (**d/d'**) PE-P2VP-HEP; (**e**) blood reference. Each image represents a 10 × 10 µm area

ideally represents 0 % coagulation. Upon exposure to blood, PE (a') and PE-MA (b') do not exhibit anticoagulant activity, although PE-MA-EDN-P2VP (c') shows some decrease, PE-MA-EDN-P2VP-HEP (d') greatly minimizes clotting. For comparison, Fig. 4.15e shows the same Raman image of dried lagomorph blood. Similarly, Fig. 4.16 illustrates Raman images of PTFE (a), PTFE-MA (b), PTFE-MA-EDN-P2VP (c), and PTFE-MA-EDN-P2VP-HEP (d) substrates before/after blood exposure. As shown, PTFE (a'), PTFE-MA (b'), and PTFE-MA-EDN-P2VP (c') do not exhibit anticoagulant activity, whereas PTFE-MA-EDN-P2VP-HEP (d') minimizes clotting. For comparison, Fig. 4.16e shows the image of dried lagomorph blood; as seen, the effectiveness of HEP covalently attached to PE-MA-EDN-P2VP and PTFE-MA-EDN-P2VP surfaces against blood coagulation is apparent.

The majority of interactions between biologically active species and synthetic materials are inherently non-favorable. However, formation of biofilms represents an important and unwelcome exception resulting from the attachment of bacteria to a synthetic surface, leading to the formation of a complex biofilm community that is often encased on a surface of polymeric materials in contact with blood. Microbial biofilms are very resilient communities that resist removal by chemical or physical means because their residents, bacterial cells, are capable of adhering to a variety of biotic and abiotic surfaces and continue to grow biofilms as long as the nutrients become available. In the context of human health, biofilms are responsible for the vast majority of deadly infections including medical-device-associated diseases. Because they often become resistant to antibiotics or host defenses, the use of antibiotics may be ineffective to many pathogens, making conventional therapies troublesome [115].

4.5   Surface Attachments of Bioactive Species

Surface medical-device-associated infections may include simple catheters, implants, stents, monitoring devices, to name just a few. The first logical step has led to the development of novel drugs, but surface modifications with drugs bring another level of challenges associated with their attachment, long-term effectiveness, altering immuno-responses, and maintenance. Among anti-biofilm formation strategies, covalent attachments of antibiotics or antimicrobial agents to polymeric surfaces have been somewhat successful with relatively longer durations, but still long-term activities might be limited [107]. Ideally, one would like to create stimuli-responsive properties on polymeric surfaces, where a surface remains silent unless an external stimulus such as bacteria triggers desirable responses. In essence, the goal is to prevent biofilm formation from developing at its inception. Almost entirely unexplored area of inhibition of bacterial film formation is the use of bacteriophages. These living bacterial viruses are capable of selective binding to specific receptors of the target bacteria. Upon binding, they inject their own DNA, which upon reproduction inside the bacteria, kill the host, and release their progeny [116]. The use of these species has been only sporadically explored [56, 117, 118]. Although control of their progeny is critical, it seems that, under controllable conditions, bacteriophages will offer an alternative and powerful approach for inhibiting bacterial infections by attaching specifically to their target host bacteria, injecting their genetic material, reproducing inside the host, killing the host, and releasing their progeny. Although the first observations of lytic phages to cure infectious diseases go back to the end of the nineteenth [119] and early twentieth centuries [116, 120, 121], it was not until 1960s, when prophylaxis and treatment of bacterial infections appeared favorable choice with a remarkable recovery efficacy above 92 % [122]. Later on, the use of bacteriophages was exploited by physical mixing with or physisorbed on polymers (Nylon™) [56, 123], glass [117], and gold [124]. The unique attribute of bacteriophages is their host-dependent reproduction; bacteriophages will remain silent until they find a specific bacterium, thus minimizing safety concerns associated with excessive concentration levels. Furthermore, due to evolution with their hosts and host specificity, there are numerous bacteriophages for each bacterium. Taking advantage of the ability of bacteriophages to recognize a host bacterium and their ability to kill specific hosts, T1 and Φ11 phages were covalently attached onto the surfaces of polytetrafluoroethylene (PTFE) and ultra-high molecular weight polyethylene (PE) [125]. Figure 4.17a–d depicts a sequence of steps leading to the formation and subsequent destruction of the bacteria attempting to form biofilms on the phage-modified polymeric substrates. The first step (Fig. 4.17a) involves the formation of reactive acid groups on polymeric substrates [47] accomplished by simple and clean microwave plasma reactions in the presence of maleic anhydride, followed by covalent attachment of T1 phages via acid-amine reactions leading to amide linkages.

When bacteria attempt to adhere to the surface of the phage-modified polymer substrate, the phage attaches specifically to an external structure of the bacterium (e.g. lipopolysaccharide or protein) and injects its genetic material into its target (Fig. 4.16c). Upon completion of this process, the phage DNA is replicated by

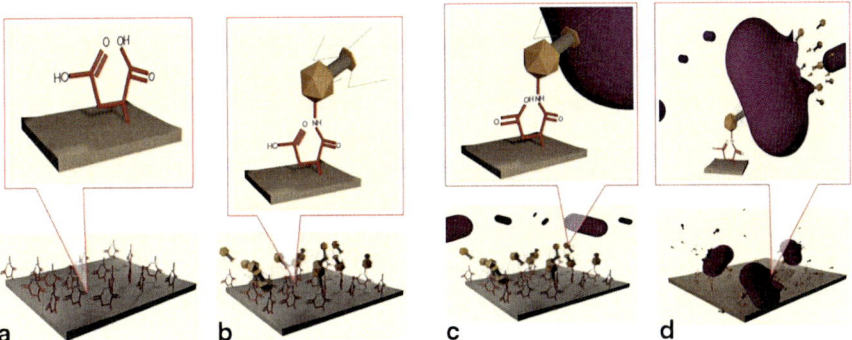

Fig. 4.17 a Covalent attachment of acid groups to polymeric surfaces. **b** Reactions of NH_2 groups of T1 with polymer surface acid groups. **c** Attachment of phages to bacteria and injection of DNA. **d** Replication of DNA and destruction of bacteria [125]

the bacterial host machinery, several capsid proteins are produced and assembled, phage DNA is packaged, and the bacteriophage progeny is released by lysis of the bacterial host. Depending on the particular bacteriophage, the progeny can be up to 200 for each individual infection. Perhaps, the most significant advantage of using bacteriophages comes from the fact that each member of the progeny is capable of infecting more bacteria and releasing a progeny of its own. This amplification effect continues until all bacteria cells are killed. Of course, one can envision potential challenges associated with control of the phage population, especially when utilized in therapy, which needs to be carefully adjusted.

The antimicrobial effectiveness of T1 and Φ11 phages covalently attached to PTFE and PE surfaces against *E. coli* and *S. aureus* was confirmed using plaque formation assays and showed that covalently attached Φ11 phages kill *S. aureus* bacteria very effectively. These studies established a universal approach of covalent attachment of bacteriophages to inert PE and PTFE polymeric surfaces. The reactions can be conducted on almost any surface and, although the limiting factor can be the maintenance of biological activity of phages, these studies show that covalent attachment of T1 and Φ11 does not adversely affect the phage's biological activities. Multiple venues for polymer surface modifications combined with over a thousand individual phage species, with hundreds of different strengths, may provide useful possibilities of creating surfaces capable of killing specific bacteria.

4.6   Summary and Outlook

In spite of alarming health-related problems associated with antimicrobial, anticoagulant, and antifouling properties, it is quite apparent that the scientific community, for the most part, is reluctant to the notion of eliminating and controlling microbial film formation or coagulation/fouling properties at their inception. This is directly reflected by the levels of funding often articulated by the statements that a gallon of Clorox in an operating room should eliminate deadly infections. Unfortunately, in many biomedical practical applications, this is simply impossible. Infections are not associated with specific diseases. It is the disease itself that may strike under most unpredictable conditions. Yet, compared to the disease-driven programs, the relatively marginal efforts focus on understanding molecular processes leading to the inhibition of infections before it is too late. The increased prevalence of antibiotic-resistant infections has created an alarming gap in medical care since patients are increasingly infected with multiresistant strains of bacteria and have therefore become "untreatable." The use of bacteriophage therapy has proven promising in several studies for it is not fully developed, and innovative approaches are necessary to tackle the uncharted areas. Taking advantage of tremendous advances in modifications of chemistries and physics, there are unprecedented opportunities for the development of new generations of biomedical materials. Bacteriophage therapy is emerging as a promising approach in combating antibiotic-resistant infections, which combined with signaling, will offer a new paradigm in fighting microbial film formation.

Acknowledgments The National Science Foundation (CMMI 332964) and J. E. Sirrine Foundation at Clemson University are acknowledged for a partial support of these studies.

References

1. Bauer S, Schmuki P, von der Mark K, Park J. Engineering biocompatible implant surfaces; part I: materials and surfaces. Prog Mater Sci. 2013;58:261.
2. Temenoff JS, Mikos AG. Biomaterials: the intersection of biology and materials science. Upper Saddle River: Pearson/Prentice Hall; 2008.
3. Wintermantel E, Suk-Woo H. Medizintechnik mit biokompatiblen Werkstoffen und Verfahren. Berlin: Springer; 2002.
4. Goddard JM, Hotchkiss JH. Polymer surface modification for the attachment of bioactive compounds. Prog Polym Sci. 2007;32:698.
5. Guimard NK, Gomez N, Schmidt CE. Conducting polymers in biomedical engineering. Prog Polym Sci. 2007;32:876.
6. Shen H, Hu X, Bei J, Wang S. The immobilization of basic fibroblast growth factor on plasma-treated poly(lacitide-co-glycolide). Biomaterials. 2008;29:2388.
7. Langer R, Tirrell DA. Designing materials for biology and medicine. Nature. 2004;428:487.
8. Lee KY, Mooney DJ. Hydrogels for tissue engineering. Chem Rev. 2001;101:1869.
9. Jiang T, Chang J, Wang C, Ding Z, Chen J, Zhang J, Kang E-T. Adsoprtion of plasmid DNA onto N, N'-(dimethylamino)ethyl-methacrylate graft-polymerized poly-L-lactic acid folm surface for promotion of in-situ gene delivery. Biomacromolecules. 2007;8:1951.
10. Mano JF. Stimuli-responsive polymeric systems for biomedical applications. Adv Eng Mater. 2008;10:515.
11. Zeilikin AN. Drug releasing polymer thin films: new era of surface-mediated drug delivery. ACS Nano. 2010;4:2494.
12. Guavin R, Khademhosseini A, Guillemette M, Langer R. Emerging trends in tissue engineering. Compr Biotechnol. 2011;5:251.
13. Davis JR. Handbook of materials for medical devices. Materials Park: ASM International; 2003.

14. Yu M, Urban MW. Polymeric surfaces with anticoagulant, antifouling, and antimicrobial attributes. Macromol Symp. 2009;283:311.
15. Liu F, Urban MW. Recent advances and challenges in designing stimuli-responsive polymers. Prog Polym Sci. 2010;35:3.
16. Urban MW. Stimuli-responsive colloids; from stratified to self-repairing polymeric films. Curr Opin Colloid Interface Sci. 2014;19:66–75.
17. Yang Y, Urban MW. Self-healing polymeric materials. Chem Soc Rev. 2013;42(17):7446–67.
18. Bae W-S, Urban MW. Creating patterned poly(dimethylsiloxane) surfaces with amoxicillin and poly(ethylene glycol). Langmuir. 2006;22:10277.
19. Aumsuwan N, Heinhorst S, Urban MW. Antibacterial surfaces on expanded poly(tetrafluoroethylene); penicillin attachment. Biomacromolecules. 2007;8:713.
20. Kroschwitz JI. Polymers: biomaterials and medical applications. New York: Wiley; 1989.
21. Kenawy ER, Worley SD, Broughton R. The chemistry and applications of antimicrobial polymers: a state-of-the-art review. Biomacromolecules. 2007;8:1359.
22. Kidane AG, Salacinski H, Tiwari A, Bruckdorfer KR, Seifalian AM. Anticoagulant and antiplatelet agents: their clinical and device application(s) together with usages to engineering surfaces. Biomacromolecules. 2004;5:798.
23. Senaratne W, Andruzzi L, Ober CK. Self-assembled monolayers and polymer brushes in biotechnology: current applications and future perspectives. Biomacromolecules. 2005;6:2427.
24. Bae WS, Convertine AJ, McCormick CL, Urban MW. Effect of sequential layer-by-layer surface modifications on the surface energy of plasma-modified poly(dimethylsiloxane). Langmuir. 2007;23:667.
25. Kim H, Urban MW. Reactions of thrombresistant multilayered thin films on poly(vinyl chloride) (PVC) surface: a spectroscopic study. Langmuir. 1998;14:7235.
26. Zhang S. Fabrication of novel biomaterials through molecular self-assembly. Nat Biotechnol. 2003;21:1171.
27. Tanii T, Hosaka T, Miyake T, Kanari Y, Zhang G-J, Funatsu T, Ohdomari I. Hybridization of deoxyribonucleic acid and immobilization of green fluorescent protein on nanostructured organosilane templates. Jpn J Appl Phys. 2005;44:5851.
28. Roy D, Knapp JS, Guthrie JT, Perrier S. Antibacterial cellulose fiber via RAFT surface graft polymerization. Biomacromolecules. 2008;9:91.
29. Lee SB, Koepsel RR, Morley SW, Matyjaszewski K, Sun Y, Russell AJ. Permanent, non-leaching antibacterial surface.1. Synthesis by atom transfer radical polymerization. Biomacromolecules. 2004;5:877.
30. Dvoracek CM, Sukhonosova G, Benedik MJ, Grunlan JC. Antimicrobial behavior of polyelectrolyte surfactant thin film assemblies. Langmuir. 2009;25:10322.
31. Lichter JA, Vliet KJV, Rubner MF. Design of antibacterial surfaces and interfaces: polyelectrolyte multilayers as a multifunctional platform. Macromolecules. 2009;42:8573.
32. Chen W, McCarthy TJ. Layer-by-Layer deposition: a tool for polymer surface modification. Macromolecules. 1997;30:78.
33. Black FE, Hartshorne M, Davies MC, Roberts CJ, Tendler SJB, Williams PM, Shakesheff KM. Surface engineering and surface analysis of a biodegradable polymer with biotinylated end groups. Langmuir. 1999;15:3157.
34. Duwez A-S, Cuenot S, Jérôme C, Gabriel S, Jérôme R, Rapino S, Zerbetto F. Mechanochemistry: targted delivery of single molecules. Nat Nanotechnol. 2006;1:122.
35. Onard S, Martin I, Chailan J-F, Crespy A, Carriere P. Nanostructuration in thin epoxy-amine films inducing controlled specific phase ethericfication: effect on the glass transition temperatures. Macromolecules. 2011;44:3485.
36. Desmet T, Morent R, Geyter ND, Leys C, Schacht E, Dubruel P. Nonthermal plasma technology as a varsatilestrategy for polymeric biomaterials surface modification: a review. Biomacromolecules. 2009;10:2351.
37. Suituma C, Wang Y, Hupert M, Barany F, McCarley RL, Soper SA. Fabrication of DNA microarrays onto poly(methyl methacrylate) with ultravilet petterning and microfluidics for the detection of low-abundant point mutations. Anal Biochem. 2005;340:123.

38. Ma Q, Zhang H, Zhao J, Gong Y-K. Fabrication of cell outer membrane mimetic polymer brush on polysulfone surface via RAFT technique. Appl Surf Sci. 2012;258:9711.
39. Tugulu S, Klok HA. Stability and nonfouling properties of poly(poly(ethylene glycol) methacrylate) brushes under cell culture conditions. Biomacromolecules. 2008;9:906.
40. Rasmussen JR, Stedronsky ER, Whitesides GM. Introduction, modification, and characterization of functional-groups on surface of low-density polyethylene film. J Am Chem Soc. 1977;99:4736.
41. Kolb HC, Finn MG, Sharpless KB. Click chemistry: diverse chemical function from a few good reactions. Angew Chem Int Ed. 2001;40:2004.
42. Binder WH, Sachsenhofer R. 'Click' chemistry in polymer and materials science. Macromol Rapid Commun. 2007;28:15.
43. Rostovtsev VV, Green LG, Fokin VV, Sharpless KB. A stepwise Huisgen cycloaddition process: copper(I)-catalyzed regioselective "ligation" of azides and terminal alkynes. Angew Chem Int Ed. 2002;41:2596.
44. Iha RK, Wooley KL, Nystrom AM, Burke DJ, Kade MJ, Hawker CJ. Applications of orthogonal "click" chemistries in the synthesis of functional soft materials. Chem Rev. 2009;109:5620.
45. Vahdata A, Bahramia H, Ansaria N, Ziaieb F. Radiation grafting of styrene onto polypropylene fibres by a 10 MeV electron beam. Radiat Phys Chem. 2007;76:787.
46. Yasuda H. Plasma polymerization. Orlando: Academic Press; 1985.
47. Gaboury SR, Urban MW. Microwave plasma reactions of solid monomers with silicone elastomer surfaces: a spectroscopic study. Langmuir. 1993;9:3225.
48. Bae WS, Urban MW. Reactions of antimicrobial species to imidazole-microwave plasma reacted poly(dimethylsiloxane) surfaces. Langmuir. 2004;20:8372.
49. Chevallier P, Janvier R, Mantovani D, Laroche G. In vitro biological performances of phosphorylcholine-grafted ePTFE prostheses through RFGD plasma techniques. Macromol Biosci. 2005;5:829.
50. Zhang Q, Wang C, Babukutty Y, Ohyama T, Kogoma M, Kodama M. Biocompatibility evaluation of ePTFE membrane modified with PEG in atmospheric ppressure glow discharge. J Biomed Mater Res. 2002;60:502.
51. Jia G, Wang HF, Yan L, Wang X, Pei RJ, Yan T, Zhao YL, Guo XB. Cytotoxicity of carbon nanomaterials: single-wall nanotube, multi-wall nanotube, and fullerene. Environ Sci Technol. 2005;39:1378.
52. Munoz-Bonilla A, Fernandez-Garcia M. Polymeric materials with antimicrobial activity. Prog Polym Sci. 2012;37:281.
53. Rabea EL, Badawy ME-T, Stevens CV, Smagghe G, Steurbaut W. Chitosan as antimicrobial agent: applicatiopns and mode of action. Biomacromolecules. 2003;4:1457.
54. Kazemzadeh-Narbat M, Kindrachuk J, Duan K, Jenssen H, Hancock REW, Wang R. Antimicrobial peptides on calcium phosphate-coated titanium for the prevention of implant-associated infections. Biomaterials. 2010;31:9519.
55. Grunlan JC, Choi JK, Lin A. Antimicrobial behavior of polyelectrolyte multilayerfilms containing cetrimide and silver. Biomacromolecules. 2005;6:1149.
56. Markoishvili K, Tsitlanadze G, Katsarava R, Morris JGJ, Sulakvalidze A. A novel sustained-release matrix based on biodegradable poly(ester amide)s and impregnated with bacteriophages and an antibiotic shows promice in management in infected venous stasis ulcers and other poorlt healing woulds. Int J Dermatol. 2002;41:453.
57. Schierholz JM, Beuth J, Pulverer G, König D-P, Scharlack RS, Kampf G, Dietze B, Wendt C, Martiny H, Große-Siestrup C. Silver-containing polymers antimicrob. Agents Chemother. 1999;43:2819.
58. Huang J, Murata H, Koepsel RR, Russell AJ, Matyjaszewski K. Antibacterial polypropylene via surface-initiated atom transfer radical polymerization. Biomacromolecules. 2007;8:1396.
59. Aumsuwan N, Heinhorst S, Urban MW. The effectiveness of antibiotic activity of penicillin attahed to expanded poly(tetrafluoroethylene) (ePTFE) surfaces: a quantitative assessment. Biomacromolecules. 2007;8:3525.

60. Aumsuwan N, Danyus RC, Heinhorst S, Urban MW. Attachment of amicillin to expanded poly(tetrafluoroethylene): surface reactions leading to inhibition of microbial growth. Biomacromolecules. 2008;9:1712.
61. Tiller JC, Lee SB, Lewis K, Klibanov AM. Polymer surfaces derivatized with poly(vinyl-N-hexylpyridinium) kill airborne and waterborne bacteria. Biotechnol Bioeng. 2002;79:465.
62. Larm O, Larsson R, Olsson P. A new non-thrombogenic surface prepared by selective covalent binding of heparin via modified reducing terminal residue. Biomater Med Devices Artif Organs. 1983;11:161.
63. Bourin M-C, Lindahl U. Review article: glycosaminoglycans and the regulation of blood coagulation. Biochem J. 1993;289:313.
64. Phaneuf MD, Berceli SA, Bide MJ, Quist WC, Logerfo FW. Covalent linkage of recombinant hirudin to poly(ethylene terephthalate) (Dacron): creation of a novel antithrombin surface. Biomaterials. 1997;18:755.
65. Seifert B, Romaniuk P, Groth T. Covalent immobilzation of hirudin improves the haemocompatibility of polylactide-polyglycolide in vitro. Biomaterials. 1997;18:1495.
66. Rabenstein DL. Heparin and heparin sulfate: structure and function. Nat Prod Rep. 2002;19:312.
67. Hirsh J, Shaughnessy SG, Halperin JL, Granger C, Ohman EM, Dalen JE. Heparin and low-molecular weight heparin: mechanisms of action, pharmacokinetics, dosing, monitoring, efficacy, and safety. CHEST. 2001;119:64S.
68. Chandy T, Das GS, Wilson RF, Rao GHR. Use of plasma glow for surface-engineering biomolecules to enhance bloodcompatibility of Dacron and PTFE vascular prosthesis. Biomaterials. 2000;21:699.
69. Christensen K, Larsson R, Emanuelsson H, Elgue G, Larsson A. Improved blood compatibility of a stent graft by combining heparin coating and abciximab. Thromb Res. 2005;115:245.
70. Goddard JM, Hotchkiss JH. Polymer surface modification for the attachment of biactive compounds. Prog Polym Sci. 2007;32:698.
71. Tanga Y, Singh J. Controlled delivery of aspirin: effect of aspirin on polymer degradation and in vitro release from PLGA based phase sensitive systems. Int J Pharm. 2008;357:119.
72. Li F-M, Gu Z-W, Li G-W, WAang S, feng X-D. Synthesis of biocompatible polymers with asparin-moieties for asparin delivery. J Bioact Compat Polym. 1991;6:142.
73. Aldenhoff YB, Koole LH. Platelet adhesion studies on dipyridamole coated polyurethane surfaces. Eur Cell Mater. 2003;5:61.
74. Krishnan S, Ayothi R, Hexemer A, Finlay JA, Sohn KE, Perry R, Ober CK, Kramer EJ, Callow ME, Callow JA, et al. Anti-biofouling properties of comblike block copolymers with amphiphilic side chains. Langmuir. 2006;22:5075.
75. Chen H, Yuan L, Song W, Wu Z, Li D. Biocompatible polymer materials: role of protein-surface interactions. Prog Polym Sci. 2008;33:1059.
76. Currie EPK, Norde W, Stuart MAC. Tethered polymer chains: surface chemistry and their impact on colloidal and suface properties. Adv Colloid Interface Sci. 2003;100:205.
77. Ma HW, Hyun JH, Stiller P, Chilkoti A. Non-fouling oligo(ethylene glycol)-functionalized polymer brushes synthesized by surface-initiated atom transfer radical polymerization. Adv Mater. 2004;16:338.
78. Fan XW, Lin LJ, Messersmith PB. Cell fouling resistance of polymer brushes grafted from Ti substrates by surface-initiated polymerization: effect of ethylene glycol side chain length. Biomacromolecules. 2006;7:2443.
79. Wagner VE, Koberstein JT, Bryers JD. Protein and bacterial fouling characteristics of peptide and antibody decorated surfaces of PEG-poly(acrylic acid) co-polymers. Biomaterials. 2004;25:2247.
80. Du H, Chandaroy P, Hui SW. Grafted poly-(ethylene glycol) on lipid surfaces inhibits protein adsorption and cell adhesion. Biochim Biophys Acta. 1997;1326:236.
81. Thomas J, Choi SB, Fjeldheim R, Boudjouk P. Silicones containing pendant biocides for antifouling coatings. Biofouling. 2004;20:227.

82. Chen H, Brook MA, Sheardown H. Silicone elastomers for reduced protein adsorption. Biomaterials. 2004;25:2273.
83. Goda T, Konno T, Takai M, Moro T, Ishihara K. Biomimetic phosphorylcholine polymer grafting from polydimethylsiloxane surface using photo-induced polymerization. Biomaterials. 2006;27:5151.
84. Chen S, Zheng J, Li L, Jiang SY. Strong resistance of phosphorylcholine self-assembled monolayers to protein adsorption: Insights into nonfouling properties of zwitterionic materials. J Am Chem Soc. 2005;127:14473.
85. Kyomotoa M, Morob T, Saigaa K, Hashimotoe M, Itoc H, Kawaguchic H, Takatorib Y, Ishiharaa K. Biomimetic hydration lubrication with various polyelectrolyte layers on cross-linked polyethylene orthopedic bearing materials. Biomaterials. 2012;33:4451.
86. Silva R, Muniz EC, Rubira AF. Multiple hydrophobic polymer ultra-thin layers covalently anchored to polyethylene films. Polymer. 2008;49:4066.
87. Bae W-S, Urban MW. Reactions of antimicrobial species to imidazole-microwave plasma reacted poly(dimethylsiloxane) (PDMS) surfaces. Langmuir. 2004;20:8372.
88. Hensarling RM, Doughty VA, Chan JW, Patton DL. "Clicking" polymer brushes with thiol-yne chemistry: indoors and out. J Am Chem Soc. 2009;131:14673.
89. Kolb HC, Sharpless KB. The growing impact of click chemistry on drug discovery. Drug Discov Today. 2003;8:1128.
90. Chan EWL, Yousaf MN. Immobilization of ligands with precise control of density to electroactive surfaces. J Am Chem Soc. 2006;128:15542.
91. Zhu K, Zhang Y, He S, Chen W, Shen J, Wang Z, Jiang X. Quantification of proteins by functionalized gold nanoparticles using click chemistry. Anal Chem. 2012;84:4267.
92. Qin G, Santos C, Zhang W, Li Y, Kumar A, Erasquin UJ, Liu K, Muradov P, Trautner BW, Cai C. Biofunctionalization on alkylated silicon substrate surfaces via "click" chemistry. J Am Chem Soc. 2010;132:16432.
93. Sun X-L, Stabler CL, Cazalis CS, Chaikof EL. Carbohydrate and protein immobilization onto solid surfaces by sequential Diels–Alder and azide–alkyne cycloadditions. Bioconjugate Chem. 2006;17:52.
94. Li H, Cheng F, Duft AM, Adronov A. Functionalization of single-walled carbon nanotubes with well-defined polystyrene by "click" coupling. J Am Chem Soc. 2005;127:14518.
95. Wu P, Feldman AK, Nugent AK, Hawker CJ, Scheel A, Voit B, Pyun J, Frechet JMJ, Sharpless KB, Fokin VV. Efficiency and fidelity in a click-chemistry route to triazole dendrimers by the copper(I)-catalyzed ligation of azides and alkynes. Angew Chem Int Ed. 2004;43:3928.
96. Sumerlin BS, Vogt AP. Macromolecular engineering through click chemistry and other efficient transformations. Macromolecules. 2010;43:1.
97. Malkoch M, Vestberg R, Gupta N, Mespouille L, Dubois P, Mason AF, Hendrick JL, Liao Q, Frank CW, Kingsbury K, et al. Synthesis of well-defined hydrogel networks using click chemistry. Chem Commun. 2006;26:2774.
98. Li B, Martin AL, Gillies ER. Multivalent polymer vesicles via surface functionalization. Chem Commun. 2007;48:5217.
99. O'Reilly RK, Joralemon MJ, Wooley KL, Hawker CJ. Functionalization of micelles and shell cross-linked nanoparticles using click chemistry. Chem Mater. 2005;17:5976.
100. Pearson H, Manti J, Urban MW. Tethered polyelectrolyte stimuli-responsive chains on polymeris surfaces. Biomater Sci. 2014;2:512.
101. Hudalla GA, Murphy WL. Using "click" chemistry to prepare SAM substrates to study stem cell adhesion. Langmuir. 2009;25:5737.
102. Adzima BJ, Tao Y, Kloxin CJ, DeForest CA, Anseth KS, Bowman CN. Spatial and temporal control of the alkyne-azide cycloaddition by photoinitiated Cu(II) reduction. Nat Chem. 2011;3:258.

103. Nimmo CM, Shoichet MS. Regenerative biomaterials that "click": simple, aqueous-based protocols for hydrogel synthesis, suface immobilization, and 3D patterning. Bioconjugate Chem. 2011;22:2199.
104. Wang X, Liu L, Luo Y, Zhao H. Bioconjugation of biotin to the interfaces of polymeric micelles via in situ click chemistry. Langmuir. 2009;25:744.
105. Prasuhn DE, Singh P, Strable E, Brown S, Manchester M, Finn MG. Plasma clearance of bacteriophage Qβ particles as a function of surface charge. J. Am Chem Soc. 2008;130.
106. Qing G, Xiong H, Seela F, Sun T. Spatially controlled DNA nanopatterns by "click" chemistry using oligonucleotides with different anchoring sites. J Am Chem Soc. 2010;132:15228.
107. Aumsuwan N, Heinhorst S, Urban MW. Antibacterial surfaces on expanded polytetrafluoroethylene; penicillin attachment. Biomacromolecules. 2007;8:713.
108. Goddard JM, Hotchkiss JH. Polymer surface modification for the attachment of bioactive compounds. Prog Polym Sci. 2007;32:698–725.
109. Houbenov N, Minko S, Stamm M. Mixed polyelectrolyte brush from oppositely charged polymers for switching of surface charge and composition in aqueous environment. Macromolecules. 2003;36:5897.
110. Ionov L, Houbenov N, Sidorenko A, Stamm M. Inverse and reversible switching gradient surfaces from mixed polyelectrolyte brushes. Langmuir. 2004;20:9916.
111. Liu F, Urban MW. Dual temperature and pH responsiveness of poly(2-(N, N-dimethylamino)ethyl methacrylate-co-n-butyl acrylate) colloidal dispersions and their films. Macromoleucles. 2008;41:6531.
112. Tang L, Whalen J, Schutte G, Weder C. Stimuli-responsive epoxy coatings. ACS Appl Mater Interfaces. 2009;1:688.
113. Liu F, Ramachandran D, Urban MW. Colloidal films that mimic cilia. Adv Funct Mater. 2010;20:3163.
114. Hua ZD, Chen ZY, Li YZ, Zhao MP. Thermosensitive and salt-sensitive molecularly imprinted hydrogel for bovine serum albumin. Lamgmuir. 2008;24:5773.
115. Webb GF, D'Agata EMC, Magal P, Ruan S. A model of antibiotic-resistant bacterial epidemics in hospitals. Proc Natl Acad Sci. 2005;102:13343.
116. Twort FW. An investigation on the nature of ultramicroscopic viruses. Lancet. 1915;ii:1241.
117. Hosseinidoust Z, Van de Ven TGM, Tufenkji N. Bacterial capture efficiency and antimicrobial activity of phage-functionalized model surfaces Langmuir. 2011;27:5472.
118. Yang L-MC, Tam PY, Murray BJ, McIntire TM, Overstreet CM, Weiss GA, Penner RM. Virus electrodes for universal biodetection. Anal Chem. 2006;78:3265.
119. Hankin EH. L'action bactericide des eaux de la Jumma et du Gange sur le vibrion du cholera. Ann Inst Pasteur. 1896;10:511.
120. D'Herelle F. Sur un microbe invisible antagoniste des bacilles dysenteriques. C R Acad Sci. 1917;165:373.
121. Stone R. Stalin's forgotten cure. Science. 2002;298:728.
122. Sulakvelidze A, Alavidze Z, Morris JGJ. Bacteriophage therapy. Antimicrob Agents Chemother. 2001;45:649.
123. Chkhaidze JD, Imedashvili NE. The use of a novel biodegradable preparation capable of the sustained release of bacteriophages and ciprofloxacin, in the complex treatment of multidrug-resistant *Staphylococcus aureus*-infected local radiation injuries caused by exposure to Sr90. Clin Exp Dermatol. 2005;30:23.
124. Li-Mei CY, Phillip YT, Benjamin JM, Theresa MM, Cathie MO, Gregory AW, Penner RM. Virus electrodes for universal biodetection. Anal Chem. 2006;78:3265.
125. Pearson HA, Sahukhal GS, Elasri MO, Urban MW. Phage-bacterium war on polymeric surfaces: can surface-anchored bacteriophages eliminate microbial infections? Biomacromolecules. 2013;14(5):1257–61.

Marek W. Urban PhD is the Sirrine Foundation Endowed Chair and professor of materials science and engineering at Clemson University. He is the author of over 300 publications and has written 3 books. His research on antimicrobial and self-repairing materials has been featured by numerous media, including the New York Times, Forbes Magazine, BBC, NBC, Discovery Channel, USA Today, Yahoo, and many others. His current research interests focus on understanding physicochemical processes governing responsiveness in materials, development of novel polymeric materials with living-like functions, and design of self-repairing synthetic materials.

Chapter 5
Layer-by-Layer Coatings as Infection-Resistant Biomaterials

Svetlana A Sukhishvili

5.1 Introduction

One promising strategy to endow biomaterials with desired bioactivity is to coat their surfaces with functional polymer films which can regulate attachment of cells and bacteria and deliver bioactive molecules in a highly localized manner. The layer-by-layer (LbL) technique established in the early 1990s created many opportunities to engineer such biofunctional coatings [1–3]. The great promise of the LbL strategy in the biomedical arena is largely due to the ability of the technique to create conformal coatings on planar or virtually any substrate [4], to use a broad variety of components in film assembly, and to generate multicomponent stratified films of tunable functionality. Recent years have seen the development of a rich palette of LbL coatings tailored to regulate cellular functions through encapsulation, control cellular and/or bacterial adhesion, endow surfaces with antifouling properties, or deliver multiple anti-inflammatory, antibacterial, and pain-killing drugs [2]. Selection of desired biocompatible film components or incorporation of growth factors can improve cell–surface interactions, enhance cellular proliferation or selectively enrich cells in vitro [5, 6]. Moreover, LbL technology enables growth and detachment of cell sheets, thus showing strong potential in tissue repair and evaluation of novel therapeutic agents [7]. Because of the highly tunable physicochemical properties of these coatings (such as film mechanical properties and hydrophobicity) and the ease of inclusion of biological molecules, which provide biologically specific stimulation, LbL films have shown great promise in controlling cellular adhesion and differentiation [8].

Perhaps one of the most promising applications of LbL coatings takes advantage of the fact that various biologically active molecules can be incorporated within LbL films in large amounts, without loss of efficacy, and then released from these films in a controlled manner. Such use of LbL assemblies as efficient high-capacity

S. A. Sukhishvili (✉)
Department of Chemistry, Chemical Biology and Biomedical Engineering, Stevens
Institute of Technology, Hoboken, NJ 07030, USA

© Springer International Publishing Switzerland 2015   81
L. Santambrogio (ed.), *Biomaterials in Regenerative Medicine and the Immune System*,
DOI 10.1007/978-3-319-18045-8_5

reservoirs/matrices for localized delivery of multiple bioactive molecules has been demonstrated in several systems. Bioactive molecules can be incorporated within films in several ways. In one scenario, a drug can be uniformly loaded within the entire film, where the amount loaded can be simply and advantageously controlled by the number of LbL film layers, or film thickness. In some cases, bioactive molecules are complexed with oppositely charged macromolecules prior to assembly within LbL architectures [9]. In other cases, LbL assemblies can also be loaded with functional molecules at the post-assembly step [10]. There have been many successful demonstrations of release of functional agents from LbL films. One particular interesting example is the case of transdermal delivery of a protein antigen and an immunoadjuvant molecule at different time scales [11]. LbL assemblies can support either "passive" delivery of loaded compounds (occurring due to drug diffusion or hydrolytic film degradation), or release of functional compounds in response to various stimuli. Stimuli-responsive drug release can occur as a result of desorption of a drug from LbL films, enhanced permeability of the multilayers, or degradation of the entire film. A broad range of stimuli, including chemical (pH, ionic strength, solvent), physical (light, electric or magnetic field, temperature), or biological, involving responses to specific bioactive compounds, has been applied to polyelectrolyte multilayer coatings [12].

Several recent manuscripts and book chapters, including ours, reviewed the various opportunities offered by LbL coatings in tissue engineering and biomedicine [2, 8,13]. Some of these reviews discuss in detail both prolonged release as well as stimuli-responsive, on-demand delivery of a broad range of biologically active molecules from LbL films and capsules for a wide spectrum of biomedical applications [2, 13]. Different from the above reviews, this chapter focuses on recent progress in the application of the LbL technique to endow biomaterials with resistance to bacterial colonization.

Bacterial colonization of biomaterials is now recognized as the leading cause of failure in implanted biomedical devices. When bacteria develop into a biofilm [14], their resistance to antibiotics can increase by as much as 10,000 times [15], making systemic antibiotic treatments inefficient in eliminating the bacterial colonization of the implant surface. In the particular case of knee or hip replacements, for example, implant infection and the resulting revision surgeries needed to treat it increase the total time of patient recovery from ~2 to ~12 months and amplify the cost by a factor of ~10 [16]. The consequences of biomaterials-associated infection are enormous, both in terms of patient well-being and cost to society.

To render biomaterials resistant to bacterial colonization, antifouling polymer coatings that create steric and entropic barriers to bacterial adhesion have been developed [17–19]. In many cases, however, this type of coating provides insufficient support of mammalian cell adhesion required for integration of orthopedic implants. A more common approach is therefore to include antimicrobial compounds such as antibiotics, cationic polymers, cationic peptides, or silver to biomaterials. In many cases, these compounds can be blended into or grafted onto device surfaces [20–22]. Often, antibacterial coatings include positively charged functional groups (e.g., amino) which interact and efficiently disrupt negatively charged bacterial walls, killing bacteria on contact [23]. Polymer coatings of biodegradable polymers which continuously elute antibiotics have also been developed [20]. In spite of several successful demonstrations of antibacterial activity, these approaches have some limitations, such as accumulation of bacterial debris at and near surfaces and premature elution and deactivation of coatings. While LbL films do share the same limitations with non-LbL antibacterial coatings, because of their chemical and morphological versatility and possibility of film stratification in the direction perpendicular to the substrate, they afford rich opportunities in combining multiple antibiofilm defense functions. Here, we briefly discuss the use of LbL assemblies as antibacterial and antibiofilm matrices with various or multiple bacterial defense features, and will then specifically focus on films which deliver bioactive molecules in response to biological stimuli.

5.2   Approaches to Design Bacteria-Resistant Coatings

A variety of approaches to preventing bacterial colonization of biomedical devices have been proposed and explored (Fig. 5.1) [21]. Extending beyond low-adhesion surfaces is immobilization and/or delivery of bactericidal, quorum quenching or biofilm disruption biomolecules, or an approach of indirectly attacking bacteria via immobilization of immunoregulatory cytokines. Bactericide-free approaches recently have been receiving growing interest, largely due to the rapid rise of antibiotic-resistant bacteria such as methicillin-resistant *Staphylococcus aureus* (MRSA), among many others [24]. Approaches interfering with the production of bacterial biofilm based on enzymatic degradation of biofilm matrices or interference with the quorum sensing have clear advantages over the use of antimicrobial compounds because of the high specificity and non-toxicity of biofilm-disrupting and quorum sensing agents. One example of coatings which provide antibacterial protection through enzymatic degradation of a biofilm matrix include surface immobilization of DNase I or Dispersin B [25, 26]. Yet high selectivity also comprises a limita-

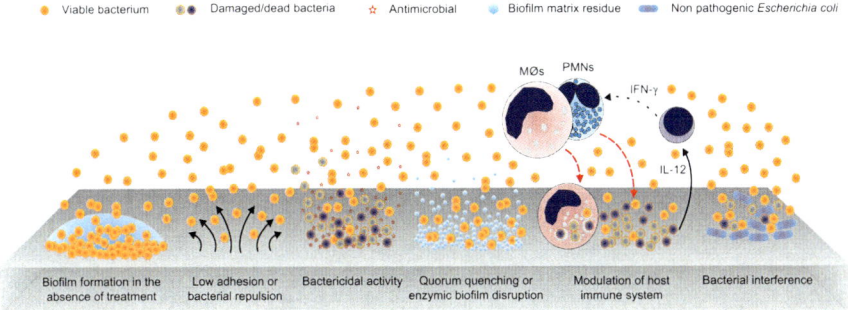

Fig. 5.1 Different strategies explored to combat bacterial colonization of biomedical devices. (From Ref. [21]). *PMN* polymorphonuclear neutrophil, *IFN-γ* interferon-γ, *IL-12* interleukin-12

tion of these approaches as compared with the use of broad-spectrum antibacterial compounds, including antibiotics. Recently, LbL films have been used to indirectly attack bacteria through activation of macrophages [27, 28]. In this approach, immunoregulatory cytokines, such as interleukin 12 (IL-12) and monocyte chemoattractant protein-1 (MCP-1) have been assembled within LbL films, and have been shown to significantly decrease infection rate in vivo in an open fracture rat model. While these new developments show significant promise, major research activities are yet focused on LbL coatings that contain broad-spectrum antimicrobials. Because of toxicity and antibiotic-resistance issues, the delivery mechanism becomes critically important in these studies. Better control of the delivery of antimicrobials from coatings, and development of novel on-demand delivery routes minimize toxicity and the contribution to antibiotic resistance of the delivered antimicrobials, therefore significantly alleviating concerns about their use. The discussion below will be focused on advances in mechanisms of delivery of antimicrobial compounds from LbL films.

5.3   Controlling Retention and Delivery of Antimicrobials Within/from LbL Coatings

Contact-killing LbL coatings. The LbL technique has been used as a convenient way to bind several types of positively charged antibacterial polymers at the surface of biomaterials. In some cases, cationic compounds are not released from the coating, but bacterial death results from a direct contact of the surface-bound positively charged bactericidal components of the film. Such contact-killing antibacterial LbL coatings have been assembled via inclusion of naturally antibacterial chitosan [29] or antimicrobial peptides (AMPs) [30, 31] within multilayer coatings. An important condition for achieving high antibacterial efficiency of the contact-killing coatings is sufficient mobility of bacteria-attacking polycations [25]. However, it is often desirable to retain cationic synthetic antibacterial polymers within coatings to minimize toxicity; other less toxic antibacterial agents, such as antibiotics or AMPs are often eluted continuously or on-demand from the LbL coatings.

Continuously eluting LbL coatings. One of the first approaches explored with LbL films is the use of these coatings as delivery matrices for silver ions. To that end, silver nanoparticles have been included within polyelectrolyte multilayers [32, 33]. In these systems, passive diffusion of silver ions from the coatings has provided antibacterial activity to the films. Although silver has been used as an antibacterial agent since ancient times, there are currently significant concerns associated with its limited efficiency and toxicity at high concentrations.

Small organic molecules—antibiotics—represent a large family of clinically used antibacterial compounds. These molecules vary in their charge and hydrophobicity, but an appropriate selection of an LbL matrix enables the advantageous use of antibiotic–matrix molecules interactions to assure inclusion of these antimicrobials within the film. For example, triclosan has been encapsulated within hydrophobic cores of LbL-assembled polymer micelles (Fig. 5.2) and was able to passively diffuse from the LbL coating, supporting continuous localized antibiotic delivery on the time scale of several days [34].

Another often-used approach is based on the use of biodegradable LbL films to continuously elute antibiotics or antimicrobial peptides from surfaces [35, 36]. A series of biocompatible cationic poly(β-amino esters) is frequently used as a degradable component for multilayer construction. Films containing these degradable polycations can incorporate and release various individual drugs [35, 37], or multiple therapeutic agents, including gentamicin, vancomycin, and/or diclofenac [38, 39]. The approach enables precise control of the dosage of the eluted drug. Moreover, the elution rate is determined by the chemistry of the assembled degradable polymers and the film stratification. Since the hydrolytic degradation rate of poly (β-amino ester) polymers depends on their hydrophobicity, charge density and strength of ionic pairing within LbL films, degradation of these coatings can be controlled by molecular parameters, as well as by these conditions of film assembly.

One problem frequently met with controlled release relying exclusively on diffusion is that loaded functional compounds leach out of LbL films in a short period of time. Moreover, small antibiotic molecules, such as gentamicin, could not be completely retained within the films designed to controllably deliver this drug exclusively via film degradation. Instead, gentamicin was eluted from LbL films through a combination of film degradation and diffusion [35]. Yet for many biomedical applications, it is desirable to extend the release time for at least several days or weeks to increase therapeutic dose and diminish risk of toxicity. A promising approach to slow down diffusion of drugs from LbL films is to insert low-permeability barrier layers (such as crosslinked polymer bilayers or graphene oxide) within LbL films [40, 41]. Another way to retain a drug within a polymer matrix is to bind it to one

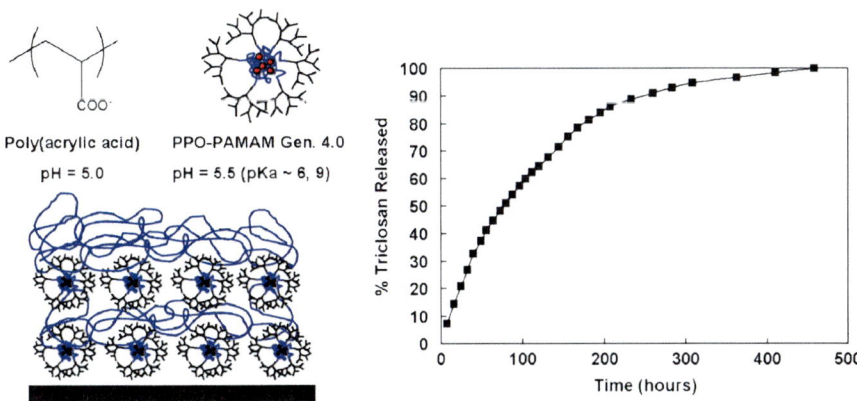

Fig. 5.2 *Left:* Schematic illustrating the formation of LbL films with positively charged PPO–PAMAM micelles encapsulating either hydrophobic drug or pyrene and negatively charged PAA. *Right:* Drug release profile of triclosan in an LbL film composed of PPO–PAMAM·triclosan/PAA. The release was performed at 37 °C in PBS. (Reprinted with permission from Ref. [34]. Copyright 2007 American Chemical Society). *PPO–PAMAM* polypropylene oxide–poly(amidoamine)

of the assembled polymers via a hydrolysable bond [42, 43]. Finally, submicron-sized hydrolysable particles of poly(D,L-lactic-co-glycolic acid) were also included within LbL films and used for multiple drug delivery [44]. These approaches, so far explored for the delivery of non-antibacterial therapeutic compounds, can be extended to controlling delivery rate of antimicrobials.

Because of the complexity of bacterial and biofilm metabolisms, new multidirectional approaches to design antibacterial coatings have been proposed to efficiently inhibit biomaterial-associated infection [45]. These LbL-based coatings have combined contact- and release-killing capabilities [46, 47], or act through a combination of a permanent protection with continuous elution from degradable LbL films [48]. In spite of significant progress in this research field, reliable antibacterial protection of implant surfaces remains challenging. In the specific case of clinically used antibiotics, continuous elution of antibiotics from surfaces contributes to the development of antibiotic-resistant bacterial strains. Therefore, delivery paths become critically important in the case of anti-biofilm coatings. Below we will discuss strategies to design infection-resistant surfaces that deliver antimicrobials on demand, that is, only when and/or where needed, thereby lowering the chances of antibiotic resistance and toxicity effects associated with continuous elution of antimicrobials.

Triggered Delivery of Antimicrobials from LbL Coatings. A variety of external stimuli, including temperature, electrochemical and redox-activation, and pH variations have been explored for triggered delivery of model or bioactive molecules from LbL surface coatings. One recent example of the use of LbL films for temperature-triggered delivery of bioactive compounds includes the use of LbL-assembled micelles with temperature-responsive cores [49]. In these systems, hydrophobic core-forming blocks of amphiphilic block copolymer micelles (BCMs) serve as depots for efficient uptake of small hydrophobic molecules, and allow for on-demand, stimuli-responsive release. Release of functional molecules from the coatings is controlled by the temperature-induced hydrophobic-to-hydrophilic phase transition of a core-forming polymer block. Responsive BCM-containing coatings of this type have been developed in several laboratories, including ours, and are reviewed in a recent publication [49]. Hydrogen-bonded LbL films containing temperature-responsive BCMs have been demonstrated to be a robust, reusable platform for localized, repeated on-demand delivery of a drug to cancer cells (Fig. 5.3) [50]. These systems are also likely to be promising for controlled delivery of hydrophobic antimicrobials.

Another up-and-coming way to deliver bioactive molecules from LbL films is to use an electric field as a trigger. The application of an electric field can disrupt binding between film-forming polymers, as in the case of biocompatible poly(L-lysine)(PLL)/heparin multilayers [51]. Alternatively, applied voltage can affect the oxidation state of LbL-assembled electroactive nanoparticles, resulting in release of film constituents [52].

The mechanism of electrochemical activation often involves pH changes which occur in the vicinity of the electrode as a result of electrolysis of water. Localized pH changes can also occur in the absence of electric field as a result of bacterial

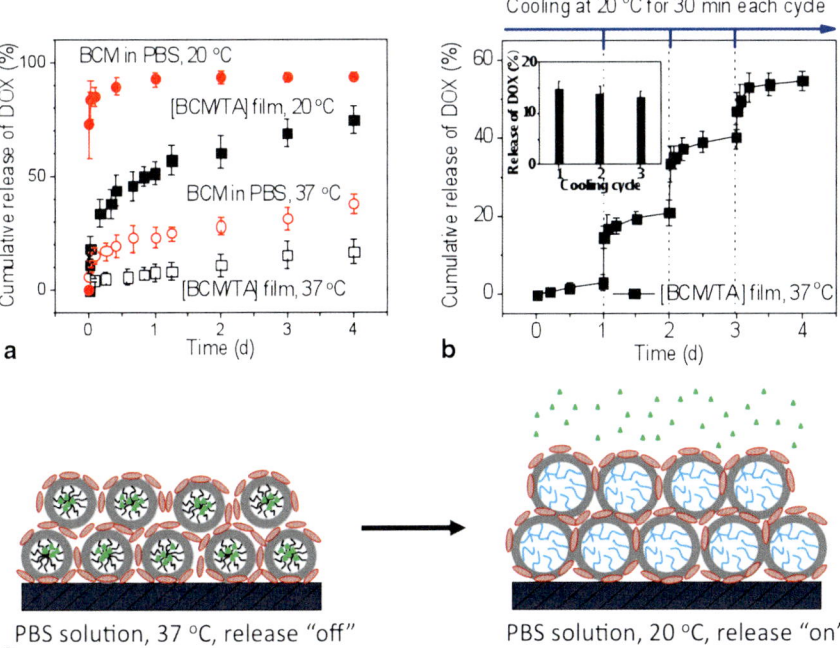

Fig. 5.3 **a** Cumulative release of doxorubicin (DOX) from free BCMs *(circles)* or from (BCM/tannic acid (TA))$_{40.5}$ films *(squares)* in PBS buffer (pH 7.4) at 20 °C *(open symbols)* and 37 °C *(filled symbols)*. DOX release was monitored by fluorometry (excitation at 480 nm), and released amounts normalized to the total amount of DOX loaded within the film. **b** Release profiles of DOX from [BCM/TA]$_{40.5}$ films immersed in PBS at a constant temperature of 37 °C, and cooled down to 20 °C for 30 min at 1, 2, and 3 days. Inset shows percentage of DOX released in each cooling cycle. **c** Schematic representation of "on-demand" DOX release from BCM-containing films. (From Ref. [50]). *PBS* phosphate buffered saline, *BCM* block copolymer micelles, *TA* tannic acid

metabolism. Significant progress has been made in designing LbL matrices which support pH-controlled localized delivery of antimicrobials. Often, these approaches include weak polyelectrolytes whose ionization state is pH-dependent. Antibiotics (cefazolin and gentamicin) have been included within LbL assemblies of poly(L-glutamic acid) and PLL, and released from the film using pH as a trigger [53]. Importantly, released antibiotics retained their antibacterial activities against *S. aureus* [53]. A model hydrophobic compound, pyrene, was also incorporated within electrostatically assembled films of amphiphilic BCMs, and films demonstrated pH-dependent release profiles of pyrene, resulting from pH response of core-forming polymer blocks [54]. Finally, antibacterial compounds can be included within cores of biodegradable micelles, and multilayers of these micelles with a hydrogen-bonding linear polymer can be destabilized by pH variations [10].

pH-decomposable hydrogen bonded films can also be stabilized within a broad pH range via chemical crosslinking, and act as reusable matrices for controlled loading and pH-induced release of antimicrobials [55]. Such crosslinked polymer network coatings retain bioactive molecules at pH 7.4 and release their contents upon pH lowering. In one recent example, films of this type were saturated with an electrostatically interacting antimicrobial peptide, and then released the peptide in response to local pH lowering associated with bacterial infection [55]. Specifically, L5 peptide was retained within the coatings at pH 7.5, and released at lower pH values. L5 peptide has demonstrated high antibacterial activity against *Staphylococcus epidermidis (S. epidermidis)* when included within crosslinked polyacid surface hydrogels, and also after release from the film [55]. This work presents the first example of smart coatings which have been named "self-defensive" [56], as such coatings remain dormant in the absence of bacterial infection, and release antibacterial agents only when bacteria arrive and locally acidify the medium. A similar bacteria-activated response has been later reported for polymer coatings which demonstrate pH-triggered conformational changes and turn bacteria-repellant when challenged with bacteria [57].

The term "self-defensive" introduced by Cado et al.[56] reflects very well the mechanism of action of these coatings. The authors have designed pathogen-responsive coatings with dual antifungal and antibacterial activity (Fig.5.4). An active AMP was incorporated within the film via grafting to hyaluronic acid (HA) chains which were then used for film construction via LbL assembly with chitosan (CHI). The films fully inhibited the development of *Staphylococcus aureus (S. aureus)* and *Candida albicans, (S. albicans)*, which are common virulent pathogens agents encountered in care-associated diseases. Importantly, the films retained their activity after three cycles of use against fresh inoculations of *C. albicans*. It is suggested that a likely trigger for localized released of AMP-modified HA degradation and release is the secretion of hyaluronidase by the pathogens [56]. Because of the limited adhesion of fibroblasts on these nontoxic films, the authors suggest the application of these coatings for surface modification of catheters or tracheal tubes rather than for coating orthopedic devices.

Recently, our group has developed several types of self-defensive coatings that contain large concentrations of antibiotics but do not release them at normal physiological conditions at pH 7.5 for at least several weeks [58]. In the specific case of

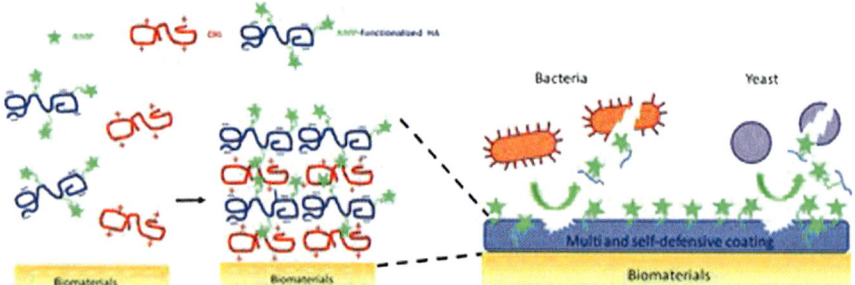

Fig. 5.4 Schematic representation of chitosan/hyaluronic acid functionalized by an antimicrobial peptide (AMP) and its activity against bacteria and yeasts based on the degradation of the film. (From Ref. [56])

antibiotics, this feature is specifically important because of the notorious development of antibiotic resistance. One type of these LbL coatings was obtained by direct assembly of tannic acid (TA) with one of several cationic antibiotics (tobromycin, gentamicin, and polymyxin B). Unlike films described above and depicted above in Fig. 5.4, the self-defensive behavior of these films is triggered by acidification of the immediate environment by pathogenic bacteria (Fig. 5.5), for example, by secretion of lactic acid by staphylococci bacteria, or acetic acid by *Escherichia coli (E. coli)*. Thus the highly localized pH-triggered burst release of antibiotics from TA/antibiotic films (Fig. 5.5c) kills bacteria only when present.

These coatings have a very high antibiotic content (300 µg/mm^3, 30 % by mass), and do not release antibiotics in phosphate buffered saline at pH 7.4 for as long as 1 month in the absence of bacteria. Therefore, in spite of the high content of antibiotics within these films, they do not contribute to the development of antibiotic resistance. Being stable and non-releasing at pH 7.5, TA/antibiotic coatings can deliver different antibiotic doses depending on the magnitude of pH decrease (Fig.5.5a). Interestingly, TA/antibiotic films assembled using spin-assisted and dip-assisted techniques show drastically different morphology, thickness and pH-/bacteria-triggered antibiotic release characteristics, and spin-assisted LbL assemblies demonstrate a transition from linear deposition of uniform 2D films to a highly developed 3D morphology for films thicker than 45 nm. Importantly, apart from the assembly technique (spin- versus dip-deposition), the rate of triggered release can be controlled through the choice of the spinning rate used during deposition. TA/antibiotic coatings as thin as 45 nm strongly inhibited growth of *S. epidermidis, S. aureus,* and *E. coli* which are known to cause infections associated with many implantable biomedical devices. The fact that TA/antibiotic coatings supported adhesion and proliferation of murine osteoblast cells makes them promising for applications in orthopedic implants, where tissue integration, in addition to prevention of bacterial colonization, is required.

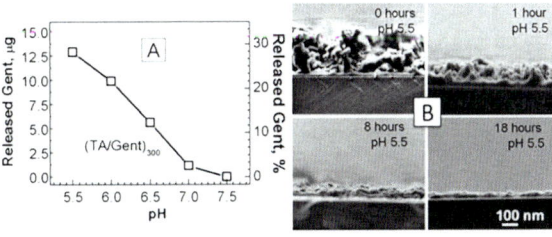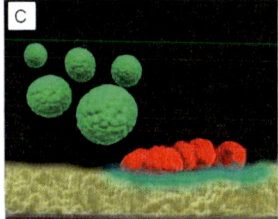

Fig. 5.5 a The amount of gentamicin (Gent) released from (TA/Gent)$_{300}$ films into 0.01 M phosphate buffer solutions containing 0.2 M NaCl at lowered pH values after 48 h, determined by mass spectrometry. Error bars are within symbol size if not shown. **b** Cross-sectional SEM images of a (TA/Gent)$_{300}$ film at different release times, indicating gradual collapse of an open 3D TA/Gent morphology during release. **c** Schematic representation of TA/antibiotic film action upon bacterial approach. A *greenish cloud* represents locally released antibiotic. Dead bacteria are shown in *red*. (Reprinted with permission from Ref. [58], Copyright 2014 American Chemical Society). *TA* tannic acid, *SEM* scanning electron microscope

To enhance the efficacy of self-defensive antibacterial coatings, they also can be designed to support multiple response functions. Very recently, nanocomposite LbL films containing montmorillonite (MMT) clay nanoplatelets and polyacrylic acid (PAA)—components of like charge at physiological conditions (pH 7.5, 0.2 M NaCl)—have been reported [59]. Before loading of antibiotics, films are highly swollen in water, and exhibit major changes in swelling as a function of pH. Yet the LbL constructs were mechanically robust because of the presence of clay nanosheets. Similarly to bulk MMT/polymer hydrogels, clay nanoplatelets can be considered as large crosslinkers, and a smaller number of crosslinks might be required to stabilize nanocomposite hydrogel structures as compared to all-polymer hydrogels. At the same time, unlike bulk hydrogels, MMT/PAA LbL films are prepared via sequential adsorption from MMT and PAA solutions on a solid, and therefore their structure and composition is governed by different (i.e., adsorption) laws than those applicable to their bulk counterparts. Under physiologic conditions, hydrogel-like MMT–PAA films can take up and sequester ~45% of dry film matrix mass of an antibiotic—gentamicin—causing dramatic film deswelling. Gentamicin remains sequestrated within the films for months under physiological conditions. When challenged with bacteria (*S. aureus, S. epidermidis,* or *E. coli*), coatings release PAA-bound gentamicin due to bacteria-induced acidification of its immediate environment, while gentamicin adsorbed to MMT nanoplatelets remains bound within the coating, affording lasting antibacterial protection (Fig. 5.6). Moreover, an increase in film swelling after gentamicin release further hinders bacterial adhesion. These multidirectional bacteria-triggered responses, enabled through a synergistic combination of inorganic (clay nanosheets) and organic (weak polyacid) components within LbL assemblies are likely to be a reason for high efficacy of the coatings. Gentamicin-loaded films demonstrated high killing efficacy against several types of bacteria relevant to infections of medical implants, such as *S. aureus, S. epidermidis,* and *E. coli.* As an example, antibacterial activity of gentamicin-loaded films was remarkably high: even when challenged with a bacterial number as high as 8×10^7 bacteria per cm^2 (close to complete bacterial monolayer coverage) in the Petrifilm test, gentamicin-loaded MMT/PAA films containing just 9.5 bilayers (matrix dry thickness of ~140 nm) demonstrated complete killing of *S. aureus.* Finally, the gentamicin-loaded nanocomposite coatings inhibited bacterial colonization while simultaneously promoting a healthy tissue response. The coatings could also be easily adapted to carry a variety of other antibiotics.

5.4   Conclusions and Outlook

LbL films with their ability to incorporate a variety of biologically active molecules and to be stratified into compartments serving as "depots" for different therapeutic agents, are attractive candidates for localized, time-scheduled delivery of multiple

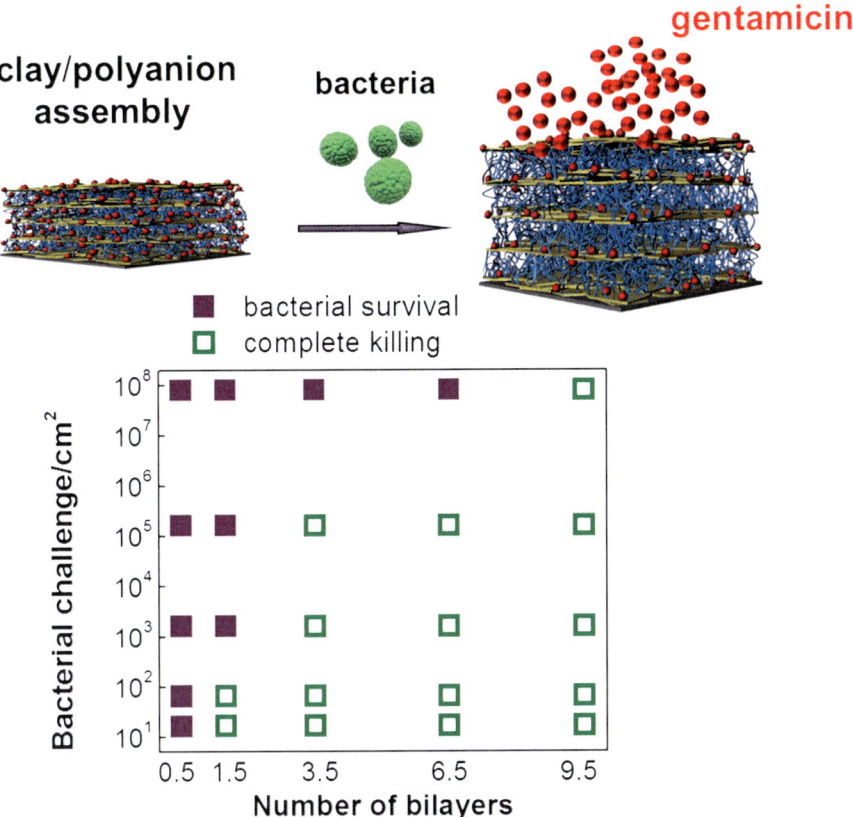

Fig. 5.6 *Above:* Schematic representation of bacterial response of nanocomposite self-defensive MMT/PAA coatings. *Below:* Survival or killing of *S. aureus* ATCC 12600 by gentamicin-loaded nanocomposite films as a function of number of bilayers within (MMT/PAA)$_n$ coatings, evaluated for different bacterial challenges. (From Ref. [59])

therapeutic agents from surfaces. This chapter has demonstrated that LbL coatings present a versatile platform for constructing multifunctional, highly efficient antimicrobial coatings at the surfaces of biomedical devices, which can control bacteria–biomaterial interactions through multiple paths. While the LbL approach has not yet made a strong appearance in everyday applications in practical biomaterials science, we believe that this technique will soon become a universal tool for engineering antibacterial coatings for the next generation of biomaterials.

Acknowledgments The author thanks Thomas Cattabiani (Stevens Institute of Technology) for his helpful comments on the chapter.

References

1. Delcea M, Möhwald H, Skirtach AG. Stimuli-responsive LbL capsules and nanoshells for drug delivery. Adv Drug Delivery Rev. 2011;63:730–47.
2. Pavlukhina S, Sukhishvili S. Polymer assemblies for controlled delivery of bioactive molecules from surfaces. Adv Drug Delivery Rev. 2011;63(9):822–36.
3. Ariga K, Lvov YM, Kawakami K, Ji Q, Hill JP. Layer-by-layer self-assembled shells for drug delivery. Adv Drug Delivery Rev. 2011;63:762–71.
4. Ariga K, Hill JP, Ji Q. Layer-by-layer assembly as a versatile bottom-up nanofabrication technique for exploratory research and realistic application. Phys Chem Chem Phys. 2007;9:2319–40.
5. Macdonald ML, Rodriguez NM, Shah NJ, Hammond PT. Characterization of tunable FGF-2 releasing polyelectrolyte multilayers. Biomacromolecules. 2010;11(8):2053–9.
6. Tsai H-A, Wu R-R, Lee IC, Chang H-Y, Shen C-N, Chang Y-C. Selection, enrichment, and maintenance of self-renewal liver stem/progenitor cells utilizing polypeptide polyelectrolyte multilayer films. Biomacromolecules. 2010;11(4):994–1001.
7. Chassepot A, Gao L, Nguyen I, et al. Chemically detachable polyelectrolyte multilayer platform for cell sheet engineering. Chem Mater. 2011;24(5):930–7.
8. Gribova V, Auzely-Velty R, Picart C. Polyelectrolyte multilayer assemblies on materials surfaces: from cell adhesion to tissue engineering. Chem Mater. 2011;24(5):854–69.
9. Dimitrova M, Affolter C, Meyer F, et al. Sustained delivery of siRNAs targeting viral infection by cell-degradable multilayered polyelectrolyte films. Proc Natl Acad Sci. 2008;105(42):16320–5.
10. Pavlukhina S, Lu Y, Patimetha A, Libera M, Sukhishvili S. Polymer multilayers with pH-triggered release of antibacterial agents. Biomacromolecules. 2010;11(12):3448–56.
11. Su X, Kim B-S, Kim SR, Hammond PT, Irvine DJ. Layer-by-layer-assembled multilayer films for transcutaneous drug and vaccine delivery. ACS Nano. 2009;3(11):3719–29.
12. De Geest BG, Sanders NN, Sukhorukov GB, Demeester J, De Smedt SC. Release mechanisms for polyelectrolyte capsules. Chem Soc Rev. 2007;36:636–49.
13. Alvarez-Lorenzo C, Concheiro A. Smart materials for drug delivery. RSC. Cambridge; 2013
14. Costerton JW, Cheng KJ, Geesey GG, Ladd TI, Nickel JC, Dasgupta M. Bacterial biofilms in nature and diseases. Ann Rev Microbiol. 1987;41:435–64.
15. Otto M. Staphylococcal biofilms. Curr Topics Microbiol Immunol. 2008;322:207–28.
16. Cataldo MA, Petrosillo N, Cipriani M, Cauda R, Tacconelli E. Prosthetic joint infection: recent developments in diagnosis and management. J Infect. 2010;61(6):443–8.
17. Banerjee I, Pangule RC, Kane RS. Antifouling coatings: recent developments in the design of surfaces that prevent fouling by proteins, bacteria, and marine organisms. Adv Mater. 2011;23(6):690–718.
18. Saldarriaga Fernández IC, van der Mei HC, Lochhead MJ, Grainger DW, Busscher HJ. The inhibition of the adhesion of clinically isolated bacterial strains on multi-component cross-linked poly(ethylene glycol)-based polymer coatings. Biomaterials. 2007;28(28):4105–12.
19. Harbers GM, Emoto K, Greef C, et al. Functionalized poly(ethylene glycol)-based bioassay surface chemistry that facilitates bio-immobilization and inhibits nonspecific protein, bacterial, and mammalian cell adhesion. Chem Mater. 2007;19(18):4405–14.
20. Price JS, Tencer AF, Arm DM, Bohach GA. Controlled release of antibiotics from coated orthopedic implants. J Biomed Mater Res. 1996;30(3):281–6.
21. Campoccia D, Montanaro L, Arciola CR. A review of the biomaterials technologies for infection-resistant surfaces. Biomaterials. 2013;34(34):8533–54.
22. Siedenbiedel F, Tiller JC. Antimicrobial polymers in solution and on surfaces: overview and functional principles. Polymers. 2012;4(1):46–71.
23. Schaer TP, Stewart S, Hsu BB, Klibanov AM. Hydrophobic polycationic coatings that inhibit biofilms and support bone healing during infection. Biomaterials. 2012;33(5):1245–54.

24. Bush K, Courvalin P, Dantas G, et al. Tackling antibiotic resistance. Nat Rev Microbiol. 2011;9(12):894–6.
25. Pavlukhina SV, Kaplan JB, Xu L, et al. Noneluting enzymatic antibiofilm coatings. ACS Appl Mater Interfaces. 2012;4(9):4708–16.
26. Swartjes JJTM, Das T, Sharifi S, et al. A functional DNase I coating to prevent adhesion of bacteria and the formation of biofilm. Adv Funct Mater. 2013;23(22):2843–9.
27. Li B, Jiang B, Boyce BM, Lindsey BA. Multilayer polypeptide nanoscale coatings incorporating IL-12 for the prevention of biomedical device-associated infections. Biomaterials. 2009;30(13):2552–8.
28. Li B, Jiang B, Dietz MJ, Smith ES, Clovis NB, Rao KMK. Evaluation of local MCP-1 and IL-12 nanocoatings for infection prevention in open fractures. J Orthop Res. 2010;28(1):48–54.
29. Fu J, Ji J, Yuan W, Shen J. Construction of anti-adhesive and antibacterial multilayer films via layer-by-layer assembly of heparin and chitosan. Biomaterials. 2005;26(33):6684–92.
30. Etienne O, Picart C, Taddei C, et al. Multilayer polyelectrolyte films functionalized by insertion of defensin: a new approach to protection of implants from bacterial colonization. Antimicrob Agents Chemother. 2004;48(10):3662–9.
31. Guyomard A, Dé E, Jouenne T, Malandain J-J, Muller G, Glinel K. Incorporation of a hydrophobic antibacterial peptide into amphiphilic polyelectrolyte multilayers: a bioinspired approach to prepare biocidal thin coatings. Adv Funct Mater. 2008;18(5):758–65.
32. Shi Z, Neoh KG, Zhong SP, Yung LYL, Kang ET, Wang W. In vitro antibacterial and cytotoxicity assay of multilayered polyelectrolyte-functionalized stainless steel. J Biomed Mater Res Part A. 2006;76A(4):826–34.
33. Lee D, Rubner MF, Cohen RE. Formation of nanoparticle-loaded microcapsules based on hydrogen-bonded multilayers. Chem Mater. 2005;17(5):1099–105.
34. Nguyen PM, Zacharia NS, Verploegen E, Hammond PT. Extended release antibacterial layer-by-layer films incorporating linear-dendritic block copolymer micelles. Chem Mater. 2007;19(23):5524–30.
35. Chuang HF, Smith RC, Hammond PT. Polyelectrolyte multilayers for tunable release of antibiotics. Biomacromolecules. 2008;9(6):1660–8.
36. Lichter JA, Van Vliet KJ, Rubner MF. Design of antibacterial surfaces and interfaces: polyelectrolyte multilayers as a multifunctional platform. Macromolecules. 2009;42(22):8573–86.
37. Moskowitz JS, Blaisse MR, Samuel RE, et al. The effectiveness of the controlled release of gentamicin from polyelectrolyte multilayers in the treatment of Staphylococcus aureus infection in a rabbit bone model. Biomaterials. 2010;31(23):6019–30.
38. Wong SY, Moskowitz JS, Veselinovic J, et al. Dual functional polyelectrolyte multilayer coatings for implants: permanent microbicidal base with controlled release of therapeutic agents. J Am Chem Soc. 2010:188–96.
39. Shukla A, Fleming KE, Chuang HF, et al. Controlling the release of peptide antimicrobial agents from surfaces. Biomaterials. 2010;31(8):2348–57.
40. Wood KC, Chuang HF, Batten RD, Lynn DM, Hammond PT. Controlling interlayer diffusion to achieve sustained, multiagent delivery from layer-by-layer thin films. Proc Natl Acad Sci. U S A. 2006;103(27):10207–12.
41. Hong J, Shah NJ, Drake AC, et al. Graphene multilayers as gates for multi-week sequential release of proteins from surfaces. ACS Nano. 2011;6(1):81–8.
42. Cao Y, He W. Synthesis and characterization of glucocorticoid functionalized poly(N-vinyl pyrrolidone): a versatile prodrug for neural interface. Biomacromolecules. 2010;11:1298–307.
43. Thierry B, Kujawa P, Tkaczyk C, Winnik FM, Bilodeau L, Tabrizian M. Delivery platform for hydrophobic drugs: prodrug approach combined with self-assembled multilayers. J Am Chem Soc. 2005;127:1626–7.
44. Soike T, Streff AK, Guan C, et al. Engineering a material surface for drug delivery and imaging using layer-by-layer assembly of functionalized nanoparticles. Adv Mater. 2010;22(12):1392–7.
45. Yuan W, Ji J, Fu J, Shen J. A facile method to construct hybrid multilayered films as a strong and multifunctional antibacterial coating. J Biomed Mater Res Part B. 2008;85B(2):556–63.

46. Li Z, Lee D, Sheng X, Cohen RE, Rubner MF. Two-level antibacterial coating with both release-killing and contact-killing capabilities. Langmuir. 2006;22(24):9820–3.
47. Fu J, Ji J, Fan D, Shen J. Construction of antibacterial multilayer films containing nanosilver via layer-by-layer assembly of heparin and chitosan-silver ions complex. J Biomed Mater Res Part A. 2006;79A(3):665–74.
48. Wong SY, Moskowitz JS, Veselinovic J, et al. Dual functional polyelectrolyte multilayer coatings for implants: permanent microbicidal base with controlled release of therapeutic agents. J Am Chem Soc. 2010;132(50):17840–8.
49. Zhu Z, Sukhishvili SA. Layer-by-layer films of stimuli-responsive block copolymer micelles. J Mater Chem. 2012;22(16):7667–71.
50. Zhu Z, Gao N, Wang H, Sukhishvili SA. Temperature-triggered on-demand drug release enabled by hydrogen-bonded multilayers of block copolymer micelles. J Control Release. 2013;171(1):73–80.
51. Boulmedais F, Tang CS, Keller B, Vörös J. Controlled electrodissolution of polyelectrolyte multilayers: a platform technology towards the surface-initiated delivery of drugs. Adv Funct Mater. 2006;16:63–70.
52. Wood KC, Zacharia NS, Schmidt DJ, Wrightman SN, Andaya BJ, Hammond PT. Electroactive controlled release thin films. Proc Natl Acad Sci U S A. 2008;105:2280–5.
53. Jiang BB, Li BY. Tunable drug loading and release from polypeptide multilayer nanofilms. Int J Nanomed. 2009;4(1):37–53.
54. Kellum MG, Harris CA, Mccormick CL, Morgan SE. Stimuli-responsive micelles of amphiphilic AMPS-b-AAL copolymers in layer-by-layer films. J Polym Sci Part A. 2010;49:1104–11.
55. Kim B-S, Park SW, Hammond PT. Hydrogen-bonding layer-by-layer-assembled biodegradable polymeric micelles as drug delivery vehicles from surfaces. ACS Nano. 2008;2(2):386–92.
56. Cado G, Aslam R, Seon L, et al. Self-defensive biomaterial coating against bacteria and yeasts: polysaccharide multilayer film with embedded antimicrobial peptide. Adv Funct Mater. 2013;23(38):4801–9.
57. Gensel J, Borke T, Pérez NP, et al. Cavitation engineered 3D sponge networks and their application in active surface construction. Adv Mater. 2012;24(7):985–9.
58. Zhuk I, Jariwala F, Attygalle AB, Wu Y, Libera MR, Sukhishvili SA. Self-defensive layer-by-layer films with bacteria-triggered antibiotic release. ACS Nano. 2014;8(8):7733–45.
59. Pavlukhina S, Zhuk I, Mentbayeva A, et al. Small-molecule-hosting nanocomposite films with multiple bacteria-triggered responses. NPG Asia Mater. 2014;6:e121.

Chapter 6
Nature-Inspired Multifunctional Host Defense Peptides with Dual Antimicrobial-Immunomodulatory Activities

Jasmeet Singh Khara and Pui Lai Rachel Ee

6.1 Introduction

Antimicrobial peptides (AMPs) have long been regarded as ancient artillery of human immunity that are indispensible components of both the adaptive and innate immune systems. The first, in a series of landmark studies, which put the spotlight on AMPs was the discovery of cecropins in 1980 from the hemolymph of the moth *Hyalophora cecropia* [1]. Further revelations of the existence of α-defensins in human neutrophils [2], and magainins in the skin of the frog *Xenopus laevis* [3], were suggestive of the ubiquitous presence of such defense molecules among eukaryotes, possessing broad-spectrum antimicrobial activity. Since then, over 2000 AMPs have been isolated and characterized to date from various plants and animal species including vertebrates such as amphibians, reptiles, birds, mammals, fish, and arthropods such as insects and crustaceans. Given the multiplicity of sources, it is no surprise that AMPs adopt diverse conformations, belonging to four main structural classes: α-helical, β-sheet, extended structures enriched with a particular amino acid, and loop peptides. Despite the exceptional disparities in sequence and configuration, AMPs are innately similar with a high proportion of hydrophobic and cationic residues, adopting an amphipathic structure upon folding [4].

Although the majority of early research was centered on the direct antimicrobial effects of AMPs, mounting evidence highlights that their broad-ranging immunomodulatory functions may be far more critical. To begin, the in vitro minimum inhibitory concentrations (MIC) of AMPs generally tend to be considerably higher than in vivo peptide concentrations under physiological conditions. Moreover, the direct bactericidal activity of AMPs is often found to be significantly compromised in the presence of monovalent (Na^+ and K^+) and divalent (Mg^{2+} and Ca^{2+}) cations, serum, and anionic polysaccharides [5]. The design of the synthetic peptide innate defense regulator 1 (IDR-1), void of direct antimicrobial activity but protective

P. L. R. Ee (✉) · J. S. Khara
Department of Pharmacy, National University of Singapore, 18 Science Drive 4,
Singapore 117543, Singapore

© Springer International Publishing Switzerland 2015
L. Santambrogio (ed.), *Biomaterials in Regenerative Medicine and the Immune System*,
DOI 10.1007/978-3-319-18045-8_6

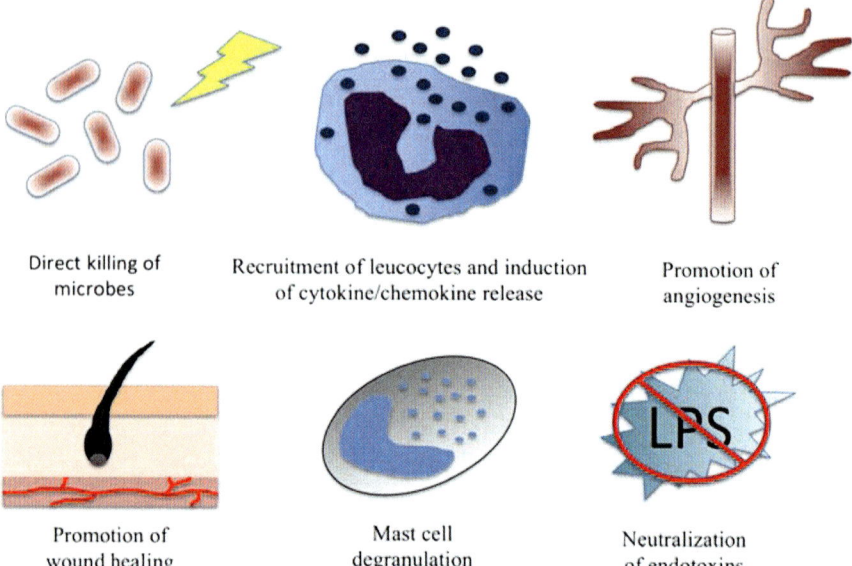

| Direct killing of microbes | Recruitment of leucocytes and induction of cytokine/chemokine release | Promotion of angiogenesis |

| Promotion of wound healing | Mast cell degranulation | Neutralization of endotoxins |

Fig. 6.1 Diverse host defense peptide-mediated protective mechanisms following exposure to infectious pathogens

against bacterial challenge in mouse models, further underlines the integral role of immunomodulatory functions in clearing bacterial infections [6].

As such, the term host defense peptide (HDP) has been increasingly applied to accurately describe peptides exhibiting immune-regulating properties in addition to direct antimicrobial activity. Figure 6.1 highlights the diverse biological functions of HDPs, with the multitude of immunomodulatory properties being found to encompass endotoxin-neutralizing ability, induction of cytokine and/or chemokine production, modulation of dendritic cell and T cell immune response, and promotion of wound healing and angiogenesis. The mechanisms behind these immune responses are discussed in greater detail in Sect. 6.2.2. The application of HDPs in human health diseases is not only restricted to inflammatory conditions such as skin disease, inflammatory bowel disease, periodontal disease, and sepsis but also includes cancer and their therapeutic potential as vaccine adjuvants. Of the countless HDPs prevalent in humans and mammals, human defensins and cathelicidins are the two most well characterized. In this chapter, we focus on the direct antibacterial activity and immunomodulatory activity of human defensins and cathelicidins, followed by a discussion on the design strategies aimed at developing novel synthetic HDPs with dual antimicrobial and immunomodulatory properties.

6.1.1 HDPs as Antimicrobials

HDPs are revered for their potential as anti-infective agents due to their broad spectrum of activity and rapid killing properties, coupled with the reduced propensity for resistance development. The ability to combat drug-resistant microbes and synergize with conventional antibiotics underscores their applicability in multidrug regimens as cost-effective and nontoxic adjuvants in infection control [7]. Though the exact mechanisms of action remain unclear, the cationic and amphiphilic regions of HDPs are widely understood to be influential in their direct microbicidal activity. These positively charged peptides initially accumulate on the bacterial surface, mediated largely by electrostatic interactions with anionic polymers, such as lipopolysaccharide and teichoic acids in Gram-negative and Gram-positive bacteria, respectively. Subsequent pore formation has been explained via several complex models including the toroidal or barrel-stave mechanisms or detergent-like membrane solubilization as espoused by the "carpet-like" mechanism [8]. Penetration of peptides into the lipid bilayer, driven by their inherent hydrophobicity, triggers membrane disruption that provokes homeostatic imbalances and loss of cellular content, culminating in cell death. However, the bacterial membrane may not necessarily be the sole target of HDPs as some translocate intracellularly and antagonize cellular division and the synthesis of proteins, DNA, RNA, and cell wall components [4, 9].

The only member of the cathelicidin family to be isolated from humans is the α-helical peptide, human cationic antimicrobial protein (hCAP)-18. Existing as a pro-peptide, hCAP-18 undergoes proteolytic cleavage at the C-terminal region to produce a 37-amino acid domain, widely known as LL-37. This cationic peptide has demonstrated potent inhibitory activity against a broad range of Gram-positive and Gram-negative bacteria in vitro [10]. Furthermore, the successful elimination of both multidrug-resistant *Acinetobacter baumannii* and methicillin-resistant *Staphylococcus aureus* (MRSA) in vitro following treatment with LL-37 highlights the potential utility of this peptide in eradicating drug-resistant infections [10, 11].

On the other hand, the two major classes of human defensins consist of α- and β-defensins, each containing six cysteine residues which form three pairs of disulfide bonds. The fundamental distinction resides in the connectivity of these disulfide bonds, with linkages between cysteine residues in positions 1–6, 2–4, and 3–5 in α-defensins, but in positions 1–5, 2–4, and 3–6 in β-defensins. In a similar fashion to cathelicidins, proteolysis of the precursor pro-peptide liberates the active element of defensins. Early work on α-defensins, or human neutrophil peptides (HNP), in the mid-1980s revealed that HNP 1–3 mixture was not only effective against Gram-positive and Gram-negative bacteria but also displayed antifungal and antiviral properties [2]. Follow-on studies extended this spectrum of activity to include various mycobacterial strains. HNP-1 exhibited bactericidal activity against *Mycobacterium tuberculosis* (MTB), while HNP-1, HNP-2, and HNP-3 were found to effectively eradicate *Mycobacterium avium–Mycobacterium intracellulare complex* (MAC) in vitro [12, 13].

The human β-defensins (hBD) were also effective in eradicating Gram-positive and Gram-negative bacteria, though hBD-2 demonstrated ~tenfold greater activity than hBD-1 [14]. However, their ability to reduce bacterial load was severely undermined upon exposure to high salt concentrations of NaCl. Interestingly, hBD-3 emerged as a promising candidate for the treatment of nosocomial infections, buoyed by evidence of its rapid bactericidal activity against multidrug-resistant clinical isolates of *S. aureus, Enterococcus faecium, Pseudomonas aeruginosa, Stenotrophomonas maltophilia,* and *A. baumannii* [15]. Combination treatment with HDPs has been shown to produce synergism and improve antimicrobial efficiency as compared to treatment with a lone agent. The addition of hBD-1, hBD-2, hBD-3, and LL-37 in various permutations revealed that a combination of two or more HDPs was essential in eliciting synergistic interactions against *S. aureus* and *Escherichia coli* [16]. A separate study evaluating the effect of combining sublethal doses of HNP-1 and LL-37 on their antimicrobial potency found that combination of both agents augmented the killing of *S. aureus* and *E. coli* [17]. These findings suggest that though devoid of activity when present alone at subinhibitory concentrations, combination of defensins with other HDPs can produce synergistic modulation of antibacterial activity in order to exert a bactericidal response.

6.1.2 HDP as Immune Modulators

HDPs are increasingly known for the multifaceted role they play in the regulation of the immune system (Fig. 6.1). Facilitating the chemotaxis of a myriad of immune cells remains one of the principal functions of HDPs. This chemotactic activity is induced either directly by chemokine production or indirectly via the activation of receptors. The ability of LL-37 to promote receptor-mediated chemotaxis of various immune cells has been widely established. LL-37 has demonstrated direct chemoattractant properties for mast cells, with cell migration mediated via the Gi protein–phospholipase C signaling pathway [18]. LL-37 was also found to be chemotactic for various leucocytes such as monocytes, neutrophils, eosinophils, and T lymphocytes [19, 20]. The involvement of the formyl peptide receptor-like 1 (FPRL1) proved central to the recruitment of human leucocytes in mounting an immune response.

In pioneering research aimed at elucidating the chemotactic properties of human defensins, Yang and coworkers revealed that hBD-2 was a potent inducer of both T cell and immature dendritic cell migration through interactions with CC chemokine receptor 6 (CCR6) [21]. In a follow-on study, the team determined that α-defensins, HNP-1, and HNP-2, were similarly chemotactic for both immature human dendritic cells and T lymphocytes, though independent of CCR6 activation [22]. Both dendritic and T cells have been implicated as key modulators of adaptive immune responses. Taken together, these findings serve to iterate the critical role of α- and β-defensins in both innate and adaptive immunity. HNP-1 and HNP-3 were also found to be chemoattractants for mast cells, T lymphocytes, and macrophages, while HNP-1 and HNP-2 were implicated as essential mediators of monocyte migration [23, 24]. Macrophage recruitment by HNP-1 was initiated following signal transduction by $G\alpha_i$ proteins and mitogen-activated protein kinases (MAPK) [23].

Besides their chemotactic activities, HDPs can elicit pro- or anti-inflammatory responses, dependent on the extent of cytokine induction or inhibition in different cell types. LL-37 effectively inhibited the release of pro-inflammatory cytokine, tumor necrosis factor- α (TNF-α), from macrophages while inducing anti-inflammatory chemokine monocyte chemoattractant protein-1 (MCP-1). In epithelial cells, however, MCP-1 production was antagonized while significant induction of interleukin (IL)-8 was observed. LL-37's immunomodulatory activity was achieved partly through the upregulation of chemokine receptors including CC chemokine receptor 2 (CCR2), C-X-C chemokine receptor type 4 (CXCR-4), and IL-8RB [25]. HNP-1, HNP-2, and HNP-3 increased TNF-α and IL-1 release from human monocytes while downregulating the expression of IL-10, a known inhibitor of cytokine synthesis. Moreover, the reduction in vascular cell adhesion molecule 1 (VCAM-1) expression in endothelial cells could potentially enhance HNP-mediated chemotaxis of leucocytes [26]. The stimulation of keratinocytes by hBD-2 and hBD-3 induced an array of pro-inflammatory cytokines and chemokines including IL-6, IL-10, interferon γ-inducible protein-10 (IP-10), MCP-1, macrophage inflammatory protein-3α (MIP-3α), and RANTES [27]. The G protein and phospholipase C (PLC) signaling pathways were found to be involved in the potentiation of this stimulatory effect.

6.2   Rationally Designed Synthetic HDPs

In a bid to develop more potent antibiotics with antibacterial-immunomodulation duality, researchers are turning towards nature-derived HDPs for inspiration. These molecules serve as a reference template for systematic modification by substitution, deletion, or scrambling of amino acid sequences to produce structurally related compounds termed as congeners. When the intention is to strategically combine the vital domains of two or more individual compounds, the end product is either a conjugate (peptide–ligand) or a hybrid (peptide–peptide). Another emerging class of synthetic HDPs is peptidomimetics, designed to structurally resemble naturally occurring defense peptides while addressing their inherent drawbacks. The following section closely examines the numerous approaches that have been utilized for the purposes of designing synthetic peptides endowed with superior killing and immune-regulating properties.

6.2.1   Antimicrobial Peptide Conjugates

In essence, this approach entails the coupling of a specific ligand or drug molecule to antimicrobial peptides with the aim of imparting dual antibacterial and immunomodulating properties. Though straightforward, in principle, the application of this design strategy has been limited thus far. Cationic steroid antibiotics (CSA), designed as synthetic mimics of the antibiotic polymyxin B, consist of steroid poly-amine conjugates capable of killing both Gram-positive and Gram-negative bac-teria [28]. Spermine, a polyamine that has been shown to possess broad-spectrum antibacterial activity, is inherent positively charged and bears a marked structural resemblance to cationic antimicrobial peptides [29]. This polycation was covalent-ly linked to dexamethasone, a corticosteroid with known anti-inflammatory and immunosuppressant properties, to give a disubstituted dexamethasone–spermine (D2S). Not only was this conjugate effective against *P. aeruginosa* and MRSA, D2S was further shown to inhibit the release of IL-6 and IL-8 from neutrophils ex-posed to lipopolysaccharides (LPS) or lipoteichoic acid (LTA) [30]. The conjugate also retained glucocorticoid activity suggesting that the anti-inflammatory activity of D2S was likely mediated via the engagement of glucocorticoid receptors.

More recently, our group has reported the conjugation of thymopentin (RKD-VY), a synthetic pentapeptide corresponding to amino acid residues 32–36 from the thymus hormone thymopoietin to the N- and C-terminus of a highly cationic antimicrobial peptide consisting of six arginine residues to confer the molecule with immunomodulatory properties [31]. Thymopentin, being an immunostimu-lant, replicates the biological activity of thymopoietin and promotes thymocyte differentiation and maturation. The resultant peptides, RR-11 (RKDVYRRRRRR) and RY-11 (RRRRRRKDVY), displayed potent bactericidal activity against both drug-susceptible and -resistant *Mycobacterium smegmatis*, while preserving the pro-inflammatory functions of thymopentin. Both RR-11 and RY-11 induced sig-nificant TNF-α release from macrophages which was found to be comparable to levels following stimulation with thymopentin alone.

Lipopeptide conjugates are also being explored for their potential as novel antibiotics with improved antibacterial and immunomodulatory properties. The conjugation of lipophilic acyl chains of various lengths to 12 amino acid residue fragments of human lactoferrin, LF12, was shown to enhance antibacterial killing considerably [32]. Lipopeptides consisting of acyl chains 12 carbon units in length (LF12-C12) were deemed to be the optimal composition, with enhancements in activity of 50-fold and 75-fold against *E. coli* and *S. aureus*, respectively. Further-more, LF12–C12 possessed a 12-fold superior endotoxin-neutralizing potency as compared to the parent peptide LF12. These results highlight the applicability of lipophilic modification to augment anti-inflammatory activity of AMPs.

6.2.2   Hybrid Antimicrobial Peptides

Hybridization is a technique combining the functional domains of two or more nat-urally occurring peptides with the intention of enhancing selectivity by improving antibacterial activity while minimizing cytotoxicity against mammalian cells. Liu and coworkers successfully employed this in the development of hybrid β-hairpin peptides by merging various segments derived from porcine cathelicidin protegrin-1 (PG-1), cecropin A isolated from hemolymph of *H. cecropia,* and a 25-amino acid peptide from the N-terminus of lactoferrin called bovine lactoferricin (LfcinB) [33]. The resulting hybrid peptides, LB–PG and CA–PG, exhibited superior antibacterial efficiencies and broader antimicrobial spectra as compared to parental peptides. The enhancement in potency of the constructed hybrids was attributed in part to their greater net positive charge, amphiphilicity, and propensity to adopt β-hairpin conformation. Moreover, LB–PG and CA–PG inhibited the expression and release of pro-inflammatory cytokines, TNF-α and inducible nitric oxide synthase (iNOS), and chemokines, MIP-1α and MCP-1, from macrophages stimulated with LPS. This finding underlines their potential as anti-inflammatory agents to suppress cytokine production and mitigate overstimulation of the immune system by bacterial endotoxins.

This design strategy has also been utilized to endow previously inactive peptides with enhanced antimicrobial and antiendotoxic activities. An LPS binding motif, termed as β-boomerang motif, GWKRKRFG, was bereft of both antimicrobial and antiendotoxic properties but provided promise due to its LPS-anchoring capabilities [34]. This boomerang motif was exploited to abolish LPS-induced aggregation, which traps peptides at the outer membrane and in turn limits their bactericidal activity. In a bid to salvage inactive peptides temporin-1Ta (TA), temporin-1Tb (TB), and KL-12, synthetic peptides incorporating the β-boomerang motif at the C-terminus were constructed, and the hybrids displayed remarkable improvement in activity against both Gram-positive and Gram-negative bacteria [35]. This modification also produced hybrids with superior endotoxin-neutralizing abilities, with peptides TA and TB having previously demonstrated poor activity in neutralizing LPS.

Synthetic hybrids may also be designed with the goal of incorporating the individual benefits of each fragment into a fusion product capable of manifesting desirable traits. Scudiero and his team set out to develop a hybrid peptide possessing the high salt tolerance capabilities of hBD3 while preserving the antibacterial activity of hBD1 [36]. Three analogues, 3N, IC, and 3I consistently displayed the highest potency against *P. aeruginosa, E.coli,* and *E. faecalis,* even at high salt concentrations. The chemotactic activity of the analogues for neutrophils and monocytes was maintained relative to parent peptides hBD1 and hBD3.

6.2.3 Antimicrobial Peptide Mimetics

In the pursuit of developing novel therapeutics with comparable efficacies to antimicrobial peptides while overcoming their shortcomings, including poor in vivo stability and high manufacturing costs, investigators have focused their attention on designing non-peptidic oligomers and polymers. Drawing inspiration from key structural features of natural AMPs, these molecules possess an inherent amphiphilic backbone and present net positive charges so as to mimic the biological functions of these peptides.

Magainin, an AMP isolated from the skin of African frog *X. laevis,* provided the template for the development of peptide mimetic, *meta*-phenylene ethynylene (mPE), which was explored for its anti-biofilm activity against oral pathogens [37]. mPE effectively inhibited the growth of bacterial pathogens *S. aureus, Porphyromonas gingivalis,* and *Streptococcus mutans,* and prevented *S. mutans* biofilm formation. An assessment into the anti-inflammatory activity of this peptide mimetic revealed a considerable decline in TNF-α secretion following LPS stimulation of macrophages, indicative of preserved biological function. mPE was further evaluated for its application in periodontal disease and displayed inhibitory activity against both *Aggregatibacter actinomycetemcomitans* and *P. gingivalis* biofilms [38]. Low concentrations of 2 μg ml^{-1} adequately suppressed IL-1β and induced IL-8 release in epithelial and myeloid cells, reiterating its potential utility as an anti-inflammatory agent in the management of other diseases.

CSAs represent another class of peptide mimetics resembling AMPs, with positive charges and hydrophobic residues on opposite faces giving rise to amphiphilic secondary conformations. Antimicrobial testing of analogues CSA-13, CSA-90, and CSA-92 against a wide array of bacterial strains found that these mimetics exhibited far superior antimicrobial activities than human cathelicidin LL-37 [39]. Furthermore, all three mimetics induced IL-8 release from keratinocytes, suggestive of dual antimicrobial and pro-inflammatory responses.

In an attempt to overcome challenges associated with poor solubility and complicated synthetic pathways, ultrashort histidine-derived AMPs (HDAMPs) were designed to possess both antimicrobial and anti-inflammatory activities. Consisting of one histidine and one or two arginine moieties, these peptidomimetics proved effective against both Gram-positive and Gram-negative bacteria, including MRSA [40]. HDAMPs also demonstrated immunosuppressive properties with LPS-neutralizing ability and significantly inhibiting the secretion of nitric oxide (NO) and TNF-α in macrophages stimulated with LPS.

Cyclic peptidomimetics consisting mainly of the cyclic lipo-α-AApeptides (*N*-acylated-*N*-aminoethyl peptides), on the other hand, displayed enhanced antimicrobial activities against both Gram-positive and Gram-negative bacteria as compared to the linear analogue [41]. The ability of these peptidomimetics to suppress inflammatory responses was apparent from the inhibition of NO and TNF-α production, mediated by antagonizing toll-like receptor 4 (TLR4)-induced nuclear factor kappa-light-chain enhancer of activated B cells (NF-kB).

6.2.4 Antimicrobial Peptide Congeners

A far more direct and common approach in designing peptides with higher antimicrobial efficacy and reduced systemic toxicity involves systematic manipulation of the amino acid sequence of peptides with known activity. Many naturally occurring peptides from mammals, amphibians, and insects have served as AMP templates including cathelicidin LL-37, hBD, lactoferrin, heparin, and thrombin. Such template-based studies are carried out by either substituting specific amino acids within the parent peptide sequence or truncating the N- and C-terminus of the parent peptide, or a combination of both these steps. These systematic modifications not only allow the investigation of crucial physiochemical parameters for improved potency and efficacy but also seek to identify the shortest possible active domains of the parental peptides. Table 6.1 provides an overview of studies adopting this strategy to achieve improved dual functionality of AMPs congeners derived from synthetic and human peptides based on the abovementioned techniques.

The effect of length, charge, helicity, hydrophobicity, and amphiphilicity on antimicrobial and hemolytic activity of AMPs has been extensively evaluated and reviewed in depth [42]. However, the modulating effect of these physiochemical parameters on the immunomodulatory functions of AMPs has been limited thus far. An investigation into the relationship between hydrophobicity and LPS-neutralizing activity of synthetic bovine myeloid antimicrobial peptide-18 (BMAP-18) revealed that both factors were linearly correlated. Increments in hydrophobicity of BMAP-18 analogues resulted in greater inhibition of NO and TNF-α production [43]. These findings were echoed in a separate study assessing the amino acid-substituted analogues of IG-19, corresponding to the α-helical region of cathelicidin LL-37, which also found that enhancements in peptide hydrophobicity showed significant linear correlation with improved LPS-binding activity [44].

Apart from overall peptide hydrophobicity, the pinpoint substitution of specific hydrophobic amino acids along a peptide sequence can potentiate both antimicrobial and anti-inflammatory activities. By swapping out a single serine residue at position nine along the hydrophobic face of LL-23 with valine, the weakly bactericidal and immunosuppressive peptide was rendered more active, with a pronounced reduction in the release of pro-inflammatory cytokines, TNF-α and MCP-1 [45]. This point mutation directed the formation of a continuous hydrophobic face, which had previously been segregated by the hydrophilic serine residue. Another work comparing two tryptophan-substituted analogues uncovered that substitution of this hydrophobic amino acid at the amphipathic interface, rather than at the center of the hydrophobic face, produced analogues with greater LPS-neutralizing activity [44]. Taken together, these results highlight the importance of site-specific tryptophan substitutions and a continuous hydrophobic surface in directing the design of novel AMPs with improved selectivity and anti-inflammatory activity.

Another approach within this design strategy utilizes a generic sequence as a reference template to develop novel AMP congeners. Tryptophan-rich peptides, such as indolicidin, isolated from bovine neutrophils, have garnered interest owing to their broad spectrum of antimicrobial activity. Park and coworkers explored short tryptophan-rich AMPs bearing the sequence XXWXXWXXWXX-NH$_2$, where X represents Leu or Lys/Arg, for their potential antibacterial and antiendotoxin activities [46]. Several synthetic analogues possessed enhanced activity against *E. coli* and *S. aureus,* while majority displayed superior killing properties against MRSA relative to indolicidin. These analogues also proved to be potent anti-inflammatory agents, evident from their ability to strongly inhibit NO production and neutralize LPS. In a similar study, the generic sequence Ac-C-HBHB(P)HBH-GSG-HBHB(P)

Table 6.1 Summary of various synthetically and human-derived AMP congeners possessing dual antimicrobial and immunomodulatory functions and their respective modifications

AMPs	Description	Modifications	Antibacterial activity spectrum	Immunomodulatory properties	Ref. no.
1. Synthetically derived peptides					
Short Trp-rich AMPs	XXWXXXWXXWXXWXX-NH$_2$ (X = Leu or Lys)	Substitution	Gram positive Gram negative	Inhibition of NO release Direct LPS binding activity	[46]
V1–V7	Ac-C-HBHB(P)HBH-GSG-HBHB(P) HBH-C-NH$_2$	Substitution	Gram positive Gram negative	Direct LPS binding activity	[47]
TP10	Deletion analogue of the chimeric cell penetrating peptide (CPP) transportan	Truncation	*Neisseria meningitidis*	Inhibition of TNF-α and IL-6 release	[51]
K$_9$L$_8$W	Ideal amphipathic α-helical Leu/Lys-rich model peptide	Substitution	Gram positive Gram negative	Inhibition of TNF-α and NO production	[49]
2. Human-derived peptides					
IG-19	Residues 13–31 of human cathelicidin LL-37	Truncation Substitution	Gram positive Gram negative	Inhibition of NO and TNF-α release Inflammation control in rheumatoid arthritis	[44, 52]
KR-12	Residues 18–29 of human cathelicidin LL-37	Truncation Substitution	Gram positive Gram-negative MRSA	Inhibition of TNF-α release Direct LPS binding activity	[53]
LL-23V9	N-terminal 23-amino acid fragment of human cathelicidin LL-37	Truncation Substitution	Gram positive Gram negative	Inhibition of TNF-α and MCP-1 release	[45]
Fr 106 and 110	N-terminal fragments of human cathelicidin LL-37	Truncation	Gram positive Gram negative and *C. albicans*	Chemoattractant for neutrophils	[54]
GKE	C- and N-terminal fragments of human cathelicidin LL-37	Truncation	Gram positive Gram negative	Inhibition of NO release Chemoattractant for neutrophils	[55]
31–50v44w	Amino acid residues 31–50 of granulysin	Truncation Substitution	*P. acnes*	Inhibited production of chemokines MCP-1, IP-10, and MDC Reduced IL-12p40 release	[56]

Table 6.1 (continued)

AMPs	Description	Modifications	Antibacterial activity spectrum	Immunomodulatory properties	Ref. no.
ApoE23	Derived from the low-density lipoprotein receptor-binding region of Apolipoprotein E (apoE)	Truncation	Gram positive Gram negative Multidrug-resistant *A. baumannii*	Reduced secretion of TNF-α, IL-6, and IL-10	[57]
GKY25	C-terminal peptide of thrombin	Truncation	Gram positive Gram negative	Inhibition of NO and TNF-α release Reduction of pro-inflammatory cytokines IL-6, IFN-g, IL-12p70	[58]
GKS26	A domain of heparin cofactor II (HCII)	Truncation	Gram positive Gram negative	In vitro decrease in IL-6 and TNF-α release In vivo reduction of IL-6 TNF-α, MCP-1, IFN-g, and IL-10 secretion	[59]

HBH-C-NH$_2$, where Ac is an acetyl moiety, H is a hydrophobic residue, P is a polar residue, and B is a cationic residue, served as a framework to design cyclic β-hairpin peptides with LPS and lipid A-binding properties [47]. Besides being antiendotoxic, these synthetic AMPs were highly selective for microbes over mammalian cells.

The substitution of D-amino acids into the peptide sequence is a practice usually aimed at decreasing the inherent cytotoxicity of AMPs as well as improving their stability to proteolytic degradation. Interestingly, however, a study investigating the effect of swapping the entire amino acid sequence from L- to D-isomers found that the D-analogue exhibited the greatest antimicrobial potency among all the peptides tested [48]. Notably, the D-analogue proved to be a stronger inhibitor of NO production and pro-inflammatory cytokines, TNF-α and MIP-2, at all concentrations. In a separate work, several D-amino acid substituted analogues of an amphipathic α-helical peptide K$_9$L$_8$W were found to possess superior cell selectivity while maintaining potent anti-inflammatory activity [49]. Given the paucity of research in this regard, the benefits of such an approach to develop more potent anti-inflammatory agents while concurrently improving their antibacterial activity warrant further investigation.

6.2.5   Genomic Mining Strategies

Recently, the first report on mycobacteriophage-derived AMPs found to exhibit potent antibacterial and immunoregulatory property has emerged from a team of scientists based out of China [50]. The researchers systematically screened the complete genome of over 70 mycobacteriophages for peptides capable of inhibiting both the growth of MTB and activity of trehalose-6,6'-dimycolate (TDM), an immunostimulant produced by virulent MTB that incites inflammatory responses and pulmonary granuloma formation. Out of 200 short-listed candidates, a 34-amino acid peptide, PK34, was found to possess the strongest antimycobacterial activity with an MIC of 50 μg/ml. PK34 also hampered the in vitro release of pro-inflammatory cytokines IFN-γ, TNF-α, MCP-1, IL-6, IL-10, and IL-12 in a concentration-dependent manner. In vivo, PK34's mitigating effect on MTB granuloma formation and TDM-induced inflammatory response was likely mediated by blocking MAPK activation. Given the diversity of bacteriophages and the abundance of information nestled within their genomic sequences, such innovative approaches could provide the key to unlocking novel compounds in our quest to expand our arsenal of antibiotics.

6.3   Clinical Applications of HDPs

Though many peptide-based therapeutics have been evaluated clinically, none have been approved for use in the clinic to date. Table 6.2 shows HDPs that are currently being pursued in clinical development for their immunomodulatory and antimicrobial activities. Three of the six peptides listed are being tested for treatment of skin infections such as acne, impetigo, and acute bacterial skin and skin structure infections (ABSSSI) caused by *S. aureus*. The development of three other peptides, iseganan, pexiganan, and omeganan, undergoing late-stage phase III trials was hampered either due to a lack of efficacy or regulatory barriers. Iseganan was unable to meet its primary end points in the treatment of oral mucositis in patients subjected to radiation therapy for cancer, which impelled IntraBiotics to halt further product development (NCT00022373). Researchers behind the development of a promising antimicrobial peptide, pexiganan, for the treatment of diabetic foot ulcers were dealt a major blow back in 1999 when the US Food and Drug Administration (FDA) deemed the drug not approvable [60]. Citing ethical concerns over the trial design and lack of evidence demonstrating superiority over conventional therapies, the FDA's decision forced Magainin Pharmaceuticals to cease operations. However, Dipexium Pharmaceuticals, bent on reversing this earlier decision, has recently initiated phase III trials for the pexiganan cream, Locilex (Table 6.2). Another drug hindered by regulatory hurdles, omiganan, was also unsuccessful in achieving FDA approval for the treatment of catheter infections following phase III trials. This failure prompted Cutanea Life Sciences to explore other potential indications, with encouraging results in phase II trials, bolstering the push towards phase III studies for the treatment of rosacea (Table 6.2).

6.4   Conclusions and Future Perspectives

Naturally occurring HDPs and their synthetic derivatives not only exhibit direct antimicrobial activity but also diverse immunomodulatory properties, underscoring their potential utility in infectious and inflammatory human diseases. High manufacturing costs and poor in vivo stability have directed the development of shorter analogues with D-amino acid substitutions or synthetic peptidomimetics aimed at enhancing resistance to proteolysis. Yet, clinical development of peptide therapeutics discussed in this chapter has been limited to topical applications as anti-infectives in skin diseases, rosacea, oral candidiasis, and diabetic foot ulcers. With shrinking antibiotic pipelines and declining approvals of new antibiotics, it is no wonder that pexiganan is the sole synthetic HDP, possessing dual antimicrobial-immunomodulating properties, that is entering phase III trials in 2014. The problem, exacerbated by limited funding sources and the rapid emergence of drug-resistant infections, warranted action from the US Senate, which responded by approving the Generating Antibiotics Incentives Now (GAIN) Act in mid-2012. The FDA Antibacterial Drug Development Task Force was soon launched to address the declining antibiotic pipeline, to evaluate and update clinical trial guidelines, and to expedite the regulatory approval of new antibiotics. With the authorities taking positive steps to facilitate the development of new antibiotics, it may be sooner rather than later that the first HDP makes it from the lab to the clinic.

Table 6.2 Synthetic host defense peptides with dual antimicrobial and immunomodulatory properties undergoing active clinical development

Peptide[a]	Description	Clinical application	Phase	Status	Company	Ref/Reg no.[b]
Brilacidin (PMX30063)	Synthetic defensin mimetic	Treatment for acute bacterial skin and skin structure infections (ABSSSI) caused by S. aureus	II	In progress	Cellceutix Corporation	NCT02052388
HB1345	Synthetic lipohexapeptide	Anti-infective for skin infections such as acne	Pre-phase I	In progress	Helix BioMedix	http://helixbio-medix.com/
LTX-109	Synthetic antimicrobial peptido-mimetic (SAMP)	Treatment of nasal MRSA/MSSA infection	I/II	Completed	Lytix Biopharma	NCT01158235
		Impetigo	II	Completed		NCT01803035
Omiganan (CLS001)	Synthetic 12-mer peptide derived from indolicidin	Treatment of inflammatory papules and pustules associated with rosacea	II	Completed	Cutanea Life Sciences	NCT01784133
PAC-113	Synthetic 12-mer peptide from histatin	Treatment of oral candidiasis in HIV-seropositive patient	IIb	Completed	Pacgen Biopharmaceuticals	NCT00659971
Pexiganan (MSI-78)	Synthetic 22-amino acid peptide isolated from the skin of the African clawed frog	Treatment of mild infections associated with diabetic foot ulcers	III	In progress	Dipexium Pharmaceuticals	NCT01590758 NCT01594762

a Research programs that were terminated or suspended indefinitely have been excluded
b Reference or registration numbers are obtained from http://clinicaltrials.gov

References

1. Steiner H, Hultmark D, Engström A, Bennich H, Boman H. Sequence and specificity of two antibacterial proteins involved in insect immunity. Nature. 1981;292(5820):246–8.
2. Ganz T, Selsted ME, Szklarek D, et al. Defensins. Natural peptide antibiotics of human neutrophils. J Clin Invest. 1985;76(4):1427.
3. Zasloff M. Magainins, a class of antimicrobial peptides from Xenopus skin: isolation, characterization of two active forms, and partial cDNA sequence of a precursor. Proc Natl Acad Sci U S A. 1987;84(15):5449–53.
4. Zasloff M. Antimicrobial peptides of multicellular organisms. Nature. 2002;415(6870):389–95.
5. Bowdish DM, Davidson DJ, Lau YE, Lee K, Scott MG, Hancock RE. Impact of LL-37 on anti-infective immunity. J Leukocyte Biol. 2005;77(4):451–9.
6. Scott MG, Dullaghan E, Mookherjee N, et al. An anti-infective peptide that selectively modulates the innate immune response. Nat Biotechnol. 2007;25(4):465–72.
7. Khara JS, Wang Y, Ke X-Y, et al. Anti-mycobacterial activities of synthetic cationic α-helical peptides and their synergism with rifampicin. Biomaterials. 2014;35(6):2032–8.
8. Hancock RE, Sahl H-G. Antimicrobial and host-defense peptides as new anti-infective therapeutic strategies. Nat Biotechnol. 2006;24(12):1551–7.
9. Peschel A, Sahl H-G. The co-evolution of host cationic antimicrobial peptides and microbial resistance. Nat Rev Microbiol. 2006;4(7):529–36.
10. Travis SM, Anderson NN, Forsyth WR, et al. Bactericidal activity of mammalian cathelicidin-derived peptides. Infect Immun. 2000;68(5):2748–55.
11. Feng X, Sambanthamoorthy K, Palys T, Paranavitana C. The human antimicrobial peptide LL-37 and its fragments possess both antimicrobial and antibiofilm activities against multidrug-resistant *Acinetobacter baumannii*. Peptides. 2013;49:131–7.
12. Ogata K, Linzer B, Zuberi R, Ganz T, Lehrer R, Catanzaro A. Activity of defensins from human neutrophilic granulocytes against *Mycobacterium avium-Mycobacterium intracellulare*. Infect Immun. 1992;60(11):4720–5.
13. Sharma S, Verma I, Khuller G. Biochemical interaction of human neutrophil peptide-1 with *Mycobacterium tuberculosis* H37Ra. Arch Microbiol. 1999;171(5):338–42.
14. Singh PK, Jia HP, Wiles K, et al. Production of β-defensins by human airway epithelia. Proc Natl Acad Sci U S A. 1998;95(25):14961–6.
15. Maisetta G, Batoni G, Esin S, et al. In vitro bactericidal activity of human β-defensin 3 against multidrug-resistant nosocomial strains. Antimicrob Agents Chemother. 2006;50(2):806–9.
16. Chen X, Niyonsaba Fß, Ushio H, et al. Synergistic effect of antibacterial agents human β-defensins, cathelicidin LL-37 and lysozyme against *Staphylococcus aureus* and *Escherichia coli*. J Dermatol Sci. 2005;40(2):123–32.
17. Nagaoka I, Hirota S, Yomogida S, Ohwada A, Hirata M. Synergistic actions of antibacterial neutrophil defensins and cathelicidins. Inflamm Res. 2000;49(2): 73–9.
18. Niyonsaba F, Iwabuchi K, Someya A, et al. A cathelicidin family of human antibacterial peptide LL-37 induces mast cell chemotaxis. Immunology. 2002;106(1):20–6.
19. Tjabringa GS, Ninaber DK, Drijfhout JW, Rabe KF, Hiemstra PS. Human cathelicidin LL-37 is a chemoattractant for eosinophils and neutrophils that acts via formyl-peptide receptors. Int Arch Allergy Imm. 2006;140(2):103–12.
20. Yang D, Chen Q, Schmidt AP, et al. LL-37, the neutrophil granule- and epithelial cell-derived cathelicidin, utilizes formyl peptide receptor-like 1 (FPRL1) as a receptor to chemoattract human peripheral blood neutrophils, monocytes, and T cells. J Exp Med. 2000;192(7):1069–74.
21. Yang D, Chertov O, Bykovskaia S, et al. β-Defensins: linking innate and adaptive immunity through dendritic and T cell CCR6. Science. 1999;286(5439):525–8.
22. Yang D, Chen Q, Chertov O, Oppenheim JJ. Human neutrophil defensins selectively chemoattract naive T and immature dendritic cells. J Leukocyte Biol. 2000;68(1):9–14.

23. Grigat J, Soruri A, Forssmann U, Riggert J, Zwirner J. Chemoattraction of macrophages, T lymphocytes, and mast cells is evolutionarily conserved within the human α-defensin family. J Immunol. 2007;179(6):3958–65.
24. Territo M, Ganz T, Selsted M, Lehrer R. Monocyte-chemotactic activity of defensins from human neutrophils. J Clin Invest. 1989;84(6):2017.
25. Scott MG, Davidson DJ, Gold MR, Bowdish D, Hancock RE. The human antimicrobial peptide LL-37 is a multifunctional modulator of innate immune responses. J Immunol. 2002;169(7):3883–91.
26. Chaly YV, Paleolog E, Kolesnikova T, Tikhonov I, Petratchenko E, Voitenok N. Neutrophil α-defensin human neutrophil peptide modulates cytoline production in human monocytes and adhesion molecule expression in endothelial cells. Eur Cytokine Netw. 2000;11(2):257–66.
27. Niyonsaba Fß, Ushio H, Nakano N, et al. Antimicrobial peptides human β-defensins stimulate epidermal keratinocyte migration, proliferation and production of proinflammatory cytokines and chemokines. J Invest Dermatol. 2006;127(3):594–604.
28. Salunke DB, Hazra BG, Pore VS. Steroidal conjugates and their pharmacological applications. Curr Med Chem. 2006;13(7):813–47.
29. Rozansky R, Bachrach U, Grossowicz N. Studies on the antibacterial action of spermine. J Gen Microbiol. 1954;10(1):11–6.
30. Bucki R, Leszczyńska K, Byfield FJ, et al. Combined antibacterial and anti-inflammatory activity of a cationic disubstituted dexamethasone-spermine conjugate. Antimicrob Agents Chemother. 2010;54(6):2525–33.
31. Wang Y, Ke X-Y, Khara JS, et al. Synthetic modifications of the immunomodulating peptide thymopentin to confer anti-mycobacterial activity. Biomaterials. 2014;35(9):3102–9.
32. Majerle A, Kidrič J, Jerala R. Enhancement of antibacterial and lipopolysaccharide binding activities of a human lactoferrin peptide fragment by the addition of acyl chain. J Antimicrob Chemoth. 2003;51(5):1159–65.
33. Liu Y, Xia X, Xu L, Wang Y. Design of hybrid β-hairpin peptides with enhanced cell specificity and potent anti-inflammatory activity. Biomaterials. 2013;34(1):237–50.
34. Bhunia A, Mohanram H, Domadia PN, Torres J, Bhattacharjya S. Designed β-boomerang antiendotoxic and antimicrobial peptides: structures and activities in lipopolysaccharide. J Biol Chem. 2009;284(33):21991–2004.
35. Mohanram H, Bhattacharjya S. Resurrecting inactive antimicrobial peptides from the lipopolysaccharide trap. Antimicrob Agents Chemother. 2014;58(4):1987–96.
36. Scudiero O, Galdiero S, Cantisani M, et al. Novel synthetic, salt-resistant analogs of human beta-defensins 1 and 3 endowed with enhanced antimicrobial activity. Antimicrob Agents Chemother. 2010;54(6):2312–22.
37. Beckloff N, Laube D, Castro T, et al. Activity of an antimicrobial peptide mimetic against planktonic and biofilm cultures of oral pathogens. Antimicrob Agents Chemother. 2007;51(11):4125–32.
38. Hua J, Scott R, Diamond G. Activity of antimicrobial peptide mimetics in the oral cavity: II. Activity against periopathogenic biofilms and anti-inflammatory activity. Mol Oral Microbiol. 2010;25(6):426–32.
39. Leszczyńska K, Namiot D, Byfield FJ, et al. Antibacterial activity of the human host defence peptide LL-37 and selected synthetic cationic lipids against bacteria associated with oral and upper respiratory tract infections. J Antimicrob Chemoth. 2013;68(3):610–8.
40. Murugan RN, Jacob B, Ahn M, et al. De novo design and synthesis of ultra-short peptidomimetic antibiotics having dual antimicrobial and anti-Inflammatory activities. PloS One. 2013;8(11):e80025.
41. Padhee S, Smith C, Wu H, et al. The development of antimicrobial α-AApeptides that suppress proinflammatory immune responses. Chem Bio Chem. 2014;15(5):688–94.
42. Matsuzaki K. Control of cell selectivity of antimicrobial peptides. BBA-Biomembranes. 2009;1788(8):1687–92.

43. Lee EK, Kim Y-C, Nan YH, Shin SY. Cell selectivity, mechanism of action and LPS-neutralizing activity of bovine myeloid antimicrobial peptide-18 (BMAP-18) and its analogs. Peptides. 2011;32(6):1123–30.

44. Nan YH, Bang J-K, Jacob B, Park I-S, Shin SY. Prokaryotic selectivity and LPS-neutralizing activity of short antimicrobial peptides designed from the human antimicrobial peptide LL-37. Peptides. 2012;35(2):239–47.

45. Wang G, Elliott M, Cogen AL, Ezell EL, Gallo RL, Hancock RE. Structure, dynamics, and antimicrobial and immune modulatory activities of human LL-23 and its single-residue variants mutated on the basis of homologous primate cathelicidins. Biochemistry. 2012;51(2):653–64.

46. Park KH, Nan YH, Park Y, et al. Cell specificity, anti-inflammatory activity, and plausible bactericidal mechanism of designed Trp-rich model antimicrobial peptides. BBA-Biomembranes. 2009;1788(5):1193–203.

47. Frecer V, Ho B, Ding J. De novo design of potent antimicrobial peptides. Antimicrob Agents Chemother. 2004;48(9):3349–57.

48. Lee E, Kim J-K, Shin S, et al. Enantiomeric 9-mer peptide analogs of protaetiamycine with bacterial cell selectivities and anti-inflammatory activities. J Pept Sci. 2011;17(10):675–82.

49. Wang P, Nan YH, Yang S-T, et al. Cell selectivity and anti-inflammatory activity of a Leu/Lys-rich α-helical model antimicrobial peptide and its diastereomeric peptides. Peptides. 2010;31(7):1251–61.

50. Wei L, Wu J, Liu H, et al. A mycobacteriophage-derived trehalose-6, 6'-dimycolate-binding peptide containing both antimycobacterial and anti-inflammatory abilities. FASEB J. 2013;27(8):3067–77.

51. Eriksson OS, Geörg M, Sjölinder H, et al. Identification of cell-penetrating peptides that are bactericidal to *Neisseria meningitidis* and prevent inflammatory responses upon infection. Antimicrob Agents Chemother. 2013;57(8):3704–12.

52. Chow LN, Choi K-YG, Piyadasa H, et al. Human cathelicidin LL-37-derived peptide IG-19 confers protection in a murine model of collagen-induced arthritis. Mol Immunol. 2014;57(2):86-92.

53. Jacob B, Park I-S, Bang J-K, Shin SY. Short KR-12 analogs designed from human cathelicidin LL-37 possessing both antimicrobial and antiendotoxic activities without mammalian cell toxicity. J Pept Sci. 2013;19(11):700–7.

54. Ciornei CD, Sigurdardóttir T, Schmidtchen A, Bodelsson M. Antimicrobial and chemoattractant activity, lipopolysaccharide neutralization, cytotoxicity, and inhibition by serum of analogs of human cathelicidin LL-37. Antimicrob Agents Chemother. 2005;49(7):2845–50.

55. Sigurdardottir T, Andersson P, Davoudi M, Malmsten M, Schmidtchen A, Bodelsson M. In silico identification and biological evaluation of antimicrobial peptides based on human cathelicidin LL-37. Antimicrob Agents Chemother. 2006;50(9):2983–9.

56. McInturff JE, Wang S-J, Machleidt T, et al. Granulysin-derived peptides demonstrate antimicrobial and anti-inflammatory effects against *Propionibacterium acnes*. J Invest Dermatol. 2005;125(2):256–63.

57. Wang C-Q, Yang C-S, Yang Y, Pan F, He L-Y, Wang A-M. An apolipoprotein E mimetic peptide with activities against multidrug-resistant bacteria and immunomodulatory effects. J Pept Sci. 2013;19(12):745–50.

58. Papareddy P, Rydengård V, Pasupuleti M, et al. Proteolysis of human thrombin generates novel host defense peptides. PLoS Pathog. 2010;6(4):e1000857.

59. Papareddy P, Kalle M, Singh S, Mörgelin M, Schmidtchen A, Malmsten M. An antimicrobial helix A-derived peptide of heparin cofactor II blocks endotoxin responses *in vivo*. BBA-Biomembranes. 2014;1838(5):1225–34.

60. Fox JL. Antimicrobial peptides stage a comeback. Nat Biotechnol. 2013;31(5):379–82.

Jasmeet Singh Khara is a postgraduate researcher under the joint NUS-Imperial College PhD Programme. His research is focused on biomaterials and macromolecular therapeutics against infectious diseases such as tuberculosis.

Pui Lai Rachel Ee is assistant professor in the Department of Pharmacy at the National University of Singapore (NUS). Her research group is actively involved in the synthesis and evaluation of biofunctionalized scaffolds for stem cell/drug delivery for tissue engineering as well as infectious diseases.1

Chapter 7
Biomaterials-Based Strategies in Blood Substitutes

Anirban Sen Gupta

7.1 Introduction

A biomaterial is broadly defined as any material that interacts with living biological tissues for medical purposes. Biomaterials can be natural, semisynthetic, or synthetic. Natural biomaterials are derived from autogenic (same individual), allogeneic (same-species donor), or xenogeneic (animal) tissues, as well as, from plants and insects. Synthetic biomaterials are developed using chemical and biotechnological methods. Semisynthetic biomaterials are the combinations of natural and synthetic material components to integrate their beneficial properties. Blood products and blood substitutes span over all three categories of biomaterials. Natural blood products for transfusion applications have been around for close to two centuries. The first recorded human blood transfusion occurred in 1818, almost 100 years prior to identification of blood-type antigens. The research in efficient preservation and transportation of donated blood began during World War I to treat wounded soldiers, but it was not until World War II that the refinement of technologies made blood transfusions widely and safely available. By the 1950s, several blood banks were established in the USA, and blood donation was promoted to the public as a form of civic responsibility. Subsequent development of many technologies, for example, freezing of red blood cells, separation and isolation of blood components via centrifugation, isolation of a specific blood component while returning the rest to the donor (apheresis), etc., have enhanced the spectrum of utilization of whole blood and its components. Currently, these products are clinically used in transfusions for traumatic injuries (e.g., accidents and battlefield wounds) and in surgical settings (e.g., transplants), for treatment of chronic and acute anemias, and for

A. Sen Gupta (✉)
Department of Biomedical Engineering, Case Western Reserve University,
Cleveland, OH 44106, USA
e-mail: axs262@case.edu

© Springer International Publishing Switzerland 2015   113
L. Santambrogio (ed.), *Biomaterials in Regenerative Medicine and the Immune System,*
DOI 10.1007/978-3-319-18045-8_7

treating drug-induced or congenital bleeding disorders [1–5]. However, only about 50% of the US population is eligible to donate blood, which is insufficient to meet the high demand. Furthermore, natural blood-based products suffer from limited shelf life due to risks of pathological contamination. For example, natural red blood cell (RBC) suspensions have a storage life of 20–40 days, while natural platelet suspensions have a storage life of 3–5 days [2, 5, 6]. In addition, natural blood products present risks of refractoriness, graft-versus-host disease, transfusion-associated immunosuppression, and acute lung injury [7–10]. To address these issues, in recent years, several approaches have been developed, for example, pathogen reduction technologies like psoralen-based or riboflavin-based UV irradiation, extensive serological testing of donor blood, leukoreduction, and specialized storage protocols [11, 12]. While these methods have enhanced the quality of natural blood products somewhat, they have also resulted in substantial increase of cost in the products. Biomaterials-based synthetic (and semisynthetic) blood substitutes can potentially address these issues, while allowing convenient production and reasonable cost.

The interest in synthetic blood substitutes developed during the 1980s as a result of the human immunodeficiency virus (HIV) crisis [13]. The crisis generated a fear of contaminated blood products, which in turn drove the research towards artificial blood substitutes that can avoid contamination. Also, the shortage of donor blood, the limited shelf life, the necessity for extensive screening and typing, and the expensive storage and portability costs acted as additional driving factors of research. Besides trauma, surgery, and drug-induced or congenital bleeding disorders, other application areas for blood substitutes are niche patients like Jehovah's Witnesses whose religious belief forbids them from utilizing natural blood transfusions. Based on these needs, several artificial blood products have undergone research in preclinical and some in early clinical stages, but currently no product has been clinically approved by the Food and Drug Administration (FDA) for human applications. One product is currently approved for veterinary applications and a few others are in early approval stages in Russia, Mexico, and South Africa. A 2008 report on a meta-analysis of 16 clinical trials of five different artificial blood products suggested increased health risks in patients treated with such products [14]. Although such analyses have resulted in some apprehension in clinical utility of these products, the current state of research in artificial blood substitutes has significant emphasis on understanding and resolving the problems posed by these products. The following sections provide descriptions of these products and blood substitute technologies currently under research, under separate areas of RBC substitutes and oxygen carriers, platelet substitutes and hemostats, and WBC substitutes and artificial plasma compounds. In each area, the sections discuss the biomaterials utilized, and the benefits and challenges.

7.2   RBC Substitutes and Oxygen Carriers

The RBCs are primarily responsible for carrying oxygen and carbon dioxide to and from tissues, respectively, by virtue of binding of the gases to hemoglobin (Hb) within the RBCs. Hb is a tetrameric protein of two α- and two β-polypeptide chains, each bound to an iron-containing heme group that can bind one oxygen molecule. The binding of oxygen to Hb is positively cooperative [15], such that a small change in oxygen partial pressure (pO_2) results in a large change in oxygen bound or released by Hb. The oxygen-carrying iron in Hb is in its reduced "ferrous" (Fe(II)) state. This Fe(II)-containing Hb can be oxidized to form methemoglobin, which has iron in the oxidized "ferric" state (Fe(III)), which is unable to bind oxygen [16, 17]. In natural RBCs, the oxygen-carrying reaction of Hb is coupled to redox cycles (e.g., driven by enzyme nicotinamide adenine dinucleotide (NAD)-cytochrome b5 reductase), such that the Fe(II)-containing Hb is always maintained in its oxygen-binding state [16, 17]. Irreversible conversion of Hb to methemoglobin not only inhibits its oxygen-carrying capacity but can also lead to dysregulated vascular tone and inflammatory responses. Therefore, engineering of an RBC substitute poses a formidable challenge with regard to maintaining the oxygen-carrying capacity of Hb, while minimizing irreversible methemoglobin formation [17]. To this end, three types of RBC substitutes or oxygen carriers have been investigated, namely, hemoglobin-based oxygen carriers (HBOCs), perfluorocarbon (PFC)-based emulsions, and iron (Fe^{2+})-containing porphyrins. The HBOCs fall under the category of natural and semisynthetic biomaterial systems, while the PFC and porphyrin systems are essentially synthetic. Table 7.1 shows design schematics of the various biomaterials-based oxygen-carrying systems.

7.2.1 Cell-Free HBOC Systems

HBOCs use natural Hb as the oxygen-carrying component, in either a cell-free suspension or encapsulated within biomaterials-based carrier vehicles, or as a complex with protective enzymes [2, 18–20]. Table 7.1 provides an up-to-date list of HBOCs investigated so far. Compared to natural RBCs, cell-free Hb suspension presents numerous advantages such as minimum antigenicity, sterilizability by ultrafiltration or low heat, and the ability to transport oxygen in plasma more efficiently because of the lack of interference by cell membrane [20]. However, unmodified cell-free Hb has very short circulation half-life because of Hb tetramer dissociation and binding to plasma immunoglobulins, that leads to their rapid clearance by the reticuloendothelial system (RES) into spleen, liver, and kidneys [21]. Additionally, cell-free Hb was found to precipitate in the loop of Henle causing severe renal toxicity. Natural RBCs can regulate the oxygen affinity of Hb via the action of an allosteric effector molecule, 2,3-diphosphoglycerate (2,3-DPG), present inside the RBCs. The lack of 2,3-DPG in cell-free Hb leads to its unfavorably high oxygen affinity, which can make oxygen release difficult. Furthermore, oxygen molecules in cell-free HB suspensions come in direct contact with blood vessel walls, often resulting in abnormally high oxygen delivery [22]. Additionally, cell-free Hb can bind nitric oxide (NO), thereby affecting normal vasodilation and vasoconstriction, leading to hypertension [23]. Also, cell-free Hb changes blood osmolarity that can alter blood volume and cause side effects [20].

Table 7.1 Biomaterials-based designs of RBC-inspired oxygen carrier systems

Oxygen carrying systems	System components	References
	Cell-free Hb	[20–23]
	Cross-linked Hb	24, 25
	Peg-stabilized Hb	29
	Polymerized Hb	26, 27, 30
	Hb encapsulated within vesicles and particles	33-38
	Polymerized Hb linked to superoxide dismutase and catalase enzyme	39
	"Picket fence" Fe-porphyrin systems binding oxygen	40
	Fe-porphyrin systems encapsulated in particles	41
	Fe-porphyrin encapsulated in cyclodextran	42
	Oxygen-dissolving perfluorocarbon molecules	43–45

Chemical modification of Hb has been carried out to address the Hb-tetramer dissociation issue. For example, chemical cross-links between the α and β chains of Hb can be created with bis-(3,5-dibromosalicyl) fumarate [24]. The product HemAssist (Baxter Healthcare Corporation, USA) uses diaspirin cross-linked Hb, which showed an increase in circulation time up to 12 h compared to <6 h for un-modified Hb [20], but the cross-linked Hb unfortunately showed a 72% increase in mortality rates in human patients compared to saline, and clinical trials were discontinued [25]. Polymerized Hb has also been created using bifunctional cross-linking reagents like glutaraldehyde and o-raffinose. However, in these approaches, it is hard to control polymer molecular weight. Purification processes have partially increased product quality with promising clinical trial results [26]. For example, PolyHeme (Northfield Laboratories, Inc., USA), a human Hb product polymerized with glutaraldehyde, progressed into phase III clinical trials in the USA in treating trauma-associated blood loss, and it showed a decreased need of natural blood transfusions [26]. Trials with Oxyglobin (Biopure Corporation, USA), another glutaraldehyde-cross-linked bovine Hb product, were conducted in the 1990s and has been since approved for veterinary use in anemia. HemoPure (Biopure Corporation, USA), a similar bovine hemoglobin polymer, has shown a reduced need of additional blood transfusions in phase III clinical trials for cardiac surgery. Although HemoPure has not been approved by the FDA in the USA yet, it currently has clinical applications in South Africa for acutely anemic patients [27]. HemoLink (Hemosol BioPharma, Canada), another human Hb product cross-linked with o-raffinose, advanced to phase III clinical trials but was discontinued in 2003 when patients receiving treatment experienced adverse cardiac events [28]. Besides polymerization of Hb, bioconjugation of other molecules onto Hb has also been investigated as a way to stabilize the tetramer dissociation and reduce immune recognition. One prime example of this is the product Hemospan (Sangart Inc., USA), which utilizes polyethylene glycol (PEG) bioconjugation onto Hb [29]. This PEG-ylated Hb system is designed to prolong its circulation half-life and also increase oxygen affinity to prevent premature oxygen off-loading. This product has been studied clinically to deliver oxygen to hypoxic tissue, and the studies showed risks of bradycardia and elevation of hepatic pancreatic enzymes even at low doses. Nonetheless, the phase I and phase II clinical trials have shown that Hemospan is well tolerated in humans for efficient oxygen delivery, and phase III trials in orthopedic surgery patients are under way in Europe. In another recent approach, a product named HemoTech has been developed that uses purified bovine Hb cross-linked intramolecularly with adenosine triphosphate (ATP) and intermolecularly with adenosine, and conjugated with reduced glutathione (GSH) [30]. The preclinical and early-phase clinical studies have shown that HemoTech works as an effective oxygen carrier in treating blood loss, anemia, and ischemic vascular conditions, and further studies are currently under way. Pyridoxalated hemoglobin polyoxyethylene (PHP, Apex Bioscience, USA) is a nitric oxide scavenger that has undergone phase II clinical trials in patients with distributive shock, but resulted in a rapid increase in blood pressure and decrease in heart rate [31], thereby showing limited clinical benefit. In another approach, genetically engineered recombinant hemoglobin (rHb) was raised in *Escherichia Coli,* and the oxygen affinity of such rHb systems can be efficiently modulated [32]. These systems are currently undergoing preclinical studies.

7.2.2   Encapsulated HBOC Systems

Due to the issues of immune recognition, toxicity, and loss of function of cell-free Hb in circulation, research has been carried out in encapsulating Hb within shells or membranes made from biocompatible materials. In natural situation, Hb stays encased within the membrane of RBCs that results in its protection and the preservation of suitable redox environment for Hb function [1]. Based on this rationale, Hb has been encapsulated within liposomal vehicles, and these Hb-containing vesicles (HbVs) have shown longer circulation times (~24 h) than free Hb, especially when the surface is modified by PEG [33]. Since lipidic shells of liposomes are similar to cell membranes, oxygen binding and transportation activity in HbVs were found to somewhat mimic natural RBCs, with comparable P_{50} values and Hill coefficients. Also, because of liposomal encapsulation, the Hb in HbVs has minimal effect on NO and vasoactivity. HbVs have been investigated in treating massive hemorrhage and hemodilution incidents in animal models as well as treating ischemia and wounds in vivo [34]. Although these studies show promise of HbVs, these systems suffer from similar issues as other liposomal formulations, for example, broad size distribution of the vehicles, variations in Hb-encapsulation efficiencies, variations in vesicle clearance, etc. Besides encapsulation within liposomes, Hb has also been encapsulated within polymeric particles, using polymers like poly(ε-caprolactone)/ poly(L-lactic acid) (PCL/PLA) copolymers, poly(L-lysine; PLL), poly(lactic-co-glycolic acid; PLGA)/PEG copolymers, etc. [35]. These polymers have undergone FDA approval in several biomedical applications, and they are therefore considered biocompatible. One issue with polymeric particles is their higher shell thickness compared to liposomes, which results in longer time for oxygen to saturate the encapsulated Hb or to be released from the Hb to tissues. By modulating molecular weight of the shell polymer, the shell-thickness and oxygen transport can be modulated. The circulation time of the Hb-loaded particles can be enhanced by surface PEGylation. The encapsulation efficiency of Hb within the polymeric particles has high variation, which is an issue regarding scale up and translation. In order for these particles to perform similar to RBCs, they need to have Hb-loading efficiency closer to RBCs, which is ~14 g/dL. Some PLGA–PEG and PLA–PEG nanoparticles have shown Hb-loading efficiencies close to 90% and circulation half-life of ~2 days [36]. Instead of nanoparticles, Hb has also been encapsulated in microparticles by coprecipitation of Hb and $CaCO_3$, cross-linking with glutaraldehyde and dissolution of $CaCO_3$, with Hb quantity per microparticle close to natural RBCs [37]. These Hb microparticles have shown oxygen binding and release characteristics similar to free Hb, but much enhanced circulation residence compared to free Hb. Similar Hb particles, about 700 nm in diameter and bearing about 80% Hb content compared to natural RBCs, have been recently reported where Hb and $MnCO_3$ were coprecipitated, immediately followed by human serum albumin addition for encapsulation and stabilization of the particles [38]. These particles have shown limited NO scavenging and reduced effect on vasoconstriction. As mentioned previously, in natural RBCs, the activity of Hb is effectively regulated by enzyme-mediated redox systems. To closely mimic this, researchers have investigated semisynthetic biomaterial-based Hb systems that additionally incorporate enzymes and other biomolecules to regulate Hb activity. To this end, polymerized Hb has been linked to antioxidant enzymes (e.g., superoxide dismutase (SOD) and catalase (CAT)) to form PolyHb–SOD–CAT systems, that can simultaneously transport molecular oxygen and remove reactive oxygen radicals to reduce the effects of ischemia perfusion injuries [39]. The perfusion of these systems through partially obstructed vessels was facilitated by their small size. Another interesting approach is to incorporate regulatory molecules such as 2,3-DPG and methemoglobin reductase along with Hb in appropriate HBOC systems to prevent hemoglobin oxidation. Ongoing and future studies are expected to shed light on the design reproducibility and efficacy of these Hb-carrying semisynthetic systems.

7.2.3   Synthetic Porphyrin Systems

In natural Hb, the oxygen-transporting active site is the nonprotein "heme" which consists of a porphyrin tetrapyrrole ring surrounding an iron atom. Based on this, research has been carried out using Fe(II)-bearing porphyrin systems for oxygen transport [40–42]. In this context, "picket fence" Fe^{2+} porphyrin molecules showed cooperative oxygen binding similar to natural Hb, but such systems were prone to irreversible oxidation in aqueous media. To address this issue, a hydrophobic environment was created for these molecules using liposomes, where amphiphilic Fe^{2+} porphyrin bearing four alkylphosphocholine groups were efficiently embedded into the bilayer of phospholipid vesicles (e.g., LipidHeme). The resultant vesicles have exhibited reversible binding and release of oxygen similar to natural Hb. Instead of lipid vesicles, human serum albumin particles have also been used to incorporate these synthetic Fe^{2+} porphyrin systems, and these have shown oxygen-transporting efficiency similar to RBCs. Surface PEG-ylation of these particles have resulted in increased circulation time and reduced oxidation. Another interesting design is the porphyrin derivative, HemoCD, that consists of 5,10,15,20-tetrakis(4-sulfonato-phenyl)porphinatoiron(II) and a methylated β-cyclodextrin dimer with a pyridine linker (Py3CD), that has similar oxygen affinities to natural Hb but is prone to get oxidized in the blood, therefore yielding a short half-life.

7.2.4   Synthetic PFC-Based Systems

Since natural Hb-based systems present issues of limited stability, circulation time, toxicity, etc., research has been carried out in creating synthetic molecules capable of transporting oxygen. One such synthetic system is the class of porphyrins described in the previous section. The other more extensively studied system is the class of PFCs. PFCs are biologically inert liquids that can carry dissolved molecular oxygen and carbon monoxide without actually binding to it, allowing easy exchange of gases. However, since PFCs are not water-miscible, they need to be formulated into emulsions for effective in vivo applications. The main difference between oxygen binding of natural Hb versus PFCs is that, in PFCs the oxygen loading is linearly related to the pO_2, whereas in natural Hb it is non-linear due to cooperative binding [2]. It has also been established that the oxygen solubility in PFC emulsions is independent of the surfactant used, which is a significant advantage [43]. As a result, various PFC emulsions have been investigated as oxygen carriers in preclinical and clinical studies [44, 45]. For example, Fluosol-DA (Green Cross Corp, Japan), an emulsion of 10 % perfluorodecalin—PFC in albumin—has shown promising results in early clinical trials with patients who have religious objections to natural blood transfusions. A newer formulation of Fluosol was investigated in high-risk coronary balloon angioplasty patients and although this emulsion was FDA approved in 1989, it was soon recalled because of issues like low oxygen delivery (0.4 mL oxygen/100 mL), premature oxygen release, and slow excretion. Fluosol has also been implicated in reducing platelet counts (thrombocytopenia), febrile symptoms, and macrophage activation. Another PFC-based emulsion that has undergone clinical trials is Oxygent (Alliance Pharmaceutical Corp, USA), where PFC was combined with F-decyl bromide to maintain droplet size and reduce macrophage activation. Oxygent was found to render higher oxygen delivery compared to Fluosol, but was still 30% less efficient than natural RBCs. Although this product reduced the need for RBC transfusion in patients undergoing major noncardiac surgery, advanced clinical trials in patients undergoing cardiopulmonary bypass were terminated in the USA because of increased risk of stroke. Another PFC-based system is Oxyfluor (HemaGen, USA) that has shown considerable promise in animal models and in early clinical trials, but further studies were stopped by Baxter International due to several safety concerns of stroke and spurious platelet counts. Oxycyte (Oxygen Biotherapeutics, USA), a third-generation PFC emulsion in water with purified egg yolk phospholipids, is currently undergoing clinical trials in patients with traumatic brain injury. Although some safety concerns were raised regarding Oxycyte's effect on the immune system and possible hemorrhagic side effects, recent animal studies have alleviated FDA concerns and studies are under way. Perftoran (Perftoran, Russia) is another PFC emulsion of perfluorodecalin and perfluoromethylcyclopiperidine in a nonionic surfactant that has been widely investigated in Russia. Although this product has shown adverse effects such as hypotension and pulmonary complications, randomized clinical trials with Perftoran conducted in Mexico City on patients undergoing cardiac valvuloplasty showed higher intraoperative pO_2 levels as well as reduced need for allogeneic RBCs. Further clinical studies are expected to establish the clinical benefit of this product.

7.2.5 *Synthetic Biomaterial Systems Mimicking Physico-Mechanical Properties of RBCs*

The biological functions of RBCs are significantly influenced by their physico-mechanical properties [46]. Natural healthy RBCs are biconcave discoid in morphology, with a diameter of ~8 μm and a thickness of ~2 μm. RBCs are also highly flexible (Young's Modulus 0.1–0.2 kPa), such that, when subjected to fluid mechanical forces, for example, passing through small capillaries, they can change into a bullet or parachute-like shape. Consequently, RBCs can undergo repeated large-amplitude deformations in microvascular circulation [47]. RBCs maintain their mechanical integrity using a two-dimensional spectrin network attached to the cytosolic side of their membrane, which gives RBCs their viscoelastic properties. Oxygenated Hb renders RBCs significantly more deformable than deoxygenated Hb, and this enables the deformation of RBCs to transport oxygen in the microvasculature. The physical size and shape of RBCs are also known to influence their hemodynamic distribution in the blood flow field. For example, in mid-to-large vessels, the RBC geometry causes them to flow in the center of the parabolic flow field, while in small vessels RBCs can become distributed throughout, allowing them to come sufficiently close to the vessel wall for efficient oxygen exchange [48]. Rationalizing from such cues, biomaterials-based mimicry of RBCs has also focused on incorporating the size, shape, and flexibility attributes of RBCs into synthetic constructs. To this end, polyelectrolyte-driven layer-by-layer assembly has been used to create polymeric particles that mimic the shape and deformability of natural RBCs [49]. In this design, Hb and bovine serum albumin (BSA) were electrostatically deposited on the surface of RBC-shaped PLGA particles of ~7-µm diameter and 400-nm shell thickness, and then the PLGA core was selectively dissolved to yield RBC-shaped Hb-loaded particles. The deformation of these particles through narrow glass capillaries demonstrated a stretching aspect ratio of $170 \pm 20\%$ (RBCs can stretch up to 250 %) and high elastic deformation. Similar RBC-mimetic particles have been fabricated using PEG hydrogel system in a stop-flow-lithography (SFL) approach, and the mechanical properties of these particles could be controlled by modulating hydrogel cross-linking [50]. In another approach, acrylate hydrogel-based RBC-mimetic particles were fabricated using a "particle replication in nonwetting templates" (PRINT) technology [51]. These particles were made in 2–3 µm molds, such that, upon hydration, the particles swelled to disks with diameters between 5.2 and 5.9 µm and heights between 1.22 and 1.54 µm. Filling of the molds resulted in a meniscus, leaving the particles thinner in the middle and thicker at the edges, resembling the biconcave nature of RBCs. As before, the modulus of the particles can be controlled by varying the cross-linking extent of the hydrogel. In vitro, these particles demonstrated the capability of undergoing efficient elastic deformation for transport through narrow channels, and, in vivo, the circulation time of the particles was found to be dependent on their elastic modulus. Hb has also been loaded into these particles, at ratios ~five times the amount of polymer by weight, and the Hb has been covalently tethered via ethyl(dimethylaminopropyl) carbodiimide/N-hydroxysuccinimide (EDC/NHS) chemistry. Yet another technology has been reported involving liposome-encapsulated actin-hemoglobin (LEAcHb) synthetic RBCs with similar shape properties as natural human RBCs, using a polymerized actin core [52]. Although these particles were much smaller (~136.8 nm) than RBCs, the biconcave shape along with the mechanical support of the membrane improved the half-life from ~8 to >72 h. In natural RBCs, the surface charge is negative that electrostatically prevents the RBCs from aggregation over a distance of 20 nm. Consequently, there has been some research in determining whether mimicking surface charge is important for design of synthetic RBCs. To this end, the surface charge of Hb-encapsulating PEG–PLA–PEG nanoparticles (<200 nm in diameter) was varied by using cationic cetyltrimethylammonium bromide (CTAB) or anionic sodium

Table 7.2 Technologies for fabricating particles of RBC or platelet-mimetic morphologies

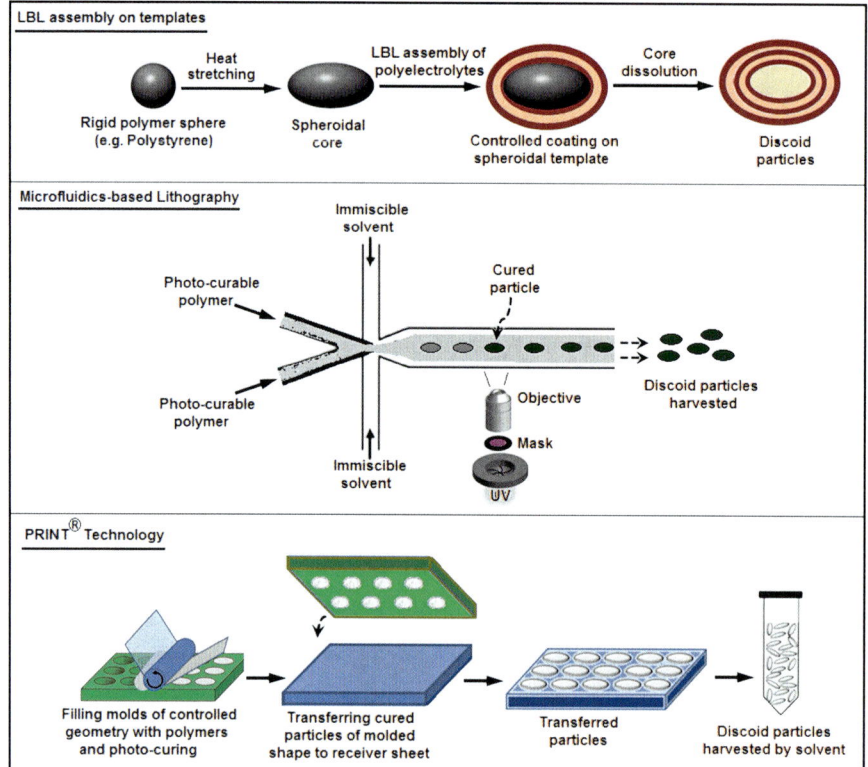

dodecyl sulfate (SDS), and the resultant half-lives of these particles were studied in mice [53]. Cationized particles were found to have a half-life of ~11 h (eightfold higher than untreated particles), while the anionized particles were quickly elimi-nated, giving a half-life of <1 h. Surface PEG-ylation has also been investigated to mask the negative charge of Hb–PLA nanoparticles, showing that such masking de-creases macrophagic uptake and clearance. Therefore, along with oxygen-binding biochemical properties, incorporation of physico-mechanical properties can lead to significant refinement in synthetic RBC design. Table 7.2 shows the various particle fabrication technologies that have been studied for engineering RBC-mimicking constructs.

7.2.6   RBCs from Stem Cells

In parallel to biomaterials-based strategies for RBC mimicry, recent advancements in regenerative medicine approaches have opened the doors for generating RBCs from stem cells and precursor cells [54, 55]. In a seminal work, researchers were able to produce erythroblasts that matured in mice into erythrocytes, and thereby led to the way for production of RBCs ex vivo [54]. During the past decade, significant advancement has been made in regulated expansion and maturation of erythroid cells towards generating enucleated RBCs. To this end, RBCs have been produced from cluster of differentiation (CD34)+ hematopoietic stem cells (HSCs) isolated from cord blood. The two major issues of this process are that, (i) the HSCs had to be cocultured with murine or human stromal cells that posed regulatory and scaling-up concerns and (ii) the yield was very low. In recent years, the coculture issues have been resolved with procedures that allow HSC expansion and differentiation in serum. Recently, RBCs were generated ex vivo from autologous HSCs of a consent-ing patient and injected into the patient without adverse effects, thereby indicating the clinical potential of this approach [55]. Using analogous approaches, research-ers have also investigated generation of RBCs from human embryonic stem cells (hESCs) and induced pluripotent stem cells (iPSCs), due to the potential of these cells as an unlimited source of HSCs. In Scotland, the iPSC-generated artificial blood is poised to undergo extensive clinical studies. In another recent approach, researchers have reported on reprogramming of mature blood cells from mice into blood-forming new HSCs, using a cocktail of eight transcription factors [55]. The reprogrammed induced HSCs (iHSCs) have shown renewal and expansion proper-ties similar to that of normal HSCs, and therefore this research opens the door for enhancing the production of HSCs for subsequent generation of blood cells. Irre-spective of the stem cell or precursor cell source, the principal challenges in all of these approaches are limited culture technologies for large-scale production, low yield of generated RBCs, and high cost of production [55]. These issues need to be aptly resolved in order to make such "regenerative blood products" universally ap-plicable in the clinical context.

7.3   Platelet Substitutes and Synthetic Hemostats

Platelets are anucleated blood cells produced from mature megakaryocytes and are primarily responsible for rendering hemostasis. Platelets render hemostasis by (i) marginating across the flowing blood volume to the site of vascular injury, (ii) actively adhering at the injury site by binding to specific proteins like von Wil-lebrand factor (vWF) and collagen, (iii) undergoing activation and aggregation via inter-platelet bridging (primary hemostasis), and (iv) presenting phosphoserine-rich negatively charged surface on their membrane for triggering coagulation cascade mechanisms [56]. Loss of platelets due to trauma or surgery as well as drug-induced or congenital defects in platelet number and function can lead to various bleed-ing complications. Transfusion of natural platelet products is routinely used in the clinic for treating such scenarios [57–59]. These transfusions primarily use alloge-neic platelet concentrates (PCs) suspended in autologous plasma. These PCs have a very short shelf life ($\sim$3–5 days) and high risk of pathologic contamination [60, 61]. Pathogen reduction technologies (e.g., psoralen-based UV treatment) can reduce contamination risks, but so far these have extended the PC shelf life only up to ~7 days [11, 12, 62]. Other platelet-derived products under investigation, for example, frozen (−80 °C) platelets, cold-stored (4 °C) platelets, lyophilized platelets, platelet-derived microparticles, and infusible platelet membranes (IPM) suffer from similar contamination risks. Also, since all these products are derived from pooled blood of multiple donors, they present high risk of blood-borne infection that can be only partly prevented by serological testing and expensive leukoreduction techniques [9, 63]. Due to these multiple issues of natural platelet-based products, there is a significant clinical interest in synthetic platelet substitutes that can render efficient hemostasis while allowing advantages of large-scale preparation, minimum contamination risk via effective sterilization, longer shelf life, and absence of biologic or pathologic side effects [64]. To this end, platelet-inspired systems have focused on mimicking platelet's mechanisms of injury site-specific adhesion, aggregation, and coagulation promotion. In addition, due to the role of platelet's morphology and physico-mechanical properties on their margination behavior under blood flow, recent biomaterials-based research has also been directed towards capturing platelet's physico-mechanical parameters and integrating them with its biointeraction parameters. Table 7.3 shows various biomaterials-based design approaches in developing platelet-inspired synthetic hemostat systems.

7.3.1 Biomaterials-Based Synthetic Systems Inspired by Platelet Adhesion

The injury site-specific adhesion of natural platelets is mediated by several mechanisms. The injured endothelium secretes multimeric vWF molecules that bear the A1-domain with specific binding capability to platelet surface receptor component glycoprotein Ibα (GPIbα) of the GPIb–IX–V complex. This vWF binding is shear dependent and reversible, and this leads to platelet attachment and rolling at the injury site [65]. Injury sites also often have exposed subendothelial matrix, with collagen type I and III being the major components. Platelet surface glycoproteins, GPIa-IIa and GPVI, have binding capability to collagen [65], and this collagen binding synergistically with vWF binding results in reducing the platelet rolling, leading to their stable adhesion. Based on these mechanisms, biomaterials-based research has focused on developing surface-modified particle systems that have vWF-binding and collagen-binding capabilities. To this end, the earliest design called "plateletosomes" involved isolation of platelet membrane glycoprotein via detergent-based extraction techniques and their subsequent incorporation within the membrane of liposomes [66]. Such extraction and incorporation make this strategy cumbersome and expensive to scale up. An alternative strategy was realized with the advent of recombinant technologies, when recombinant GPIbα (rGPIbα) and GPIa-IIa (rGPIa-IIa) could be generated. These rGPIbα and rGPIa-IIa fragments have been conjugated on the surface of liposomes, latex beads, and albumin particles to create synthetic systems that have vWF-binding and collagen-binding

Table 7.3 Biomaterials-based designs of platelet-inspired synthetic hemostat systems

Hemostat systems	System components	References
	Platelet membrane glyco-proteins extracted and incorporated in liposome membrane surface	66
	Recombinant GPIa-IIa and GPIb-alpha fragments conjugated to liposomal or polymeric particle surface	67–69
	Lipid vesicles surface decorated with VWF-binding and collagen binding peptide motifs	70, 71
	Natural (e.g., albumin) and synthetic (e.g., latex) particles surface-coated with Fibrinogen (Fg)	74–77
	Low-crosslinked synthetic microgel particles surface-decorated with fibrin-specific binding motifs	79
	Liposomes, natural polymeric and synthetic polymeric particles surface-decorated with Fg-derived RGD and H12 peptide motifs	76, 77, 81, 83
	Lipidic and polymer-based particles decorated hetero-multivalently on the surface with adhesion-promoting and aggregation-promoting peptide motifs	86, 90

capabilities [5, 67, 68]. The vWF molecule is unraveled at high shear to expose the GPIbα-binding A1 domain, and therefore the rGPIbα-modified particle systems have shown effective shear-dependent vWF binding. The collagen binding of platelets is essentially shear independent, and this property was efficiently mimicked by the rGPIa-IIa-modified particles. Since for natural platelets, the adhesion is mediated by a combination of vWF binding and collagen binding, therefore in advancement of the platelet adhesion-mimicking design, both rGPIbα and rGPIa-IIa have been co-conjugated on the surface of liposomes and albumin particles [69]. These designs have shown efficient binding to collagen surfaces in the presence of soluble vWF, with more binding evident at higher shear rates, thereby closely mimicking natural platelet adhesion.

The recombinant technologies to make rGPIba and rGPIa–IIa can be highly expensive for scaling up and clinical translation. Also, because of the large size of these recombinant proteins, there may be issues of steric hindrance regarding decoration of particles with multiple copies of such moieties. These issues can be potentially addressed by using small molecular weight peptides to impart vWF-binding and collagen-binding capabilities [70]. To this end, recently, research-ers have utilized a vWF-binding peptide (VBP) sequence TRYLRIHPQSWVHQI derived from the C2 domain (residues 2303–2332) of the coagulation factor FVIII and a collagen-binding peptide (CBP), which is a 7-mer repeat of the Glycine(G)–Proline(P)–Hydroxyproline(O) tri-peptide (i.e., e[GPO]7-) with helicogenic affinity to fibrillar collagen but minimal affinity to platelet collagen receptors GPIa/IIa and GPVI (hence no risk of systemically activating platelets) [70]. Using liposomes as model particles, VBPs and CBPs were conjugated onto the particle surface. When decorated with only the VBP moieties, the liposomes exhibited shear-dependent enhancement of adhesion onto vWF-coated surfaces or on collagen surfaces in the presence of soluble vWF. Studies have indicated that this vWF-binding mechanism of VBP is possibly occurring at the D'–D3 domain of vWF and not the platelet GPIbα-interactive A1 domain [71]. This implies that the VBP-decorated particles will be capable of binding to vWF without competing or interfering with the natural platelet binding. Similarly, liposomes decorated with only CBP moieties exhibited significant shear-independent binding to collagen-coated surfaces under flow. With the rationale for synergistic adhesion mechanisms, the VBP and CBP moieties were combined on a liposome surface, and the resultant particles showed significantly higher adhesion to vWF/collagen surfaces at low-to-high shear ranges, compared to liposomes bearing VBP or CBP moieties alone [70]. Furthermore, compared to liposomes modified heteromultivalently with rGPIbα and CBP, those modified with VBP and CBP showed higher levels of adhesion, suggesting that the issues of mu-tual steric interference in decorating large protein fragments on a synthetic particle surface can be resolved with small peptide decorations. These results established the feasibility of mimicking the dual adhesion mechanisms of platelets on synthetic particle platforms using small peptides.

7.3.2   Biomaterials-Based Synthetic Systems Inspired by Platelet Aggregation

The adhesion of natural platelets to vWF and collagen at the injury site triggers activation mechanisms, by virtue of which platelets change their morphology from biconvex discoid to stellar. This activation also results in stimulation of platelet sur-face integrin GPIIb–IIIa from a resting to a ligand-binding conformation. Stimulated GPIIb–IIIa is capable of binding the blood protein fibrinogen (Fg) via Fg's multiple peptide domains, namely tripeptide Arg–Gly–Asp (RGD) and HHLGGAKQAGDV (also known as H12) in its α and γ chains, respectively, on both termini of Fg [72, 73]. As a result, Fg can bridge multiple platelets together resulting in active platelet aggregation at the injury site. The activated platelets also secrete agonists-like ad-enosine diphosphate (ADP) which can further activate local platelets, adding to the aggregating population. Based on these processes, research has been carried out to mimic platelet's aggregation mechanisms by decorating synthetic particle surfaces by Fg itself, Fg fragments or Fg-based RGD, and H-12 peptides [5, 74–77]. The earliest designs of this approach used surface engineering of RBCs with Fg and Fg-derived RGD peptides [74]. These designs, termed "thromboerythrocytes," were envisioned to be a super-fibrinogen system where the surface decoration with Fg or Fg-derived RGDs was to enhance active platelet aggregation by interacting with the stimulated GPIIb–IIIa integrins. These RBC-based designs demonstrated hemostatic efficacy in vivo by reducing tail-bleeding times in thrombocytopenic rats. However, for clinical translation, these designs would require a large number of autologous antigen-matched RBCs or sufficient RBCs from universal donors, both of which are of limited availability. To address such limitations, some research efforts have been directed towards creating "universal donor-like RBCs" by masking RBC surface antigens with synthetic polymer coatings, or enzymatically converting the A and B antigens into O type, or stimulating stem cells into becoming universal donor-type RBCs [78]. The scale-up efficacy of such approaches is yet to be determined. Another Fg-based approach that has undergone preclinical and clinical evaluation in the USA and Europe is Fibrocaps™ (ProFibrix, The Netherlands), where a solution of fibrinogen and thrombin are separately spray dried and then combined to produce a mixture of suspendable microparticles. Fibrocaps have been used in clinical trials for treatment of bleeding during surgery and trauma-related injury, and they have shown significant reduction in time to hemostasis after direct administration to surgical wound within a gelatin sponge, compared to the sponge alone. Advanced phase clinical trials for this product is currently under way. One limitation of this product is that it needs to be directly applied to the wound (sprayed or applied in a sponge) and therefore may have limitations for intravenous (IV) administration. A few other Fg-based products that are undergoing extensive research as semisynthetic hemostats are Synthocytes™, Thrombospheres™, and Fibrinoplate™, all of which are essentially made of human albumin microparticles surface coated with Fg [5]. These products have undergone preclinical in vivo studies and some early-phase clinical trials, but more detailed clinical evaluations are needed to establish their safety and treatment efficacies. In a similar approach, liposomes have been coated with Fg, and these have demonstrated increased platelet recruitment and aggregation on collagen-covered surfaces in vitro [5]. Another recent interesting approach is the use of fibrin-based microgel systems that can respond to hemodynamic environment at an injury site to potentially enhance platelet recruitment and aggregation as well as directly seal the bleeding site. To this end, recent research has been reported on poly(N-isopropylacrylamide)-based low-cross-linked mircogel particles surface decorated with fibrin-specific recognition motifs that can enable binding to injury site by specific interaction with nascent fibrin protofibrils [79]. This is a promising design, with the caveat that the binding of these constructs require prior presence of sufficient fibrin at the injury site, which in turn requires prior sufficient propagation of the coagulation cascade. This property is similar to the fibrin-specific constructs developed previously by a different group of researchers [80], and this may make such constructs more applicable towards clot-targeted drug delivery, rather than clot-promoting hemostats.

Using particle surface coatings with Fg can present issues of immunogenicity, biocompatibility, and storage stability. Therefore, several approaches with the same design rationale have utilized Fg-derived small molecular weight peptides for particle surface decoration. To this end, polymeric nano- and microparticles, liposomes, and albumin particles have been decorated with linear RGD peptides (e.g., GRGDS), cyclic RGD peptides (e.g., cyclo-CNPRGDY[-OEt]RC), and H-12 peptides (i.e., HHLGGAKQAGDV) [5, 76, 77, 81]. A potential issue with using linear RGD peptides like GRGDS for such applications is that this peptide does not have interactive capability specifically with active platelet GPIIb–IIIa, but this can also interact with a variety of other integrins on other cell types (e.g., integrin αVβ3 on endothelial cells) [5]. Therefore, for in vivo applications, these RGD moieties may not provide functional specificity towards hemostasis. Nonetheless, recent research has demonstrated that GRGDS-decorated polymeric nanoparticles can render bleeding reduction in vivo [82]. It is important to note that these studies used models of acute bleeding in traumatic injury where most active platelets would localize at the injury site by default due to heavy bleeding out and therefore aggregation enhancement of these platelets via any RGD motif would indicate some extent of hemostatic efficacy. Functional specificity of the GRGDS motifs towards active platelet GPIIb–IIIa versus other cell integrins was not investigated in these studies. In comparison, the cyclic RGD peptide cyclo-CNPRGDY[-OEt]RC has been established to have high binding affinity specifically to stimulated platelet GPIIb–IIIa compared to other integrins [83], and therefore may provide higher functional specificity in platelet-mimetic hemostatic action. Similarly, the H-12 peptide also has binding specificity to stimulated platelet GPIIb–IIIa and can provide similar platelet-relevant functional specificity in hemostasis. The H-12 peptides have also been used to surface decorate disc-shaped "nanosheets" made of PLGA polymer, as another design approach for synthetic platelets [84]. All of these peptide-decorated designs have shown the capability of inducing platelet aggregation in vitro, and some have shown resultant hemostatic effect to reduce bleeding time in vivo in a variety of small animal models like tail-vein bleeding, tail-transection bleeding, femoral artery bleeding, liver resection bleeding, and blast trauma. Scaling up to bleeding models in larger animals and demonstrating efficacy in both chronic and acute bleeding scenarios would be necessary towards establishing the clinical potential of these designs.

7.3.3   Integrating Platelet-Inspired Adhesion and Aggregation Properties

Natural platelets render primary hemostasis by a synergistic interplay between the adhesive and aggregatory functionalities [56, 65]. Therefore, in recent years, there has been significant interest in developing biomaterials-based systems that can combine these two functions. To this end, latex beads have been developed that are surface decorated simultaneously with rGPIbα protein fragments and H-12 peptides [85]. Studies with beads bearing such combined surface modifications revealed that the two motifs need to be presented at two different canopy levels from the particle surface by using different spacer lengths, in order to minimize steric interference of their respective bioactivity. Such steric interference issues can be potentially resolved by combining the adhesive and aggregatory functions using small peptides instead of large proteins. This has been demonstrated recently by combining adhesion-promoting VBP and CBP peptides and aggregation-promoting cyclic RGD on liposomes as model particles [71, 86]. These "functionally integrated" platelet-mimetic systems showed higher hemostatic efficacy in vitro (in flow chamber setup) as well as in vivo (in mouse tail-transection bleeding model), compared to systems that had adhesion functionality only or aggregation functionality only. These studies suggest that capturing multiple platelet functions using heteromultivalent ligand modifications on biocompatible particle platforms may provide more efficient designs for synthetic platelet analogs.

7.3.4   Synthetic Biomaterial Systems Mimicking Physico-Mechanical Properties of Platelets

Platelet's morphology and physico-mechanical properties play important role in their interaction with other blood cells, especially RBCs, and in effect determine platelet's margination capabilities in hemodynamic flow [87, 88]. Compared to RBC's discoid biconcave shape and 8–10 μ diameter size, the "resting" platelet has a biconvex discoid shape and 2–3 μ diameter. Also, platelets are comparatively stiffer than RBCs. These properties result in platelets marginating to an RBC-free zone in close proximity to the vascular wall under hemodynamic flow, in mid-to-large sized vessels. Based on this rationale, there is recent research interest to engineer particles of platelet-mimetic shape, size, and flexibility as "physical" design parameters of synthetic platelets. In the context of platelet modulus and flexibility, studies have been carried out with rGPIbα-decorated liposomes having various lipid membrane fluidities [67]. The degree of membrane fluidity influenced the membrane flexibility and therefore the deformability of the particles. This was found to be directly correlative to the particle behavior on vWF-coated surfaces, whereby "soft" liposomes rolled slowly suggesting better adhesion, while "hard" liposomes rolled faster. In another recent report, platelet-relevant size, shape, mechanical properties, and vWF adhesion functionalities were integrated on an albumin particle platform [89]. Compared to stiff spherical particles, platelet-mimetic soft discoid particles exhibited better margination and adhesion (lower k_{off}) as measured by the duration of particle contact with the surface. In further advancement of this research, platelet-mimetic soft discoid particles were surface decorated heteromultivalently with VBP, CBP, and RGD peptides, and these functionally integrated "platelet-like nanoparticles" (PLNs) exhibited enhanced hemostatic efficacies in vitro and in vivo, compared to stiff spherical particles [90]. These results establish the feasibility of incorporating platelet-inspired multiple design parameters on a single platform towards engineering a more effective synthetic platelet construct.

7.3.5   Platelets from Stem Cells

Similar to the case with RBCs, regenerative medicine approaches have also focused on generating platelets from precursor cells and stem cells. To this end, one strategy has been to produce platelets in vitro from mature megakaryocytes by action of thrombopoietin, various interleukins, bone morphogenic protein, and stem cell factors [91]. Megakaryocytes, in turn, have been reported to be effectively generated from CD34+ hematopoietic stem cells (HSCs) derived from bone marrow, cord blood, and peripheral blood. Alternatively, recent research has also demonstrated the ability to generate HSCs and megakaryocytes from hESCs or iPSCs using specialized culture conditions [91]. In an exciting evolution of these approaches, researchers have recently demonstrated the establishment of stable immortalized megakaryocyte progenitor cell lines (imMKCLs) from iPSC-derived hematopoietic progenitor cells, which in turn can lead to a potentially inexhaustible source of platelets [92]. In parallel work, researchers have reported on a microscale bioreactor system that integrates the major chemical and physical components of the bone marrow to produce platelets from iPSC-derived megakaryocytes [93]. These approaches emphasize the translational potential of obtaining platelets from precursor cells and stem cells for clinical applications in the future. The current challenge is to maintain robust high yield of the platelets using the current culture conditions and bioreactor designs.

7.4   WBC Substitutes

The major function of WBCs is to maintain immune surveillance in blood circulation against foreign bodies and diseased cells and render cytotoxic and phagocytic incapacitation of these objects. This function is regulated by complex signaling events, cell–cell interactions, and secretions. Such multifactorial processes have not been effectively mimicked yet on synthetic platforms. Nevertheless, there has been some interesting experimental development in functional modulation or mimicry of WBCs, for example, (i) "Immunoengineering" approaches where synthetic materials (e.g., nanoparticles) are designed to deliver specific antigens to target cells/tissues to prod site-selective immune responses [94], (ii) "Leuko-polymerosomes" where polymeric vesicles are heteromultivalently surface decorated with selectin- and integrin-binding motifs to render leukocyte-mimetic interactions with "diseased" cells, followed by release of hydrogen peroxide (H_2O_2) from the vesicles to kill the target cells [95], and (iii) "leukolike vectors (LLVs)" where nanoporous silicon particles are surface coated with cellular membranes isolated from freshly harvested leukocytes to render properties like avoidance of opsonization and macrophagic uptake, preferential binding to inflamed endothelium, and facilitated transport of therapeutics across the endothelium while avoiding the lysosomal pathway [96]. Table 7.4 shows various biomaterials-based design approaches in developing leukocyte-inspired synthetic systems. These approaches are still in their early evaluation phases and will require efficient reproducibility, characterization, and scale up for translational applications.

Table 7.4 Biomaterials-based designs of leukocyte-mimetic systems

WBC mimicking systems	System components	References
	Antigen-presenting nanoparticle systems for specific immunomodulation	94
	Leuko-polymersomes with polymer vesicles surface-decorated by Sialyl Lewis motifs and other cell adhesion molecules	95
	Leuko-like Vectors with porous silica nanoparticles coated by leukocyte-derived membrane	96

7.5   Plasma Substitutes and Expanders

Surgery, trauma, and other critical conditions often lead to heavy blood volume loss (hypovolemia) and effective restoration of intravascular volume using plasma substitutes remains a clinically significant strategy. Hypovolemia leads to suboptimal nutritive effect of the circulation, and this results in the body trying to compensate for this deficit by redistribution of blood flow to vital organs (e.g., heart and brain). However, this results in under-perfusion of other tissues of the body, and this may ultimately lead to multiple organ dysfunctions. Two main classes of biomaterials, namely, crystalloids and colloids, have been investigated as plasma substitutes/expanders in treating hypovolemia [97, 98]. Crystalloids form crystals in their solid state and can dissolve in water as separate small molecules (e.g., sucrose, dextrose, etc.) or ions (e.g., sodium chloride, sodium bicarbonate, etc.) that can pass through semipermeable membrane (e.g., cell membrane). The most well-known crystalloids for plasma substitute applications are 0.9% saline (sodium chloride), dextrose (5–50% glucose in water), and Hartman's solution or Ringer's lactate (1–2% sodium bicarbonate). Colloids are fine particles that can form a dispersion or suspension in a continuous medium, and they persist in the intravascular compartment for a longer time than crystalloids since colloid particles are too large to cross the vascular endothelium. Colloidal suspensions can also draw water in from the interstitial and intracellular fluid compartments because of their high osmolarity. The most well-known colloids for plasma substitute applications are natural materials like human albumin suspensions (HAS) and cross-linked gelatin polypeptide suspensions (e.g., products like Gelofusine® and Haemaccel®), as well as, semisynthetic polymers like hydroxyethyl starch (HES, e.g., the product Voluven®) and dextran. All the above crystalloid and colloid-based plasma expanders have shown various degrees of side effects. For example, saline-based systems have shown effects of acidosis, hyperchloremia, and negative effects on renal functions. Hartmann's solution has negative effect in patients with impaired lactate metabolism and is also avoided in patients undergoing administration of citrate-containing blood products due to the risk of coagulation. Large-volume infusions of dextrose solutions can increase the risk of hyponatremia, edema, acidosis, and thrombophlebitis. Among colloidal suspensions, albumin has been associated with partial hypotension and coagulopathic side effects due to the presence of vasoactive peptides. Similar effects on platelet functions and coagulation cascade have been reported for gelatin suspensions. Dextran and other cross-linked starch suspensions have been implicated in anaphylactic reactions as well as coagulopathy and impairment of renal functions. Nonetheless, these crystalloid and colloid products can effectively render intravascular volume resuscitation, and they have therefore progressed into clinical applications.

7.6   Conclusion

The quest for "artificial blood" components continues to be clinically significant, to overcome the limitations of scarce supply, expensive isolation and storage, pathologic contamination risks, and biologic side effects posed by natural blood products. To this end, biomaterials have provided unique avenues of designing synthetic and semisynthetic systems to mimic functions of natural RBCs, platelets, leukocytes, and plasma. While only a few of these products have progressed into clinical applications so far, several others are currently under rigorous preclinical and clinical evaluations. Hence, refinement and optimization of their designs can be expected to lead to effective clinical translation in the future. In parallel, in the field of regenerative medicine, there is ongoing research regarding generation of blood cells ex vivo, from precursor cells and stem cells using biomaterials-based reactor and culture technologies, and some of these systems are progressing into preclinical or clinical investigation. Continued research at the interface of biomaterials and regenerative medicine can ultimately lead to the optimized design and production of cellular and noncellular components of natural blood and their mimics, that can be potentially integrated in the future to fine-tune the composition of "artificial blood."

References

1. Sakai H, Sou K, Horinouchi H, Kobayashi K, Tsuchida E. Review of hemoglobin-vesicles as artificial oxygen carriers. Artif Organs. 2009;33:139–45.
2. Modery-Pawlowski CL, Tian LL, Pan V, Sen Gupta A. Synthetic approaches to RBC mimicry and oxygen carrier systems. Biomacromolecules. 2013;14:939–48.
3. Mohanty D. Current concepts in platelet transfusion. Asian J Transfus Sci. 2009;3:18–21.
4. Lee D, Blajchman MA. Novel treatment modalities: new platelet preparations and substitutes. Br J Haematol. 2001;114:496–505.
5. Modery-Pawlowski CL, Tian LL, Pan V, McCrae KR, Mitragotri S, Sen Gupta A. Approaches to synthetic platelet analogs. Biomaterials. 2013;34:526–41.
6. Hess JR. An update on solutions for red cell storage. Vox Sang. 2006;91:13–9.
7. Hillyer CD. Blood banking and transfusion medicine. Philadelphia: Churchill Livingstone Elsevier; 2007.

8. Smith JW, Gilcher RO. Red blood cells, plasma, and other new apheresis-derived blood products: improving product quality and donor utilization. Transfus Med Rev. 1999;13:118–23.
9. Corash L. Inactivation of viruses, bacteria, protozoa, and leukocytes in platelet concentrates: current research perspectives. Transfus Med Rev. 1999;13:18–30.
10. Carson JL, Hill S, Carless P, Hébert P, Henry D. Transfusion triggers: a systematic review of the literature. Transfus Med Rev. 2002;16:187–99.
11. Solheim BG. Pathogen reduction of blood components. Transfus Apher Sci. 2008;39:75–82.
12. Seghatchian J, de Sousa G. Pathogen-reduction systems for blood components: the current position and future trends. Transfus Apher Sci. 2006;35:189–96.
13. Giangrande PLF. The history of blood transfusion. Br J Haematol. 2000;110:758–67.
14. Natanson C, Kern SJ, Lurie P, Banks SM, Wolfe SM. Cell-free hemoglobin-based blood substitutes and risk of myocardial infarction and death: a meta-analysis. JAMA. 2008;299:2304–12.
15. Goutelle S, Maurin M, Rougier F, Barbaut X, Bourguignon L, Ducher M, Maire P. The Hill equation: a review of its capabilities in pharmacological modelling. Fundam Clin Pharmacol. 2008;22:633–48.
16. Umbreit J. Methemoglobin—It's not just blue: a concise review. Am J Hematol. 2007;144:134–44.
17. Dorman SC, Kenny CF, Miller L, Hirsch RE, Harrington JP. Role of redox potential of hemoglobin-based oxygen carriers on methemoglobin reduction by plasma components. Artif Cells Blood Substit Immobil Biotechnol. 2002;30:39–51.
18. Chang TMS. Blood substitutes based on nanobiotechnology. Trends Biotechnol. 2006;24:372–77.
19. Winslow RM. Red cell substitutes. Semin Hematol. 2007;44:51–59.
20. Squires JE. Artificial blood. Science. 2002;295:1002–5.
21. Buehler PW, D'Agnillo F, Schaer DJ. Hemoglobin-based oxygen carriers: from mechanisms of toxicity and clearance to rational drug design. Trends Mol Med. 2010;16:447–57.
22. Winslow RM. Cell-free oxygen carriers: scientific foundations, clinical development, and new directions. Biochim Biophys Acta. 2008;1784:1382–6.
23. Rioux F, Drapeau G, Marceau F. Recombinant human hemoglobin (rHb1.1) selectively inhibits vasorelaxation elicited by nitric oxide donors in rabbit isolated aortic rings. J Cardiovasc Pharmacol. 1995;25:587–94.
24. Stowell CP, Levin J, Spiess BD, Winslow RM. Progress in the development of RBC substitutes. Transfusion. 2001;41:287–99.
25. Sloan EP, Koenigsberg MD, Philbin NB, Gao W. DCLHb Traumatic Hemorrhagic Shock Study Group. European HOST Investigators. Diaspirin cross-linked hemoglobin infusion did not influence base deficit and lactic acid levels in two clinical trials of traumatic hemorrhagic shock patient resuscitation. J Trauma. 2010;68:1158–71.
26. Gould SA, Moore EE, Hoyt DB, Burch JM, Haenel JB, Garcia J, DeWoskin R, Moss GS. The first randomized trial of human polymerized hemoglobin as a blood substitute in acute trauma and emergent surgery. J Am Coll Surg. 1998;187:113–20.
27. Jahr JS, Moallempour M, Lim JC. HBOC-201, hemoglobin glutamer-250 (bovine), Hemopure (Biopure Corporation). Expert Opin Biol Ther. 2008;8:1425–33.
28. Cheng DC, Mazer CD, Martineau R, Ralph-Edwards A, Karski J, Robblee J, Finegan B, Hall RI, Latimer R, Vuylsteke A. A phase II dose-response study of hemoglobin raffimer (Hemolink) in elective coronary artery bypass surgery. J Thorac Cardiovasc Surg. 2004;127:79–86.
29. Vandegriff KD, Winslow RM. Hemospan: design principles for a new class of oxygen therapeutic. Artif Organs. 2009;33:133–8.
30. Simoni J, Simoni G, Moeller JF, Feola M, Wesson DE. Artificial oxygen carrier with pharmacologic actions of adenosine-5'-triphosphate, adenosine, and reduced glutathione formulated to treat an array of medical conditions. Artif Organs. 2014;38:684–90.
31. Kinasewitz GT, Privalle CT, Imm A, Steingrub JS, Malcynski JT, Balk RA, DeAngelo J. Multicenter, randomized, placebo-controlled study of the nitric oxide scavenger pyridoxalated hemoglobin polyoxyethylene in distributive shock. Crit Care Med. 2008;36:1999–2007.

32. Varnado CL, Mollan TL, Birukou I, Smith BJ, Henderson DP, Olson JS. Development of recombinant hemoglobin-based oxygen carriers. Antioxid Redox Signal. 2013;18:2314–28.
33. Sakai H, Sou K, Horinouchi H, Kobayashi K, Tsuchida E. Hemoglobin-vesicle, a cellular artificial oxygen carrier that fulfils the physiological roles of the red blood cell structure. Adv Exp Med Biol. 2010;662:433–8.
34. Kaneda S, Ishizuka T, Goto H, Kimura T, Inaba K, Kasukawa H. Liposome-encapsulated hemoglobin, TRM-645: current status of the development and important issues for clinical application. Artif Organs. 2009;33:146–52.
35. Rameez S, Alosta H, Palmer AF. Biocompatible and biodegradable polymersome encapsulated hemoglobin: a potential oxygen carrier. Bioconjug Chem. 2008;19:1025–32.
36. Sheng Y, Yuan Y, Liu C, Tao X, Shan X, Xu F. In vitro macrophage uptake and in vivo biodistribution of PLA-PEG nanoparticles loaded with hemoglobin as blood substitutes: effect of PEG content. J Mater Sci Mater Med. 2009;20:1881–91.
37. Duan L, Yan X, Wang A, Jia Y, Li J. Highly loaded hemoglobin spheres as promising artificial oxygen carriers. ACS Nano. 2012;6:6897–904.
38. Xiong Y, Liu ZZ, Georgieva R, Smuda K, Steffen A, Sendeski M, Voigt A, Patzak A, Bäumler H. Nonvasoconstrictive hemoglobin particles as oxygen carriers. ACS Nano. 2013;7:7454–61.
39. Chang TM, Powanda D, Yu WP. Analysis of polyethylene-glycol-polylactide nano-dimension artificial red blood cells in maintaining systemic hemoglobin levels and prevention of methemoglobin formation. Artif Cells Blood Substit Immobil Biotechnol. 2003;31:231–47.
40. Tsuchida E, Sou K, Nakagawa A, Sakai H, Komatsu T, Kobayashi K. Artificial oxygen carriers, hemoglobin vesicles and albumin-hemes, based on bioconjugate chemistry. Bioconjug Chem. 2009;20:1419–40.
41. Collman JP, Brauman JI, Rose E, Suslick KS. Cooperativity in O_2 binding to iron porphyrins. Proc Natl Acad Sci U S A. 1978;75:1052–5.
42. Kano K, Kitagishi H. HemoCD as an artificial oxygen carrier: oxygen binding and autoxidation. Artif Organs. 2009;33:177–82.
43. Freire MG, Gomes L, Santos LM, Marrucho IM, Coutinho JA. Water solubility in linear fluoroalkanes used in blood substitute formulations. J Phys Chem B. 2006;110:22923–9.
44. Gould SA, Rosen AL, Sehgal LR, Sehgal HL, Langdale LA, Krause LM, Rice CL, Chamberlin WH, Moss GS. Fluosol-DA as a red-cell substitute in acute anemia. N Engl J Med. 1986;314:1653–6.
45. Riess JG. Perfluorocarbon-based oxygen delivery. Artif Cells Blood Substit Immobil Biotechnol. 2006;34:567–80.
46. Wang X, Gao W, Peng W, Xie J, Li Y. Biorheological properties of reconstructed erythrocytes and its function of carrying-releasing oxygen. Artif Cells Blood Substit Immobil Biotechnol. 2009;37:41–4.
47. Goldsmith HL, Marlow J. Flow behaviour of erythrocytes. I. Rotation and deformation in dilute suspensions. Proc R Soc Lond B. 1972;182:351–84.
48. Charoenphol P, Mocherla S, Bouis D, Namdee K, Pinsky DJ, Eniola-Adefeso O. Targeting therapeutics to the vascular wall in atherosclerosis–carrier size matters. Atherosclerosis. 2011;217:364–70.
49. Doshi N, Zahr AS, Bhaskar S, Lahann J, Mitragotri S. Red blood cell-mimicking synthetic biomaterial particles. Proc Natl Acad Sci U S A. 2009;106:21495–9.
50. Haghgooie R, Toner M, Doyle PS. Squishy non-spherical hydrogel microparticles. Macromol Rapid Commun. 2010;31:128–34.
51. Merkel TJ, Jones SW, Herlihy KP, Kersey FR, Shields AR, Napier M, Luft JC, Wu H, Zamboni WC, Wang AZ, Bear JE, DeSimone JM. Using mechanobiological mimicry of red blood cells to extend circulation times of hydrogel microparticles. Proc Natl Acad Sci U S A. 2011;108:586–91.
52. Li S, Nickels J, Palmer AF. Liposome-encapsulated actin-hemoglobin (LEAcHb) artificial blood substitutes. Biomaterials. 2005;26:3759–69.

53. Xu F, Yuan Y, Shan X, Liu C, Tao X, Sheng Y, Zhou H. Long-circulation of hemoglobin-loaded polymeric nanoparticles as oxygen carriers with modulated surface charges. Int J Pharm. 2009;377:199–206.
54. Douay L, Andreu G. Ex vivo production of human red blood cells from hematopoietic stem cells: what is the future in transfusion? Transfus Med Rev. 2007;21:91–100.
55. Rousseau GF, Giarratana MC, Douay L. Large-scale production of red blood cells from stem cells: what are the technical challenges ahead? Biotechnol J. 2014;9:28–38.
56. Hoffman M, Monroe DM III. A cell-based model of hemostasis. Thromb Haemost. 2001;85:958–65.
57. Heal JM, Blumberg N. Optimizing platelet transfusion therapy. Blood Rev. 2004;18:149–65.
58. Rubella P. Revisitation of the clinical indications for the transfusion of platelet concentrates. Rev Clin Exp Hematol. 2001;5:288–310.
59. Kauvar DS, Lefering R, Wade CE. Impact of hemorrhage on trauma outcome: an overview of epidemiology, clinical presentations, and therapeutic considerations. J Trauma. 2006;60:S3–11.
60. Murphy S. Platelet storage for transfusion. Semin Hematol. 1985;22:165–77.
61. Liumbruno G, Bennardello F, Lattanzio A, Piccoli P, Rossetti G. Recommendations for the transfusion of plasma and platelets. Blood Transfus. 2009;7:132–50.
62. Janetzko K, Hinz K, Marschner S, Goodrich R, Klüter H. Pathogen reduction technology (Mirasol®) treated single-donor platelets resuspended in a mixture of autologous plasma and PAS. Vox Sanguinis. 2009;97(3):234–39.
63. Schreiber GB, Busch MP, Kleinman SH, Korelitz JJ. The risk of transfusion- transmitted viral infections. The retrovirus epidemiology donor study. N Engl J Med. 1996;334:1685–90.
64. Blajchman MA. Substitutes for success. Nat Med. 1999;5:17–18.
65. Ruggeri ZM, Mendolicchio GL. Adhesion mechanisms in platelet function. Circ Res. 2007;100:1673–85.
66. Rybak ME, Renzulli LA. A liposome based platelet substitute, the plateletsome, with hemostatic efficacy. Biomater Artif Cells Immobil Biotechnol. 1993;21:101–18.
67. Takeoka S, Teramura Y, Okamura Y, Tsuchida E, Handa M, Ikeda Y. Rolling properties of rGPIbalpha-conjugated phospholipid vesicles with different membrane flexibilities on vWf surface under flow conditions. Biochem Biophys Res Commun. 2002;296:765–70.
68. Nishiya T, Kainoh M, Murata M, Handa M, Ikeda Y. Reconstitution of adhesive properties of human platelets in liposomes carrying both recombinant glycoproteins Ia/IIa and Ib alpha under flow conditions: specific synergy of receptor-ligand interactions. Blood. 2002;100:136–42.
69. Nishiya T, Kainoh M, Murata M, Handa M, Ikeda Y. Platelet interactions with liposomes carrying recombinant platelet membrane glycoproteins or fibrinogen: approach to platelet substitutes. Artif Cells Blood Substit Immobil Biotechnol. 2001;29:453–64.
70. Ravikumar M, Modery CL, Wong TL, Sen Gupta A. Mimicking adhesive functionalities of blood platelets using ligand-decorated liposomes. Bioconjug Chem. 2012;23:1266–75.
71. Haji-Valizadeh H1, Modery-Pawlowski CL, Sen Gupta A. A factor VIII-derived peptide enables von Willebrand factor (VWF)-binding of artificial platelet nanoconstructs without interfering with VWF-adhesion of natural platelets. Nanoscale. 2014;6:4765–73.
72. Pytela R, Piersbacher MD, Ginsberg MH, Plow EF, Ruoslahti E. Platelet membrane glycoprotein IIb/IIIa: member of a family of Arg-Gly-Asp-specific adhesion receptors. Science. 1986;231:1559–62.
73. Plow EF, D'Souza SE, Ginsberg MH. Ligand binding to GPIIb-IIIa: a status report. Semin Thromb Hemost. 1992;18:324–32.
74. Coller BS, Springer KT, Beer JH, Mohandas N, Scudder LE, Norton KJ, West SM. Thromboerythrocytes. In vitro studies of a potential autologous, semi-artificial alternative to platelet transfusions. J Clin Invest. 1992;89:546–55.
75. Levi M, Friederich PW, Middleton S, de Groot PG, Wu YP, Harris R, Biemond BJ, Heijnen HF, Levin J, ten Cate JW. Fibrinogen-coated albumin microcapsules reduce bleeding in severely thrombocytopenic rabbits. Nat Med. 1999;5:107–11.

76. Okamura Y, Takeoka S, Teramura Y, Maruyama H, Tsuchida E, Handa M, Ikeda Y. Hemostatic effects of fibrinogen gamma-chain dodecapeptide-conjugated polymerized albumin particles in vitro and in vivo. Transfusion. 2005;45:1221–28.
77. Bertram JP, Williams CA, Robinson R, Segal SS, Flynn NT, Lavik EB. Intravenous hemostat: nanotechnology to halt bleeding. Sci Transl Med. 2009;1:11–22.
78. Gelderman MP, Vostal JG. Current and future cellular transfusion products. Clin Lab Med. 2010,30.443–52.
79. Brown AC, Stabenfeldt SE, Ahn B, Hannan RT, Dhada KS, Herman ES, Stefanelli V, Guzzetta N, Alexeev A, Lam WA, Lyon LA, Barker TH. Ultrasoft microgels displaying emergent platelet-like behaviours. Nat Mater. Epub ahead of print; 2014 doi:10.1038/nmat4066.
80. Yu X, Song SK, Chen J, Scott MJ, Fuhrhop RJ, Hall CS, Gaffney PJ, Wickline SA, Lanza GM. High-resolution MRI characterization of human thrombus using a novel fibrin-targeted paramagnetic nanoparticle contrast agent. Magn Reson Med. 2000;44:867–72.
81. Ravikumar M, Modery CL, Wong TL, Sen Gupta A. Peptide-decorated Liposomes Promote arrest and aggregation of activated platelets under flow on vascular injury relevant protein surfaces in vitro. Biomacromolecules. 2012;13:1495–502.
82. Lashof-Sullivan MM, Shoffstall E, Atkins KT, Keane N, Bir C, VandeVord P, Lavik EB. Intravenously administered nanoparticles increase survival following blast trauma. Proc Natl Acad Sci U S A. 2014;111:10293–98.
83. Cheng S, Craig WS, Mullen D, Tschopp JF, Dixon D, Piersbacher MD. Design and synthesis of novel cyclic RGD-containing peptides as highly potent and selective integrin αIIbβ3 antagonists. J Med Chem. 1994;79:659–67.
84. Okamura Y, Fukui Y, Kabata K, Suzuki H, Handa M, Ikeda Y, Takeoka S. Novel platelet substitutes: disk-shaped biodegradable nanosheets and their enhanced effects on platelet aggregation. Bioconjug Chem. 2009;20:1958–65.
85. Okamura Y, Handa M, Suzuki H, Ikeda Y, Takeoka S. New strategy of platelet substitutes for enhancing platelet aggregation at high shear rates: cooperative effects of a mixed system of fibrinogen gamma-chain dodecapeptide- or glycoprotein Ibalpha-conjugated latex beads under flow conditions. J Artif Organs. 2006;9:251–58.
86. Modery-Pawlowski CL, Tian LL, Ravikumar M, Wong TL, Sen Gupta A. In vitro and in vivo hemostatic capabilities of a functionally integrated platelet-mimetic liposomal nanoconstruct. Biomaterials. 2013;34:3031–41.
87. Al Momani T, Udaykumar HS, Marshall JS, Chandran KB. Micro-scale dynamic simulation of erythrocyte-platelet interaction in blood flow. Ann Biomed Eng. 2008;36:905–20.
88. Tokarev AA, Butylin AA, Ermakova EA, Shnol EE, Panasenko GP, Ataullakhanov FI. Finite platelet size could be responsible for platelet margination effect. Biophys J. 2011;101:1835–43.
89. Doshi N, Orje JN, Molins B, Smith JW, Mitragorti S, Ruggeri ZM. Platelet mimetic particles for targeting thrombi inflowing blood. Adv Mater. 2012;24:3864–69.
90. Anselmo AC, Modery-Pawlowski CL, Menegatti S, Kumar S, Vogus DR, Tian LL, Chen M, Squires TM, Sen Gupta A, Mitragotri S. Platelet-like nanoparticles: mimicking shape, flexibility, and surface biology of platelets to target vascular injuries. ACS Nano. Epub ahead of print; 2014.
91. Avanzi MP, Mitchell WB. Ex vivo production of platelets from stem cells. Br J Haematol. 2014;165:237–47.
92. Nakamura S, Takayama N, Hirata S, Seo H, Endo H, Ochi K, Fujita K, Koike T, Harimoto K, Dohda T, Watanabe A, Okita K, Takahashi N, Sawaguchi A, Yamanaka S, Nakauchi H, Nishimura S, Eto K. Expandable megakaryocyte cell lines enable clinically applicable generation of platelets from human induced pluripotent stem cells. Cell Stem Cell. 2014;14:535–48.
93. Thon JN, Mazutis L, Wu S, Sylman JL, Ehrlicher A, Machlus KR, Feng Q, Lu S, Lanza R, Neeves KB, Weitz DA, Italiano JE Jr. Platelet bioreactor-on-a-chip. Blood. Epub ahead of print; 2014.
94. Swartz MA, Hirosue S, Hubbell JA. Engineering approaches to immunotherapy. Sci Transl Med. 2012;4:148rv9.

95. Hammer DA, Robbins GP, Haun JB, Lin JJ, Qi W, Smith LA, Ghoroghchian PP, Therien MJ, Bates FS. Leuko-polymersomes. Faraday Discuss. 2008;139:129–41.
96. Parodi A, Quattrocchi N, van de Ven AL, Chiappini C, Evangelopoulos M, Martinez JO, Brown BS, Khaled SZ, Yazdi IK, Enzo MV, Isenhart L, Ferrari M, Tasciotti E. Synthetic nanoparticles functionalized with biomimetic leukocyte membranes possess cell-like functions. Nat Nanotechnol. 2013;8:61–8.
97. Booth C, Highley D. Crystalloids, colloids, blood, blood products and blood substitutes. Anaesth Intensive Care Med. 2010;11:50–5.
98. McCahon R, Hardman J. Pharmacology of plasma expanders. Anaesth Intensive Care Med. 2007;8:79–81.

Dr. Anirban Sen Gupta is a primary faculty (associate professor) in Biomedical Engineering at Case Western Reserve University, in Cleveland Ohio, the USA. Dr. Sen Gupta's education and training is in chemistry, chemical engineering and biomedical engineering, and his research focus is on bio-inspired materials for advanced therapy applications (www.senguptalab.com).

Chapter 8
Biomaterials-Based Immunomodulation of Dendritic Cells

Evelyn Bracho-Sanchez, Jamal S. Lewis and Benjamin G. Keselowsky

Biomaterials are synthetic or naturally derived materials suitable for incorporation into the human body, and are meant to perform, augment, or replace a natural physiological function [1]. In conjunction with clinical benefit, the reproducibility, stability, availability, and relatively cheap cost of manufacture has driven the demand for numerous biomaterials on a global scale in the past quarter century [1]. Biomaterials are central to many strategies for regenerative medicine providing structural support and facilitating interactions with host cells and tissues. Components such as biologically active proteins, nucleotides, or small drug molecules may be incorporated into either a biomaterial scaffold that maintains structural integrity or biomaterial particles that target delivery to specific cellular and subcellular compartments. Material properties are able to be tuned to provide beneficial interactions with the host, for example, through biodegradation and/or the controlled release of factors. While biomaterials encompass many forms (e.g., metals, ceramics, protein-based polymers), the vast majority of biomaterial immunomodulation approaches to date utilize synthetic polymers.

The immune system provides a formidable barrier to the clinical application of many biomaterials, and controlling interactions with components of the immune system has recently become a major focus in the field [2]. The immune system is intricately organized and composed of multiple layers of protection that work in unison for the defense of the host against would-be invaders. Historically, biomaterials scientists have developed materials that simply try to limit chronic aggravation

E. Bracho-Sanchez (✉) · J. S. Lewis · B. G. Keselowsky
J. Crayton Pruitt Family Department of Biomedical Engineering, University of Florida,
1275 Center Dr. BMS J291, PO Box 116131, Gainesville, FL, USA
e-mail: e.bracho@ufl.edu

J. S. Lewis
e-mail: jamalslewis@ufl.edu

B. G. Keselowsky
e-mail: bkeselowsky@bme.ufl.edu

© Springer International Publishing Switzerland 2015
L. Santambrogio (ed.), *Biomaterials in Regenerative Medicine and the Immune System*,
DOI 10.1007/978-3-319-18045-8_8

of the immune system. However, recent biomaterials approaches to actively engage and modulate immune responses to achieve specific outcomes hold great promise. Key cellular players in immunological defense and homeostasis are dendritic cells (DCs), professional antigen-presenting cells (APCs) that have a crucial role in dictating T cell-mediated immunity. DCs naturally encounter, uptake, process, and present antigen to naïve T cells, and appropriately shape the resulting T and B cell responses. These features make DCs a major target of strategies to manipulate the immune system. Pioneering cellular-based therapies are ongoing using DCs, where DCs are generated from peripheral blood cells, exposed to antigen and stimuli *ex vivo,* and then reintroduced into the patient's circulatory system. Successes from these cellular therapy approaches highlight the importance of DC modulation. However, widespread application of DC cellular therapies is limited for most treatments, as the costs are tremendous. Recent accomplishments using implanted/injected biomaterials as a means to direct DC function while remaining *in situ* are encouraging, and they also address the high costs of cell-based therapies.

Biomaterial-based methodologies of modulating DCs can be broadly categorized as directing DC responses toward either inflammatory or suppressive phenotypes. Numerous approaches have aimed to augment the inflammatory response of DCs for their use in infectious diseases and cancer applications. Fewer, but a growing number of approaches are pursuing suppressive DC phenotypes, which is of interest for autoimmunity and transplant rejection therapies. Strategies currently employed to direct DCs through the introduction of biomaterials into the body include the use of both implanted scaffolds and injected particulates. The modulation of material properties, such as hydrophobicity, surface chemistry, and degradation rate, as well as the biomaterial-based delivery of proteins, nucleic acids, and small drug molecules, are approaches actively being investigated for directing immune responses. Suitable outcomes for strategies promoting inflammatory DCs are typically increased expression of stimulatory [3] and co-stimulatory molecules [3], and release of pro-inflammatory cytokines [4]. Conversely, desired outcomes of strategies promoting suppressive DCs are often decreased expression of stimulatory and co-stimulatory molecules [5], release of regulatory cytokines [4], and increased expression of inhibitory molecules and tolerogenic mediators [6]. The ability to manipulate DC function via biomaterials is proving to be enabling for numerous immune-related applications.

In this chapter, we discuss tools of biomaterials-based DC immunomodulation currently being developed, and suggest that they are broadly applicable to regenerative medicine. Given that regenerative medicine applications often require a careful balance between inflammatory and suppressive immunological processes involved in wound healing and regeneration, the field may substantially benefit from the tools and principles uncovered from such biomaterial-based immunomodulation approaches. To begin, major principles of immunology are introduced and relevant specifics of DC biology briefly described.

8.1 Innate and Adaptive Immunity Overview

The immune system is a complex network of cellular interactions and processes that has evolved as an organized and lethal defense mechanism against invaders, as well as providing homeostatic regulation of self and nonself. Recognition of foreign entities initiates a series of events that can lead to isolation and antigen-specific elimination in the case of a pathogen, or tolerance, as in the case of food proteins. In vertebrates, immunity can be divided into two main categories: the innate and the adaptive immune systems [7].

Innate immunity is considered the first line of defense, consisting of anatomical barriers such as the skin and mucosal membranes, soluble factors, and key cellular players [8]. Members of this system include the complement cascade (made up of a large number of distinct plasma proteins reacting with one another to opsonize pathogens and induce a series of inflammatory responses), neutrophils, natural killer (NK) cells, macrophages, and DCs. If a pathogen has penetrated beyond physical skin and mucosal barriers, then the cellular elements of innate immunity are typically engaged. Innate cellular defense mechanisms are triggered by signaling through pattern recognition receptors (PRRs), which are encoded in the germline and can be secreted or expressed on the host cellular surfaces. These PRRs can recognize and rapidly respond to a wide variety of common molecular features known as pathogen-associated molecular patterns [7, 8], facilitating uptake and clearance of the foreign material. Upon uptake and degradation of a pathogen, cellular machinery presents peptides from degraded proteins on the cell surface bound to specialized stimulatory molecules known as human leukocyte antigens (HLA) in humans and major histocompatibility complex (MHC) in mice [7]. This process is known as antigen presentation, and it can be carried out by a number of cells collectively known as APCs.

The adaptive arm of the immune system provides specific responses to newly encountered antigens, and is mainly driven by the interactions between APCs and lymphocytes [9]. Two major classes of lymphocytes exist, B and T cells, both originating from a common progenitor in the bone marrow. These lymphocytes patrol the body, routing through the bloodstream and lymphatic system to travel between peripheral and lymphatic tissues. The basic functions of B cells include antigen presentation, antibody production, and immunological memory. B lymphocytes are equipped with antigen-recognizing, membrane-bound immunoglobulins termed B cell receptors that allow it to interact with and uptake a vast array of antigens. When a naïve B cell encounters an antigen with affinity for its B cell receptor, the B cell will then differentiate into either a plasma or a memory B cell, which proliferates to generate a pool of these antigen-specific cells to combat the invasion. This process is known as clonal selection and expansion. Plasma cells produce soluble antibodies, which are the secreted forms of the B cell receptor. Antibodies that bind to surface antigens on pathogens will attract the first components of the complement cascade leading to complement activation. Clonally expanded memory B cells possess the same antigen-specific B cell receptor and will persist *in vivo* for an extended period following the resolution of the infection. These cells form the basis of an immunological memory response in the event of secondary invasion by the same pathogen.

Much like B lymphocytes, T lymphocytes express a unique, membrane-bound protein termed the T cell receptor, which recognizes antigenic peptides that are bound to stimulatory molecules on the surface of APCs. T cells also express associated co-receptors that coordinate with the T cell receptor. Different types of these co-receptors (e.g., cluster of differentiation 4; CD4) will correspond to the function of a T cell. Lymphocytes expressing CD4 are generally categorized as helper T cells. The CD4 molecule recognizes stimulatory molecules HLA-D/MHC II on the surface of professional APCs. Main subtypes of CD4$^+$ helper T cells include Th1, Th2, and Th17. The Th1 T cell subset supports pro-inflammatory cell-mediated immunity, which is intended to eliminate intracellular pathogens and inducing B cell production of opsonizing IgG [10]. Th1 T cells are characterized by the release of effector cytokines such as interferon (IFN)-γ, tumor necrosis factor (TNF)-α, and interleukin (IL)-2. Th2 T cells on the other hand are associated with less inflammatory immune responses and B cell production of IgG, IgA, and IgE [10]. More recently, the Th17 subset was found to abundantly produce the pro-inflammatory cytokine IL-17 [11] and is associated with antimicrobial responses in epithelial and mucosal tissues. T lymphocytes expressing CD8 receptors are generally categorized as cytotoxic T cells and recognize intracellular antigen presented on stimulatory molecules HLA-A/B/C (human) or MHC I (mouse), which are expressed by all nucleated cells including both professional and nonprofessional APCs. CD8$^+$ T cells directly bind to infected cells and eliminate them by releasing cytotoxins.

Professional APCs (DCs, macrophages, B cells) serve as the bridge between the innate and adaptive immune system. These cells patrol the body and can capture pathogens in their immature state via several uptake mechanisms. Once engulfed by an APC, antigen is loaded into acidic vesicles where proteins are degraded into peptide fragments, and the antigen is presented in the context of the stimulatory HLA/MHC molecules. The binding of a T cell receptor to antigen-loaded HLA/MHC molecules is commonly termed signal 1 of the T cell–APC interaction. Once bound to a T cell receptor, APCs can also provide signal 2 by presenting co-stimulatory molecules, which bind cognate T cell receptors that activate the T cell and trigger clonal T cell expansion. Furthermore, APCs can produce a signal 3, which refers to the release of cytokines. Through these mechanisms, APCs shape T cell responses. APCs can also sometimes engage in the secondary mechanism of cross presentation. During this process, an extracellularly supplied antigen is loaded on to HLA-A/B/C or MHC I molecules (instead of the primary route via HLA-D/MHC II) and presented to CD8$^+$ T cells. The role of macrophages as APCs is primarily to stimulate immunity to previously encountered antigens, and the role of B cells as APCs is primarily to engage helper T cell support to cognate antigen. DCs, in contrast, are of particular interest due to their unique ability to initiate and tune naïve lymphocyte responses to new antigens. As such, DCs are a central regulator of immunity and

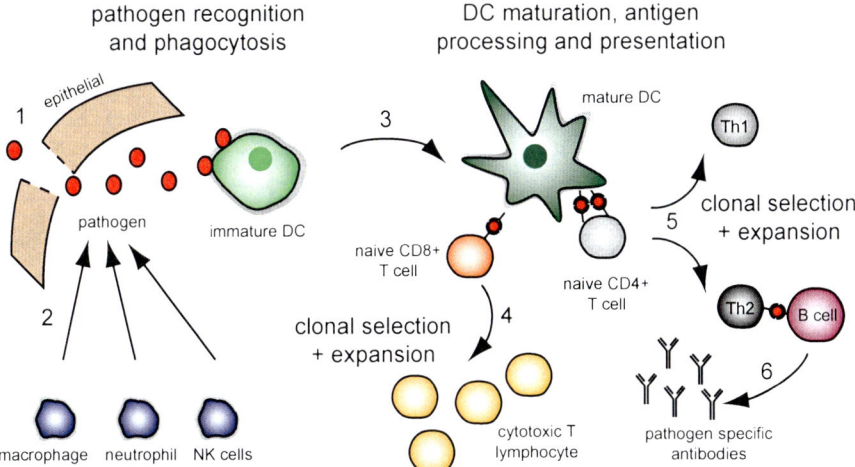

Fig. 8.1 Simplified immune response to foreign pathogen. (*1*) Pathogen encounters epithelial barrier which serves as first line of defense. (*2*) If the barrier is broken, the innate immune system will attempt to clear it by attacks mediated largely by macrophages, neutrophils, and NK cells. (*3*) To initiate adaptive immune system, dendritic cells (*DCs*) encounter antigen, undergo maturation during the antigen processing and present it on their surface for recognition by T cells. (*4*) CD8$^+$ T cells then undergo clonal expansion and become antigen-specific cytotoxic T cells. (*5*) CD4$^+$ T cells also undergo clonal expansion and interact with B cells leading to (*6*) the production of pathogen specific antibodies

have become a major target for immunomodulation. In summary, Fig. 8.1 depicts a simplified schematic of the typical immune response to a foreign pathogen.

The successful application of immunological principles has been most widely implemented through the use of vaccines. Vaccination engages innate and adaptive immunity and has been able to manipulate immunological responses for the treatment of numerous pathologies, particularly infectious diseases, and more recently cancer. Vaccines have been primarily successful in eliciting humoral (antibody-mediated) responses, but have had limited ability to induce cellular immunity (which provides clearance of intracellular pathogens in infected cells). Incorporation of biomaterials tools into vaccine strategies holds potential to address this and other limitations and will be discussed further.

8.1.1   DC Subsets

Two main subsets of DCs are plasmacytoid DC (pDCs) and conventional DCs. Both cell types express low levels of stimulatory (e.g., HLA-D/MHC II) and co-stimulatory molecules (e.g., CD40, CD80, and CD86) at homeostatic resting state, which are upregulated upon activation and are distinguished by high expression of the integrin receptor $\alpha_x\beta_2$ (the alpha subunit of which is also known as CD11c). Plasmacytoid DCs arise from lymphoid committed precursors and constitute a small subset of DCs mainly found in the blood and lymphoid tissue, and enter the lymph nodes through peripheral blood circulation [7]. Upon activation, pDCs are characterized by the copious release of type 1 IFN (IFN-α and IFN-β), particularly in response to viral infections. Plasmacytoid DCs, however, do not play a crucial role in the activation of naïve T lymphocytes, which is in contrast to conventional DCs. Conventional DCs originate from myeloid precursors and form a small subset of hematopoietic cells that populate most lymphatic and non-lymphatic tissue. Conventional DCs are the most potent APCs of the immune system as they have an enhanced ability to sense injury, capture antigen, process, and present it to T lymphocytes. Conventional DCs can also further differentiate into phenotypes that are either inflammatory or tolerogenic [12] (Fig. 8.2). Another major type of DC, the Langerhans cell, resides in the skin and mucosa, and unlike conventional DCs, during homeostasis they originate from a unique pool of skin-localized precursors [13]. While these (and more) distinct DC subsets have been identified, the field of biomaterials-based immune modulation has primarily focused on conventional DCs due to their ability to initiate *de novo* antigen response and their relative abundance.

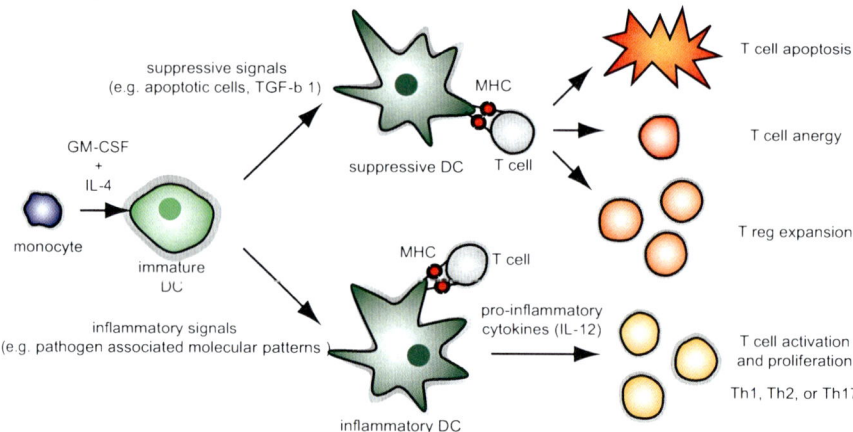

Fig. 8.2 Dendritic cell phenotype dictates T cell function. Immature dendritic cells can be derived from monocytes in the presence of IL-4 and GM-CSF. Dendritic cells can then become tolerogenic or inflammatory depending on the environmental stimulus, and subsequently direct T cell polarization. Inflammatory danger signals and pathogen-associated molecular patterns (*PAMPs*) can activate iDCs into classically mature (inflammatory) DCs which express high levels of MHC class II and co-stimulatory molecules, release pro-inflammatory cytokines, migrate to lymph nodes and stimulate naïve T cells to differentiate into Th1, Th2, or Th17 effector T cells. Suppressive DCs can be induced in response to signals such as apoptotic cells and TGF-β1. Suppressive DCs can induce antigen-specific T cell apoptosis, anergy and Treg expansion. *GM-CSF* granulocyte macrophage colony stimulating factor, *TGF* transforming growth factor, *MHC* major histocompatibility complex

8.1.1.1   Inflammatory DC Phenotype

In the immature state, DCs are constantly monitoring their microenvironment, sampling, processing, and presenting antigens, and ready to respond to signals of danger. They capture antigens through several mechanisms including pinocytosis [7] (an actin-driven mechanism in which the plasma membrane ruffles to form a vesicle incorporating foreign material), receptor-mediated endocytosis [7] (either via C-type lectin receptors, or CD64 and CD32), and phagocytosis [7] (internalization of larger particulates usually aided by specific membrane receptors). Once immunogenic antigen is captured in conjunction with activation signals (e.g., from PRRs), DCs undergo phenotypical changes rendering their mature state. These changes are characterized by morphology alteration (e.g., loss of cell surface adhesion molecules, cytoskeleton reorganization, and increased cellular motility [7]), increased expression of HLA/MHC stimulatory molecules, upregulation of co-stimulatory molecules (such as CD40, CD80, and CD86), and release of pro-inflammatory cytokines (e.g., IL-12). DC maturation is closely linked with migration from peripheral tissue to the secondary lymphoid organs (spleen, lymph nodes, tonsils, appendix, and Peyer's patches) where they interact with T cells and influence their function.

8.1.1.2   Suppressive DC Phenotype

In the periphery, DCs encounter self-derived antigens and present them to T cells as a normal process of homeostasis. Rarely, this presentation results in the pathological activation of the adaptive immune system and can result in autoimmunity against self-antigen. Typically, the normal homeostatic reaction involves DC initiation of suppressive immune networks that can lead to T cell anergy, deletion, or expansion of a regulatory phenotype. For instance, transforming growth factor-β1 (TGF-β1) and/or IL-10 secreted by suppressive DCs can induce regulatory T cells (Tregs). These Tregs suppress inflammatory immune responses, and are characterized by the expression of surface receptors CD4 and CD25 and the transcription factor, forkhead box P3 (Foxp3). Subset of Tregs can also secrete IL-10 and have suppressive roles when in contact with effector T cells. Generally, characteristics of tolerance-promoting DCs have been reported to include low expression of stimulatory and co-stimulatory molecules, low production of inflammatory cytokines (e.g., IL-12), increased production of regulatory cytokines (e.g., IL-10, TGF-β1), high levels of inhibitory molecules (e.g., programmed death ligand 1; PD-L1), as well as high expression of tolerogenic mediators such as indoleamine 2,3-dioxygenase (IDO) [12, 14]. There is tremendous potential in targeting DCs for therapeutic applications across the spectrum of immune diseases.

8.2   Biomaterials-Based Immunomodulation

Numerous synthetic polymers have been investigated for their use as biomaterials in regenerative medicine applications, including polyesters (e.g., poly(lactic acid), poly (glycolic acid), and their copolymers), polyorthoesters, polyanhydrides, and polycarbonates. Of these, polymers fabricated using biodegradable copolymer poly(lactic co-glycolic) acid (PLGA) have been the most researched for the delivery and controlled release of immunotherapeutics, and their interaction with DCs is well characterized [15]. By altering the composition of the PLGA copolymer, microparticles can be fabricated to provide an initial burst of encapsulated immunomodulatory molecule followed by sustained release. This feature has notably been identified as enabling the development of therapeutics which can administer both prime and boost doses in a single injection [5]. This section details efforts in the manipulation of immune responses through the use of polymeric systems as controlled release carriers of immunomodulatory molecules, as well as the potential for biomaterials to provide direct interactions with DCs, for example, through proteins adsorbed onto the biomaterial surface. A conceptual summary of biomaterials-based immunomodulation of DCs is illustrated in Fig. 8.3.

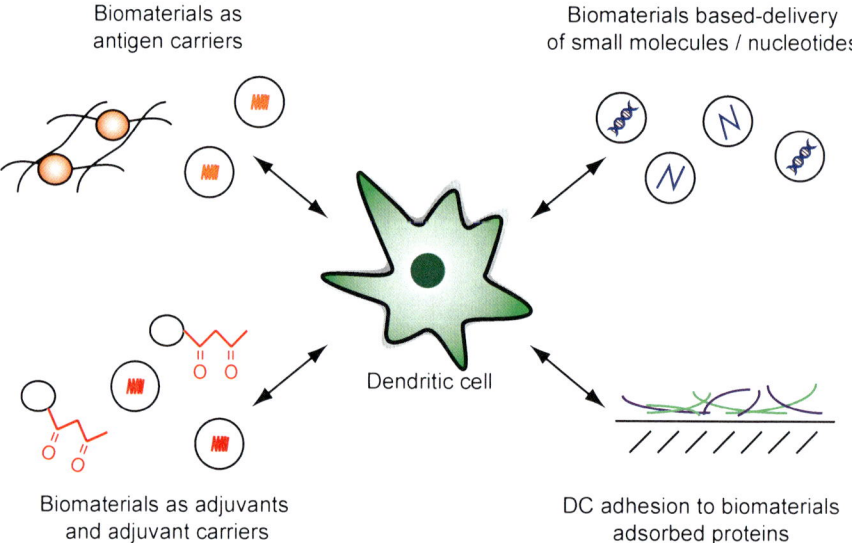

Fig. 8.3 Overview of biomaterials-based dendritic cell modulation. Common strategies can be broadly divided into those employing biomaterials as antigen carriers for targeted delivery of small molecules and nucleotides and controlled release of adjuvants. Biomaterials can furthermore directly function as an adjuvant, possibly through dendritic cell *(DC)* interactions with surface-adsorbed proteins

8.2.1   Biomaterials-Based DC Manipulation Toward an Inflammatory Phenotype

The quintessential scenario in which an inflammatory DC phenotype would be desired is vaccine applications. Vaccines are generally composed of an antigen, adjuvant, and a delivery system (e.g., a liquid solvent able to be injected). Adjuvants can be materials or molecules which stimulate innate immunity by activating APCs. This engagement of the innate branch of immunity results in the development of long-lasting, antigen-specific immune responses toward the co-delivered antigen [16]. The only Food and Drug Administration (FDA)-approved agents for use as adjuvants in the USA, interestingly, are materials consisting of inorganic particulates of calcium phosphate, aluminum hydroxide, and aluminum phosphate, which are collectively referred to as alum [16]. Alum causes aggregation of antigen onto the particulates, providing an antigen depot, and eliciting a strong inflammatory response [16]. Although these materials have been used for decades, their mechanism of action is still being uncovered. Recent reports have shown that alum engages the NALP (Nacht-domain, NAIP (neural apoptosis inhibitor protein), C2TA (MHCII transctiptor activator), HET-E (incompatibility locus protein from Podospora anserina) and TP1 (telomerase-associated protein), leucine-rich repeat (LRR), and pyrin domain-containing protein)-3 inflammasome, resulting in the production of pro-inflammatory cytokines IL-1β and IL-18 [17]. However, alum has limitations in that it is not optimal for different types of antigens, and is ineffective at inducing cellular immunity. This motivates development of new adjuvant technologies where biomaterials can be implemented as both antigen carriers and serve as adjuvants or adjuvant carriers.

8.2.2   Biomaterials as Antigen Carriers

The use of biomaterials is an attractive approach as they can be formulated to provide an antigen depot with long-term protection, persistence, and controlled release of antigens, potentially eliminating the need for booster doses, while increasing the likelihood for antigen uptake and processing by APCs. In contrast to alum, synthetic biodegradable polymers can be used to provide a depot for more antigens than just proteins, including, for example, small antigenic peptides. Antigens can be loaded into, or conjugated onto biomaterial particles or scaffolds, protecting the cargo from enzymatic degradation. For example, studies conducted by Ali et al. sought to create an inflammatory microenvironment through the use of implanted porous PLGA scaffolds as a cancer vaccine [18]. This approach aimed at not only using biomaterials as a delivery vehicle but also as a structural support in which DCs are attracted through release of a recruiting molecule, granulocyte macrophage colony-stimulating factor (GM-CSF), and in which the DCs can reside while receiving co-encapsulated antigen (tumor lysate) and activation signal cytosine triphosphate, phosphodiester link, guanine triphosphate (CpG). The investigators reported that after 14 days *in vivo*, the cellular infiltration of the scaffolds containing GM-CSF was significantly higher when compared to the matrix containing no recruiting

agent. Not only were the cells primed in a controlled microenvironment but activated DCs were also able to leave the PLGA matrix and travel to draining lymph
nodes for further interaction with T cells [18]. Notably, these immune-modulating
materials were able to develop specific and protective antitumor immunity report-
ing a 90% survival rate in melanoma mouse models.

Biomaterials in particulate form can promote targeted delivery of antigen to DCs
and intracellular DC compartments, facilitating specific processing and presentation on either MHC I or MHC II complexes. For example, Lewis et al. demonstrated
that DC uptake was markedly increased when PLGA microparticles were surface
conjugated with either anti-DEC-205 antibody, anti-CD11c antibody, or the CD11c-
binding peptide, P-D2 [19]. Notably, the PD-2 peptide conjugation greatly improved
DC MHC II antigen presentation compared to controls, by prolonging presentation
to a 4-day time span, *in vitro*. It was reported that none of the modified or unmodified PLGA particles (50:50 composition) activated DCs (activating factors were not
included). The P-D2 peptide surface modification also greatly promoted *in vivo* DC
uptake and trafficking of microparticles to lymph nodes [19]. In contrast to delivering exogenous antigen for MHC II loading, promotion of cross-presentation, the
exogenous delivery of antigen to induce immunity through the MHC I pathway,
is known to be challenging. This is due to the fact that internalized particulates
are quickly trafficked to acidic endosomes where they are enzymatically degraded.
However, several groups have made the attempt with varying degrees of success.
For example, Sneh-Edri et al. investigated PLGA nanoparticles decorated with
peptides with an endoplasmic reticulum localizing motif and loaded with antigenic
materials. Although enhanced localization to the endoplasmic reticulum was not
achieved, the particle modifications did increase endocytosis and influenced intracellular trafficking, which provided low levels of cross-presentation of the antigenic
peptide, for a modestly prolonged time of 2 days, *in vitro* [20]. More successful
strategies have overcome the challenge of inducing cross-presentation through the
development of pH-responsive polymers [21] that disrupt endosomal membranes,
thereby delivering antigen to the cytoplasm (where processing occurs for MHC I
loading). For example, Flanary et al. investigated the use of poly(propylacrylic acid,
PPPA) as a protein vaccine carrier, and reported enhanced delivery of ovalbumin
(as a model antigen) into the MHC I pathway [22]. In the ionized state, PPAA maintains aqueous solubility, but when the environment becomes acidic, as is the case
of endosomes, the pendant carboxyl groups become protonated and the polymer
transitions to an insoluble state that is membrane destabilizing, delivering cargo to
the cytosol. This cytosolic delivery resulted in an increased MHC I presentation and
antigen-specific CD8$^+$ lymphocyte activation, promoting cellular immunity [22].
These studies illustrate the use of biomaterials for the delivery of antigen to cellular
and subcellular compartments.

8.2.2.1 Biomaterials as Adjuvants and Adjuvant Carriers

In addition to being able to provide an antigen depot, biomaterials can also serve adjuvant functions by either delivering adjuvant molecules [23] or by providing direct

activation through cell–material interactions [24]. Several experimental adjuvant molecules, such as unmethylated bacterial CpG DNA (CpG), polyinosinic–poly-cytidylic acid (PolyI:C), lipopolysaccharide (LPS), and monophosphoryl lipid A (MPLA) are known to activate APCs directly. The mechanism involves the binding of toll-like receptors on DCs, resulting in the upregulation of MHC II, co-stimulatory molecules CD80, CD86, and CD40 and secretion of pro-inflammatory cytokines such as IL-6. Studies by Schlosser et al. showed that CpG and PolyI:C encapsulated along with ovalbumin (model antigen) in a PLGA microparticle stimulated a potent cytotoxic T lymphocyte response. A single immunization was provided and significant levels of antigen-specific tetramer (SIINFEKL/H-2K(b))-positive cytotoxic T lymphocytes were detected, as was production of IFN-α, efficient cytolysis, and protection from vaccinia virus infection [23]. Similarly, Beaudette et al. reported that the co-encapsulation of ovalbumin and CpG resulted in higher cytotoxic T lymphocyte activity when compared to particles co-administered with the CpG adjuvant in the free form [25]. In another example, Elamanchili et al. demonstrated that DCs pulsed with MPLA-encapsulating PLGA nanoparticles showed upregulation of MHC II and CD86 molecules resulting in the activation of naïve T lymphocytes [26]. Extending upon their previous work [18] (described above), Mooney and colleagues further experimented with CpG and tumor lysate-loaded PLGA scaffold materials, implanted as a cancer vaccine, this time comparing co-encapsulation with different recruiting factors GM-CSF, Flt3L, or CCL20 [27]. The authors demonstrated a significant increase in recruited and activated DC numbers (with high levels of MHC II and CD86) infiltrated into the scaffold, leading to potent antitumor T cell responses. These vaccine formulations resulted in long-term survival in melanoma mice models [27]. Combination approaches such as the co-delivery of adjuvants, particularly of PRR-activating molecules, along with the immune-modulating nucleotides such as silencing RNA (siRNA) are promising [28, 29]. Specifically, studies conducted by Pradhan et al. conjugated IL-10 siRNA and CpG onto the surface of PLGA microparticles loaded with encapsulated pDNA antigen. Their studies demonstrated uptake, high expression of stimulatory and co-stimulatory molecules (CD80, CD86, and MHC II), low expression of inhibitory cytokines (IL-10) and high expression of pro-inflammatory cytokines (IL-12 and TNF-α) by DCs treated with microparticles when compared to its soluble formulation. This DC profile shifted the T cell activation to a Th1 subset which resulted in increased survival in lymphoma animal models [29].

In contrast to approaches where biomaterials serve as an adjuvant carrier, another strategy is designing materials that can directly act as adjuvants themselves. Studies conducted by Hudalla et al. in the Collier group developed peptides bearing *p*-nitrophenyl phosphonate ligands that self-assembled into nanofibers displaying properly folded, biologically active protein antigens [30]. These nanofibers were able to successfully present controlled amount of a model protein antigen (green fluorescent protein) and acted as a self-adjuvanting vaccine material in mice [30]. Additionally, Narasimhan and colleagues demonstrated the ability of a family of polyanhydride nanoparticles to differentially act as self-adjuvants, depending on the polymer composition, inducing potent immunomodulatory activity [31–33]. Along these lines, it has been observed that various PLGA compositions influence DC

activation differently. Studies conducted by Fischer et al. have shown microparticles with a 50:50 ratio of lactide to glycolide maintain DC immaturity [34] while
the more hydrophobic ratio of 75:25 has been reported to induce DC activation
[24]. Guiding principles for materials chemistry-based immune modulation are still
being explored, but surface immobilized adhesive proteins may play a role by directing DC adhesion [35–37]. Additionally, like alum and other particulate materials, PLGA microparticles are capable of engaging the NALP3 inflammasome [17].
Altogether, this literature supports the concept that biomaterials have tremendous
potential to improve current adjuvant technologies by either direct activation of
APCs or encapsulation of stimulating agents.

8.2.3   Biomaterials-Based DC Manipulation Toward
a Suppressive Phenotype

So far, discussion has centered on the modulation of DCs to initiate an inflammatory
immunological response, and this has been the largest focus in the field. However,
biomaterials immune modulation of DCs for suppressive applications is gaining
momentum and hinges on the fact that DCs can also support suppressive pathways
that maintain homeostasis and dictate key regulators of peripheral tolerance. It has
been reported that suppressive DCs can elicit T cell anergy, expansion of regulatory
T cells and repression of B cell differentiation and expansion [38]. T cell anergy is
a tolerogenic mechanism in which lymphocytes become hyporesponsive following
the incomplete delivery of co-stimulatory signals [39]. In addition to inducing T
cell anergy, DCs can also expand the formation of $CD4^+CD25^+FoxP3^+$ regulatory
T cells. This Treg subset can directly halt effector T cells and by inhibiting antigen
presentation from DCs [40]. Therefore, guidance of DC phenotype toward a suppressive or tolerogenic state holds great promise, particularly for the mitigation of
autoimmune diseases and transplant rejection. While the list of works to date is
shorter compared to inflammatory applications, biomaterial approaches also offer
opportunities to employ controlled release and payload protection to deliver modulatory molecules and nucleotides for suppressive applications. These biomaterial
strategies can thereby overcome barriers of limited bioavailability that often accompanies administration of these types of therapeutic molecules.

Numerous pharmacological methods have been reported to inhibit DC maturation, including the use of anti-inflammatory (e.g., corticosteroids [41], sa-
licylates [42]) or immunosuppressive (e.g., mycophenolate mofetil [43], $1\alpha,25$-
dihydroxyvitamin D_3 [44, 45]) agents, as well as certain cytokines (TGF-β [46],
IL-10 [47]).The systemic soluble administration of pharmacological agents typically has severe limitations in dosing and long-term use. Targeted delivery of these
agents to specific cells *in vivo* can be critical, given the pleiotropic nature many
possess, with potential off target effects and toxicity on bystander cells and tissues.
This limitation provides another window of opportunity for the use of biomaterials

to provide targeted delivery and controlled release of these types of factors. For example, the systemic administration of the widely used immunosuppressant rapamycin, *in vivo,* decreases DC surface expression of CD40, CD80, CD86, and MHC II molecules [48]. Taking advantage of this, Bryant et al. investigated the use of PLGA particles for the delivery of donor antigens to induce transplant tolerance in allogeneic islet transplantation model [49]. Infusion of the particles mediated tolerance in approximately 20% of recipients, and combination with a low-dose rapamycin at the time of transplant significantly improved tolerance to approximately 60% efficacy. This method offered a biomaterials-based approach to inducing long-term donor graft-specific tolerance [49]. However, long-term systemic administration of rapamycin and other such immunosuppressants at normal therapeutic levels results in systemic immunosuppression by affecting off target lymphocytes [50]. This has motivated the use of rapamycin encapsulating particles. For example, work by the Little group examined the effects of rapamycin-loaded PLGA microparticles on DCs [45]. This study established low expression of stimulatory and co-stimulatory molecules (MHC II, CD86, CD40) and decreased T cell activation [45]. Furthermore, Maldonado et al. demonstrated that biodegradable PLGA nanoparticles carrying either protein or peptide antigen, co-formulated with rapamycin induced a durable antigen-specific tolerogenic effect [51]. Specifically, treated animals showed reduced allergic hypersensitivity disorders, protection from disease relapse in a multiple sclerosis model, and prevention of antidrug antibodies responses in a hemophilia A model. Treatment with these nanoparticles resulted in the inhibition of $CD4^+$ and $CD8^+$ T cell activation, an increase in regulatory T cells, and durable B cell tolerance resistant to multiple immunogenic challenges with protein or peptide antigen. Importantly, the authors demonstrated that only the encapsulated rapamycin and not the soluble form could induce immunological tolerance at the doses used [51]. Altogether, these works show how immunomodulation can be improved through the use of biomaterials encapsulating small drug molecules.

DC adhesive interactions with biomaterials and biomaterials-adsorbed proteins have also been investigated, and these interactions have been reported to be able differentially promote suppressive DC phenotypes [35, 52]. For example, studies performed by Acharya et al. indicate that DC morphology and production of cytokines is differentially dependent upon adhesive substrates [53]. Specifically, DCs cultured on albumin and serum-coated tissue culture-treated substrates maintained low levels of CD80, CD86, and MHC II, and produced increased levels of IL-10 [53]. Using a different approach, Leclerc et al. demonstrated that DCs can receive biological cues from tethered signaling molecules [54]. The authors reported covalently immobilized GM-CSF onto planar silane and polystyrene surfaces resulted in increased DC differentiation in response to GM-CSF tethering [54].

Delivery of immune-modulating nucleotides also has the potential to modulate DC responses and further dictates the outcome of an immunological response. Biomaterials can serve as delivery vehicles for siRNA [29], gene vectors [55], messenger RNA [56], and antisense oligonucleotides [57]. Biomaterial carriers can provide a vehicle to protect nucleotides from systemic elimination, and provide intact delivery to subcellular compartments [58]. For example, Giannoukakis and colleagues demonstrated that nonobese diabetic mouse (NOD; a well-established animal model for type 1 diabetes)-derived DCs treated *ex vivo* with antisense oligonucleotides resulted in diminished expression of costimulatory molecules (CD40, CD80, and CD86) and a significant delay in the incidence of autoimmune diabetes in NOD mice [57]. Furthermore, phase I safety clinical trials concluded positively, where antisense-treated autologous DCs were manipulated *ex vivo* and administered as a cellular therapy to established type 1 diabetes adult patients [59]. Nevertheless, the limiting high cost of cellular therapies has motivated the group to pursue a biomaterials microparticle-based approach. Along this line, encapsulation of antisense oligonucleotides in a polymeric microparticle was able to effect gene knockdown in DCs [60]. The subcutaneously injected microparticles were trafficked to pancreatic draining lymph nodes, augmented Tregs, and provoked hyporesponsiveness to β cell antigen without compromising global immunity [60]. The studies presented in this section highlight the delivery of small molecules and nucleotides through the use of biomaterials as well as surface interactions specifically targeting DCs for suppressive applications, and are directly relevant to regenerative medicine.

8.3   Summary

Biomaterials can be engineered to fulfill many functions in the context of regenerative medicine. They have long been studied to perform, augment, or replace a natural function, historically attempting to avoid provoking the immune system. However, more recently, biomaterials have gained much interest as a tool with which to actively engage the immune system, particularly through the targeting of DCs. DCs play a crucial role in dictating the fate of an immune response and maintaining homeostasis and have therefore become the focus of numerous experimental approaches. This chapter has outlined key concepts for the understanding of immune responses, described relevant DC subsets and functions, and detailed illustrative biomaterials-based technologies directing DC phenotype toward either an inflammatory or a suppressive role. Techniques being developed include the use of biomaterials as antigen carriers, biomaterials as adjuvants and adjuvant carriers, the delivery of small molecules and nucleotides, and adhesive interactions with biomaterials. The field of biomaterials-driven immunotherapies will continue to grow as critical mechanisms are uncovered and more advanced material designs emerge. Notably, the concepts encompassing the field of regenerative medicine have grown over the years to include fields such as tissue engineering and wound healing. Looking forward, through the use of new tools such as biomaterials-based immunomodulation of DCs, it is conceivable that the range of regenerative medicine will be further expanded to include related disciplines such as cancer, transplant rejection, autoimmune diseases, and infectious diseases.

References

1. Sionkowska A. Current research on the blends of natural and synthetic polymers as new biomaterials: review. Prog Polym Sci. 2011;36(9):1254–76.
2. Lewis JS, Roy K, Keselowksy BG. Materials that harness and modulate the immune system. MRS Bulletin. 2014;25–34.
3. Winter M, Beer HD, Hornung V, Kramer U, Schins RP, Forster I. Activation of the inflammasome by amorphous silica and TiO2 nanoparticles in murine dendritic cells. Nanotoxicology. 2011;5(3):326–40.
4. Lutz MB, Schuler G. Immature, semi-mature and fully mature dendritic cells: which signals induce tolerance or immunity? Trends Immunol. 2002;23(9):445–9.
5. Lewis JS, Roche C, Zhang Y, et al. Combinatorial delivery of immunosuppressive factors to dendritic cells using dual-sized microspheres. J Mater Chem B Mater Biol Med. 2014;2(17):2562–74.
6. Hwu P, Du MX, Lapointe R, Do M, Taylor MW, Young HA. Indoleamine 2,3-dioxygenase production by human dendritic cells results in the inhibition of T cell proliferation. J Immunol. 2000;164(7):3596–9.
7. Banchereau J, Briere F, Caux C, et al. Immunobiology of dendritic cells. Annu Rev Immunol. 2000;18:767–811.
8. Akira S, Uematsu S, Takeuchi O. Pathogen recognition and innate immunity. Cell. 2006;124(4):783–801.
9. Iwasaki A, Medzhitov R. Toll-like receptor control of the adaptive immune responses. Nat Immunol. 2004;5(10):987–95.
10. Broere F, Apasov SG, Sitkovsky MV, van Eden W. Principles of immunopharmacology: 3rd revised and extended edition: Basel: Springer; 2011.
11. Korn T, Bettelli E, Oukka M, Kuchroo VK. IL-17 and Th17 Cells. Annu Rev Immunol. 2009;27:485–517.
12. Steinman RM, Hawiger D, Nussenzweig MC. Tolerogenic dendritic cells. Annu Rev Immunol. 2003;21:685–711.
13. Hoeffel G, Wang Y, Greter M, et al. Adult Langerhans cells derive predominantly from embryonic fetal liver monocytes with a minor contribution of yolk sac-derived macrophages. J Exp Med. 2012;209(6):1167–81.
14. Huang L, Baban B, Johnson BA, 3rd, Mellor AL. Dendritic cells, indoleamine 2,3 dioxygenase and acquired immune privilege. Int Rev Immunol. 2010;29(2):133–55.
15. Keselowsky BG, Xia CQ, Clare-Salzler M. Multifunctional dendritic cell-targeting polymeric microparticles: engineering new vaccines for type 1 diabetes. Hum Vaccin. 2011;7(1):37–44.
16. Jones KS. Biomaterials as vaccine adjuvants. Biotechnol Prog. 2008;24(4):807–14.
17. Sharp FA, Ruane D, Claass B, et al. Uptake of particulate vaccine adjuvants by dendritic cells activates the NALP3 inflammasome. Proc Natl Acad Sci U S A. 2009;106(3):870–5.
18. Ali OA, Huebsch N, Cao L, Dranoff G, Mooney DJ. Infection-mimicking materials to program dendritic cells in situ. Nat Mater. 2009;8(2):151–8.
19. Lewis JS, Zaveri TD, Crooks CP 2nd, Keselowsky BG. Microparticle surface modifications targeting dendritic cells for non-activating applications. Biomaterials. 2012;33(29):7221–32.
20. Sneh-Edri H, Likhtenshtein D, Stepensky D. Intracellular targeting of PLGA nanoparticles encapsulating antigenic peptide to the endoplasmic reticulum of dendritic cells and its effect on antigen cross-presentation *in vitro*. Mol Pharm. 2011;8(4):1266–75.
21. El-Sayed ME, Hoffman AS, Stayton PS. Smart polymeric carriers for enhanced intracellular delivery of therapeutic macromolecules. Expert Opin Biol Ther. 2005;5(1):23–32.
22. Flanary S, Hoffman AS, Stayton PS. Antigen delivery with poly(propylacrylic acid) conjugation enhances MHC-1 presentation and T-cell activation. Bioconjug Chem. 2009;20(2):241–8.
23. Schlosser E, Mueller M, Fischer S, et al. TLR ligands and antigen need to be coencapsulated into the same biodegradable microsphere for the generation of potent cytotoxic T lymphocyte responses. Vaccine. 2008;26(13):1626–37.

24. Yoshida M, Mata J, Babensee JE. Effect of poly(lactic-co-glycolic acid) contact on maturation of murine bone marrow-derived dendritic cells. J Biomed Mater Res A. 2007;80(1):7–12.

25. Beaudette TT, Bachelder EM, Cohen JA, et al. *In vivo* studies on the effect of co-encapsulation of CpG DNA and antigen in acid-degradable microparticle vaccines. Mol Pharm. 2009;6(4):1160–9.

26. Elamanchili P, Diwan M, Cao M, Samuel J. Characterization of poly(D, L-lactic-co-glycolic acid) based nanoparticulate system for enhanced delivery of antigens to dendritic cells. Vaccine. 2004;22(19):2406–12.

27. Ali OA, Tayalia P, Shvartsman D, Lewin S, Mooney DJ. Inflammatory cytokines presented from polymer matrices differentially generate and activate DCs. Adv Funct Mater. 2013;23(36):4621–8.

28. Singh A, Suri S, Roy K. In-situ crosslinking hydrogels for combinatorial delivery of chemokines and siRNA-DNA carrying microparticles to dendritic cells. Biomaterials. 2009;30(28):5187–200.

29. Pradhan P, Qin H, Leleux JA, et al. The effect of combined IL10 siRNA and CpG ODN as pathogen-mimicking microparticles on Th1/Th2 cytokine balance in dendritic cells and protective immunity against B cell lymphoma. Biomaterials. 2014;35(21):5491–504.

30. Hudalla GA, Modica JA, Tian YF, et al. A self-adjuvanting supramolecular vaccine carrying a folded protein antigen. Adv Healthc Mater. 2013;2(8):1114–9.

31. Goodman J, Vela Ramirez J, Boggiato P, et al. Nanoparticle chemistry and functionalization differentially regulates dendritic cell–nanoparticle interactions and triggers dendritic cell maturation. Part Part Syst Char. 2014;1269–80.

32. Haughney SL, Ross KA, Boggiatto PM, Wannemuehler MJ, Narasimhan B. Effect of nanovaccine chemistry on humoral immune response kinetics and maturation. Nanoscale. 2014;6(22):13770–8.

33. Vela-Ramirez JE, Goodman JT, Boggiatto PM, et al. Safety and biocompatibility of carbohydrate-functionalized polyanhydride nanoparticles. Aaps J. 2015;17(1):256–67.

34. Fischer S, Uetz-von Allmen E, Waeckerle-Men Y, Groettrup M, Merkle HP, Gander B. The preservation of phenotype and functionality of dendritic cells upon phagocytosis of polyelectrolyte-coated PLGA microparticles. Biomaterials. 2007;28(6):994–1004.

35. Acharya AP, Dolgova NV, Xia CQ, Clare-Salzler MJ, Keselowsky BG. Adhesive substrates modulate the activation and stimulatory capacity of non-obese diabetic mouse-derived dendritic cells. Acta Biomater. 2011;7(1):180–92.

36. Acharya AP, Dolgova NV, Moore NM, et al. The modulation of dendritic cell integrin binding and activation by RGD-peptide density gradient substrates. Biomaterials. 2010;31(29):7444–54.

37. Rogers TH, Babensee JE. The role of integrins in the recognition and response of dendritic cells to biomaterials. Biomaterials. 2011;32(5):1270–9.

38. Volchenkov R, Karlsen M, Jonsson R, Appel S. Type 1 regulatory T cells and regulatory B cells induced by tolerogenic dendritic cells. Scand J Immunol. 2013;77(4):246–54.

39. Schwartz RH. T cell anergy. Annu Rev Immunol. 2003;21:305–34.

40. Sakaguchi S, Sakaguchi N, Shimizu J, et al. Immunologic tolerance maintained by CD25+ CD4+ regulatory T cells: their common role in controlling autoimmunity, tumor immunity, and transplantation tolerance. Immunol Rev. 2001;182:18–32.

41. Woltman AM, de Fijter JW, Kamerling SW, Paul LC, Daha MR, van Kooten C. The effect of calcineurin inhibitors and corticosteroids on the differentiation of human dendritic cells. Eur J Immunol. 2000;30(7):1807–12.

42. Hackstein H, Morelli AE, Larregina AT, et al. Aspirin inhibits *in vitro* maturation and *in vivo* immunostimulatory function of murine myeloid dendritic cells. J Immunol. 2001;166(12):7053–62.

43. Mehling A, Grabbe S, Voskort M, Schwarz T, Luger TA, Beissert S. Mycophenolate mofetil impairs the maturation and function of murine dendritic cells. J Immunol. 2000;165(5):2374–81.

44. Penna G, Adorini L. 1 Alpha,25-dihydroxyvitamin D3 inhibits differentiation, maturation, activation, and survival of dendritic cells leading to impaired alloreactive T cell activation. J Immunol. 2000;164(5):2405–11.

45. Jhunjhunwala S, Raimondi G, Thomson AW, Little SR. Delivery of rapamycin to dendritic cells using degradable microparticles. J Control Release. 2009;133(3):191–7.
46. Yamaguchi Y, Tsumura H, Miwa M, Inaba K. Contrasting effects of TGF-beta 1 and TNF-alpha on the development of dendritic cells from progenitors in mouse bone marrow. Stem Cells. 1997;15(2):144–53.
47. Steinbrink K, Wolfl M, Jonuleit H, Knop J, Enk AH. Induction of tolerance by IL-10-treated dendritic cells. J Immunol. 1997;159(10):4772–80.
48. Hackstein H, Taner T, Zahorchak AF, et al. Rapamycin inhibits IL-4–induced dendritic cell maturation *in vitro* and dendritic cell mobilization and function *in vivo.* Blood. 2003;101(11):4457–63.
49. Bryant J, Hlavaty KA, Zhang X, et al. Nanoparticle delivery of donor antigens for transplant tolerance in allogeneic islet transplantation. Biomaterials. 2014;35(31):8887–94.
50. Powell JD, Pollizzi KN, Heikamp EB, Horton MR. Regulation of immune responses by mTOR. Annu Rev Immunol. 2012;30:39–68.
51. Maldonado RA, LaMothe RA, Ferrari JD, et al. Polymeric synthetic nanoparticles for the induction of antigen-specific immunological tolerance. Proc Natl Acad Sci U S A. 2014;12(2):E156–65.
52. Babensee JE, Paranjpe A. Differential levels of dendritic cell maturation on different biomaterials used in combination products. J Biomed Mater Res A. 2005;74(4):503–10.
53. Acharya AP, Dolgova NV, Clare-Salzler MJ, Keselowsky BG. Adhesive substrate-modulation of adaptive immune responses. Biomaterials. 2008;29(36):4736–50.
54. Leclerc C, Brose C, Nouze C, et al. Immobilized cytokines as biomaterials for manufacturing immune cell based vaccines. J Biomed Mater Res A. 2008;86(4):1033–40.
55. Yuba E, Kojima C, Sakaguchi N, Harada A, Koiwai K, Kono K. Gene delivery to dendritic cells mediated by complexes of lipoplexes and pH-sensitive fusogenic polymer-modified liposomes. J Control Release. 2008;130(1):77–83.
56. Palama IE, Cortese B, D'Amone S, Gigli G. mRNA delivery using non-viral PCL nanoparticles. Biomater Sci. 2015;3:144.
57. Machen J, Harnaha J, Lakomy R, Styche A, Trucco M, Giannoukakis N. Antisense oligonucleotides down-regulating costimulation confer diabetes-preventive properties to nonobese diabetic mouse dendritic cells. J Immunol. 2004;173(7):4331–41.
58. Wang J, Lu Z, Wientjes MG, Au JLS. Delivery of siRNA therapeutics: barriers and carriers. Aaps J. 2010;12:492–503.
59. Giannoukakis N, Phillips B, Finegold D, Harnaha J, Trucco M. Phase I (safety) study of autologous tolerogenic dendritic cells in type 1 diabetic patients. Diabetes Care 2011;34(9):2026–32.
60. Phillips B, Nylander K, Harnaha J, et al. A microsphere-based vaccine prevents and reverses new-onset autoimmune diabetes. Diabetes. 2008;57(6):1544–55.

Evelyn Bracho-Sanchez is a PhD candidate in the J. Crayton Pruitt Family Department of Biomedical Engineering at the University of Florida. Her focus has been on understanding biomaterials-based immunomodulation through the use of biomaterials and immune modulating enzymes. Particularly, Evelyn is interested in protein therapies for the treatment of type 1 diabetes.

Jamal S. Lewis received his PhD degree in biomedical engineering in 2012 from the University of Florida. As a postdoctoral associate at the University of Florida, Jamal's research is focused on the development of a dendritic cell targeting, microparticle-based vaccine for the prevention of type 1 diabetes. He is also the principal investigator of a Phase I Small Business Innovation Research grant to explore commercializing this microparticle-based vaccine formulation at OneVax, LLC, where he also serves as a senior scientist.

Benjamin G. Keselowsky is an associate professor in the J. Crayton Pruitt Family Department of Biomedical Engineering at the University of Florida. Funding is gratefully acknowledged—currently from the National Institutes of Health (R01DK091658, R01DK098589) and in the past from the National Science Foundation, the Arthritis Foundation, and the Juvenile Diabetes Research Foundation. Professor Keselowsky's research group seeks to understand immune cell interactions with biomaterials, as well as the engineering of controlled release biomaterials capable of directing immunological processes. A large focus of the group is engineering polymeric biomaterials-based microparticles as a vaccine to retrain the immune system, correcting aberrant activation toward pancreatic self-antigens for type 1 diabetes.

Chapter 9
Nanosystems for Immunotherapeutic Drug Delivery

Alex Schudel, Michael C. Bellavia and Susan N. Thomas

Abbreviations

Alum	Aluminum hydroxide gel
Da	Dalton
DTSSP	3,3'-dithiobis(sulfosuccinimidylpropionate)
NIR	Near infrared
nm	Nanometer
OVA	Chicken ovalbumin
PAM	Tri-palmitoyl-S-glyceryl cysteine lipopeptide with a pentapeptide SKKKK
PEG	Poly(ethylene glycol)
PLGA	Poly(lactic-co-glycolic acid)
SB	Transforming growth factor-β inhibitor SB505124
TGF-β	Transforming growth factor-β
TLR	Toll-like receptor
TRX	3,5-didodecyloxybenzamidine

S. N. Thomas (✉)
George W. Woodruff School of Mechanical Engineering, Parker H. Petit Institute for Bioengineering and Bioscience, Georgia Institute of Technology, 315 Ferst Drive NW, Atlanta, GA, USA
e-mail: susan.thomas@gatech.edu

A. Schudel
School of Materials Science and Engineering, Parker H. Petit Institute for Bioengineering and Bioscience, Georgia Institute of Technology, 315 Ferst Drive NW, Atlanta, GA, USA
e-mail: aschudel@gatech.edu

M. C. Bellavia
Wallace H. Coulter Department of Biomedical Engineering, Georgia Institute of Technology and Emory University, 315 Ferst Drive NW, Atlanta, GA, USA

© Springer International Publishing Switzerland 2015
L. Santambrogio (ed.), *Biomaterials in Regenerative Medicine and the Immune System*,
DOI 10.1007/978-3-319-18045-8_9

9.1   Introduction

Immunotherapy is of resurging interest due to its potential to provide benefit in the treatment of a broad array of pathologies. Since a significant portion of clinically approved immunotherapeutics are small molecule or biologic agents, strategies that enhance the delivery of these immunotherapeutic drugs to tissues enriched in immune cells important in the initiation and regulation of immune response, such as lymphoid tissues, have been developed. Hence, as in chemotherapy, formulation approaches that improve immunotherapeutic drug bioactivity while minimizing off-target effects and toxicities are of high interest and include novel co-formulation, targeting, and release schemes that have the potential to optimize agent delivery to immune cells to enhance the immunotherapeutic potential of administered agent(s). This chapter details how materials engineering, formulation design, and delivery schemes have improved immunological outcomes for a variety of therapeutic purposes. The presented literature is by no means exhaustive but is intended to provide a few illustrative examples as snapshots of emerging concepts and approaches used in the field.

9.2   Nanoformulations for Improved Delivery
of Immunotherapeutic Drugs

The therapeutic efficacy of administered drug is restricted by several barriers that reduce drug bioavailability and lifetime of action within the target site. Deleterious off-target effects and toxicities also reduce the maximum tolerable dose. Advances in biomaterial-based drug nanoformulations have facilitated marked improvements in these respects (Fig. 9.1) [1], resulting, in particular, from their capacity to prolong circulation times of intravenously infused agents as well as enhance delivery to cells within target tissues.

9.2.1   Delivery Systems for Prolonged Drug Circulation Times

A major and persistent drug delivery challenge is rapid clearance of administered agent, resulting in poor half-life and limited tissue accumulation [1]. Carriers have been designed in an array of formulations, such as liposomes, polymer micelles, nanoparticles, and fractal constructs (dendrimers), as well as various metallic particles that are of a high enough effective hydrodynamic size to resist clearance by glomerular filtration in the kidneys [2]. As one example, Miki et al. explored the benefit of free versus micelle-formulated interleukin (IL)-2, a cytokine that regulates the activity of lymphocytes and in particular T cells [3], to enhance the efficacy of a dendritic cell-based cancer vaccine [4]. Intravenously infused IL-2 is used to supplement cancer therapy [5], but has a short half-life [6] limiting its therapeutic

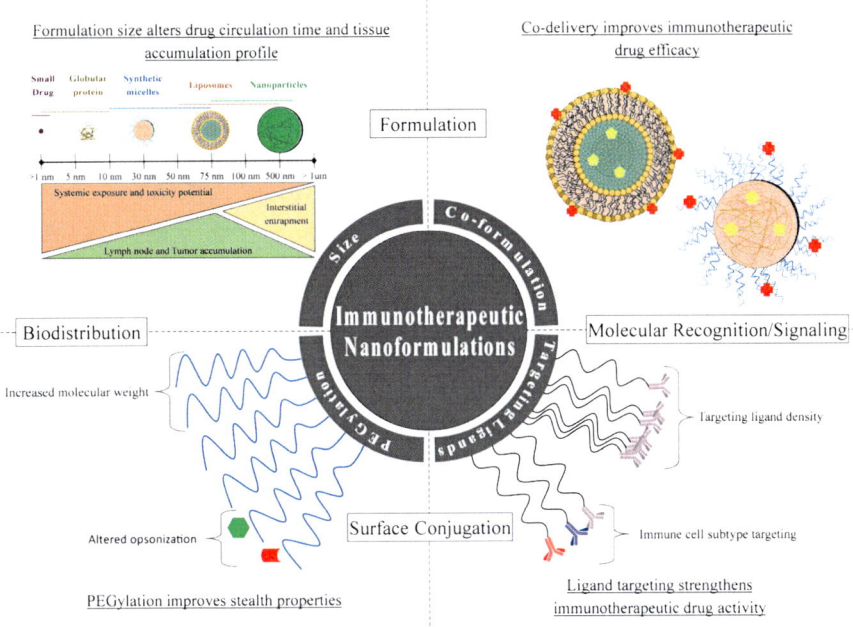

Fig. 9.1 Schematic diagram of nanoformulations that improve immunotherapeutic drug efficacy. *Top left:* nanoformulations that increase the effective hydrodynamic size of drugs improve their circulation times by resisting glomerular filtration, enhance tumor accumulation by the enhanced permeability and retention effect, and may promote lymphatic uptake from the interstitium. *Top right:* co-delivery via co-formulation improves the synergistic signaling activity of multiple immunotherapeutic agents, for example, antigen and adjuvant in vaccine applications. *Bottom left:* PEGylation increases the effective hydrodynamic size and stealth properties of nanoformulations to increase drug half-life after infusion. *Bottom right:* ligand targeting can increase uptake by immune cells in a density-dependent fashion. *PEG* poly(ethylene glycol)

window. To improve upon this limitation, IL-2 was encapsulated within a ~75-nm micelle composed of a block co-polymer of poly(ethylene glycol) (PEG) and poly amino acids by electrostatic interactions. This formulation dramatically prolonged the serum half-life of IL-2 relative to when it was administered in its free form [4]. Treatment with micelle-encapsulated IL-2 in combination with the tumor vaccine also resulted in a corresponding increase in the frequency of cytotoxic T lymphocytes, cytotoxic T lymphocyte accumulation in the tumor, and a reduction in tumor growth relative to the vaccine in combination with free IL-2 [4].

While formulation into a micelle, nanoparticle, or liposome improves the circulation half-life of small molecule and biologic drugs, further improvement has been achieved through carrier surface modifications. This is because while increased size reduces filtration, it also typically augments uptake by the reticuloendothelial system, which is composed of phagocytic cells located in reticular connective tissues in the liver, spleen, and lymph nodes that facilitate the removal of foreign bodies and physiological debris [7]. Hence, modifications that increase stealth properties by preventing uptake by the reticuloendothelial system prolong drug circulation times. The most renowned approach to accomplish this is via surface modification with PEG, termed PEGylation. PEGylation alone can in fact increase drug circulation times by virtue of increasing the agent's effective hydrodynamic size [8]. Opsonization, which increases uptake within the reticuloendothelial system, is also resisted by PEGylation [8]. In a study by Kaminskas et al., polylysine dendrimers were modified by surface derivation with PEG of three molecular weights: 200, 570, or 2000 Da [9]. Modified dendrimers administered intravenously into rats demonstrated an increase in plasma half-life with increasing PEG molecular weight [9]. Follow-up studies elaborated on the potential for PEGylated dendrimers to prolong the plasma half-life of conjugated doxorubicin [10], suggesting the potential for PEGylation in immunotherapeutic drug delivery.

9.2.2   Lymph Node Targeting

Secondary lymphoid organs are the sites in the body where organized lymphocyte accumulation occurs and function to maintain mature naive lymphocytes. Lymph nodes in particular are considered to be the tissues where the initiation of adaptive immune response occurs [11, 12]. Lymph nodes are connected in series by the lymphatics that drain interstitial tissues to maintain tissue fluid balance [11, 12]. Lymph entering the lymph node travels through the subcapsular sinus where solutes are filtered by size and opsonization [12, 13]. This ability for the lymph node to facilitate sampling of lymph-borne antigen by resident immune cells is considered central to its role in regulating adaptive immunity [11, 14].

Several immunotherapeutic schemes targeting lymph nodes have been developed, with most approaches tested to date exploiting the drainage function of lymphatics as a delivery mechanism after administration in peripheral tissues. Uptake from the interstitium by the lymphatics is most sensitive to the size of administered agent, with formulations of hydrodynamic diameter ranging from 5 to 50 nm being most efficiently taken up [12, 15, 16]. To this end, several approaches using micelles, polymer nanoparticles, and liposomes as some examples have been employed.

Micelles are composed of block copolymers that self-assemble to sizes from several nanometers to several hundred nanometers. Micelles offer the notable advantage in drug delivery applications of the ability to encapsulate hydrophobic drugs within their core and deliver this cargo to tissues and cells. For example, Dane et al. used a micellar formulation composed of block copolymer PEG-bi-poly(propylene sulfide) to promote immunosuppression and allograft survival via delivery of hydrophobic immunosuppressive drugs. Following intradermal administration, micelles appreciably accumulated within lymph nodes and were taken up by resident immune cells [17]. When animals were treated with micelles loaded with the immunosuppressive drugs rapamycin and tacrolimus, survival of tail skin from C57BL/6 donor mice transplanted onto the tails of BALB/c recipient mice was significantly prolonged [17].

In addition to delivering hydrophobic drugs, the micelle structure has also been used in the emulsion polymerization of hydrophobic polymers to create a polymer nanoparticle drug delivery vehicle [18]. Anionic polymerization of propylene sulfide can be carried out within micelles composed of pluronic F127, an amphiliphic block copolymer made from PEG-poly(propylene glycol)-PEG [18]. The resulting 30 nm diameter poly(propylene sulfide) nanoparticles have been explored in several applications for their capacity to deliver immunotherapeutic drugs to lymph nodes for immunomodulation. Recently, de Titta et al. conjugated TLR9 ligand CpG oligonucleotides (unmethylated DNA strands containing cytosine and guanine) were to these polymer nanoparticles. Upon footpad injection, nanoparticle-conjugated CpG elicited a more potent maturation response of draining lymph node-resident dendritic cells relative to free CpG [19]. Furthermore, when co-injected with nanoparticles conjugated to the chicken ovalbumin protein (OVA), treatment with CpG-conjugated nanoparticles more potently induced antigen-specific effector and memory T cell responses relative to treatment with free CpG [19]. Capitalizing on this T cell activation, vaccinated mice were challenged with OVA expressing murine lymphomas and melanomas and a significant retardation in tumor formation was found in mice vaccinated with OVA-conjugated nanoparticles co-infused with CpG-conjugated nanoparticles rather than free CpG [19]. These results emphasize the potential for lymph node delivery to improve the potency of cancer vaccines [19].

Elaborating upon the integral role that tumor-draining lymph nodes play in tumor immunity [20], Thomas et al. explored how the delivery of immunotherapeutic adjuvant drugs to the tumor-draining versus non-tumor-draining lymph node influenced tumor progression and development of antitumor immunity [21]. In this work, tumor lymphatic drainage was biased by implanting murine B16F10 melanomas in the left dorsal skin of mice, thereby focusing lymphatic drainage to the tumor-draining ipsilateral lymph nodes and creating non-tumor-draining lymph nodes on the contralateral side [21]. When fluorescently labeled, the aforementioned poly(propylene sulfide) nanoparticles were found to accumulate appreciably only in lymph nodes ipsilateral to the site of intradermal injection [21]. With daily administration in the dermis ipsilateral to the tumor, adjuvant drug-loaded nanoparticles resulted in a slowing of tumor growth not observed when administered intravenously or in the skin contralateral to the tumor [21]. The nanoparticle-mediated delivery to the lymph node was crucial to this effect since treatment with free drug with or without nanoparticle co-infusion had no influence on tumor growth [21]. These differences in efficacy were attributed to the observed increase in the maturation and activation status of dendritic and T cells resident within the tumor-draining lymph node [21]. Correspondingly, drug-loaded nanoparticle delivery to tumor-draining lymph nodes boosted the adaptive immune response to endogenously produced tumor antigen that also accumulated within tumor-draining lymph nodes, as indicated by an increased frequency of tumor infiltrating antigen-specific $CD8^+$ T cells [21]. These results illustrate the potential for tumor-draining lymph node-targeted drug delivery for tumor immunotherapy.

Liposomes also have demonstrated utility in immunotherapeutic drug delivery, with evidence that their efficacy in adaptive immunotherapy and vaccination applications is related to the extent of accumulation within lymph nodes. As one example, CpG and an antibody specific for dendritic cell maturation marker CD40 were incorporated into 80-nm PEGylated liposomes and assessed for their therapeutic effect when intratumorally injected into B16F10 melanomas [22]. The combined treatment with CpG and anti-CD40 antibody conferred a stronger inhibition of tumor growth and improved animal survival when liposomally formulated compared to the soluble free drugs [22]. Formulated drug treatment also lowered the levels of inflammatory cytokine produced systemically in response to treatment as well as prevented toxicity-associated animal weight loss [22]. Notably, the liposome formulation resulted in significant increases in drug accumulation within draining lymph nodes and increased uptake of CpG by lymph node-resident cells [22].

9.2.3   Formulation

Thus far, the passive influence of formulation design to prolong drug half-lives and improve delivery to lymph nodes has been discussed. The following section expands this topic to explore formulations for improved synergy of two or more drugs delivered in combination and to facilitate targeting and immune stimulation by receptor-mediated recognition.

9.2.3.1   Co-delivery versus Co-infusion

Immunotherapy using multiple agents with synergistic activity has been proposed for a variety of indications. The rationale is that co-treatment with drugs that exert complementary or orthogonal signaling function may amplify the desired response. In other scenarios, multiple agents are required for functionality and necessitate co-delivery to the same cell population(s). Several materials engineering approaches have been explored to address the challenges of multi-agent delivery.

Vaccines are the most effective form of immune modulation developed to date and involve the conditioning of adaptive immune response via multiple immunological signals. In its most basic form, vaccination entails exposure of the immune system to attenuated or inactivated pathogen. Antigen, the substance or part of the pathogen to which the immune response is generated against, is taken up, processed, and presented by antigen-presenting cells to antigen-specific lymphocytes that recognize the specific antigen and proliferate. The local immune microenvironment at the time of antigen presentation dictates the differentiation programs followed by proliferating lymphocytes. Nature has thus evolved antigen-presenting cells to recognize numerous pathogen-associated molecular patterns such that upon exposure to pathogen, these cells become activated, produce stimulatory cytokines, and express maturation co-stimulatory molecules on their surface [23]. Hence, antigen-specific lymphocytes proliferate as the result of antigen recognition in an immune stimulatory microenvironment, driving the induction of robust adaptive immunity.

To mimic the effects of such vaccination strategies using a technology with more finely-tuned control of dose for improved safety, engineered material systems have been developed. Formulations that enhance the delivery of both antigen and the immune stimulatory signal, termed an adjuvant, to their same target cells, and to ensure their efficient synergistic interactions, have been explored in an approach called co-delivery [24]. As one example, Bal et al. explored the effects of free versus cationic liposome-encapsulated OVA and TLR ligand as an adjuvant on anti-OVA adaptive immune response induced by intradermal injection in mice [25]. OVA and the TLR ligands CpG or PAM_3CSK4, a synthetic lipoprotein consisting of a tri-palmitoyl-S-glyceryl cysteine lipopeptide with a pentapeptide SKKKK (PAM), were passively incorporated into 130-nm liposomes [25]. Immunization with both OVA and CpG or PAM increased anti-OVA antibody titers and, in the case of CpG, decreased the ratio of anti-OVA IgG_1/IgG_{2a} antibody titers [25]. Liposomal co-encapsulation of either adjuvant with OVA also tended to more greatly bias in a decreasing fashion the ratio of anti-OVA IgG_1/IgG_{2a} antibody titers relative to co-immunization with the free form of the agents, though no statistical significance was reported [25]. However, liposomal co-encapsulation of OVA and CpG did result in a significant increase in the levels of interferon (IFN)-γ produced by splenocytes upon OVA re-stimulation after immunization relative to immunization with free OVA mixed with free CpG. These results suggest a benefit of co-formulation to ensure co-delivery to promote the robust generation of adaptive immunity [25].

The benefit of co-delivery has also been explored in other applications aimed towards harnessing the therapeutic potential of synergistic drug signaling activity. In cancer, a significant mechanism by which tumors evade the immune system is through the production of immunosuppressive cytokines, such as transforming growth factor-β (TGF-β) [26]. This pleiotropic cytokine suppresses the activity of cytotoxic T lymphocytes and natural killer cells while promoting an increase in regulatory T lymphocytes [26]. Inhibition of TGF-β therefore has the potential to amplify the T cell stimulatory activity of cytokine IL-2 that is used in the treatment of metastatic melanoma. To this end, the bioactivity of TGF-β inhibitor SB505124 (SB) delivered in conjunction with IL-2 was assessed in vivo for the capacity to enhance antitumor immunity [27]. Furthermore, the benefit of delivery via co-infusion versus a co-delivery approach whereby a 120-nm biodegradable core-shell nanogel was used to individually and co-encapsulate both agents was also explored [27]. When IL-2 and SB were co-encapsulated within the nanogel, intratumoral injection reduced B16F10 tumor growth rate and mass after 1 week as well as prolonged animal survival compared to co-infusion of individually formulated or soluble SB and IL-2 [27]. Intravenous injection of nanogel co-encapsulated drugs also resulted in prolonged survival relative to individual drug treatments in either free or nanogel-encapsulated form in a model of melanoma lung metastasis induced by intravenous infusion of B16F10 cells [27]. This demonstrated in vivo efficacy suggests that such a co-delivery approach has the potential to significantly improve the therapeutic activity of a variety of immunotherapeutic drug combinations.

9.2.3.2   Receptor-Mediated Targeting

Formulation designs utilizing surface conjugation of antibodies have been developed for what has become known as "active targeting." [1]. The utility of such approaches targeting immune cell receptors for immunotherapy applications have been explored to elaborate on the role that receptor type and density play in the quality and magnitude of the generated immune response. Cruz et al. reported the benefit of dendritic cell receptor targeting on improving the efficacy of biodegradable poly(lactic-*co*-glycolic acid) (PLGA) nanoparticle uptake and immunogenicity both in vitro and in vivo. Interestingly, however, they noted only small differences in dendritic cell uptake in vitro and capacity to induce proliferation of antigen-specific T cells in lymph nodes draining the site of injection when comparing the differential effects of targeting dendritic cell receptor subtypes CD40, DEC-205, and CD11c [28]. Otherwise, in vitro cytokine production by treated dendritic cells and the efficacy of generated antigen-specific cytotoxic T cell responses was nearly equivalent [28]. These results illustrate the potential for receptor targeting in achieving a robust immunological response using particle-mediated delivery approaches, and they suggest that several surface markers may equivalently facilitate such targeting.

To determine the effect of targeting ligand density on induced immunity, Bandyopadhyay et al. modified the surface of OVA-loaded PLGA nanoparticles with different densities of antibody to the dendritic cell lectin receptor DEC-205 [29]. After co-incubation, the amount of IL-10 produced by dendritic cells was found to increase with surface concentration of anti-DEC 205 antibody [29]. Mice immunized with anti-DEC-205 antibody-coated OVA-nanoparticles also demonstrated enhanced cytokine production by splenocytes restimulated with OVA, with the amount of IL-10 and IL-5 increasing with increasing anti-DEC-205 antibody density [29]. Differences in uptake could not explain these results; it was instead proposed that the multimeric display of anti-DEC-205 antibody results in receptor cross-linking causing observed differences in CD36-dependent IL-10 production [29]. These findings indicate that the density of surface-presented targeting ligands can influence the elicited magnitude and quality of induced immunological responses through modulation of interactions with immune cell-expressed surface receptors.

9.3   Triggered Drug Release from Immunotherapeutic Drug Nanoformulations

After delivery, nanoformulated drug must be released in order to mediate its effects. To focus drug bioactivity to its intended target as well as minimize side effects and associated toxicities, innovations in materials engineering and bioconjugation chemistry have been developed to trigger drug release specifically in response to microenvironmental or extraphysiological cues. The first generally requires some sort of physiological dichotomy to exist between the target site and the surrounding

Table 9.1 Proposed mechanisms of drug release from engineered nanoformulations for targeted delivery

Target	Potential release mechanism
Cellular endosome	pH
	Oxidation–reduction potential
Tumor	pH
	Oxidation–reduction potential
	Enzymatic
	Light-triggered

physiological system. The most common properties that can be exploited for chemical and physical release involve pH, oxidation/reduction potential, and enzymatic degradation. External stimuli (e.g., light, ultrasound, etc.) have also been explored to great effect. Here we discuss a few examples by which therapeutic drug release from drug delivery systems in a triggered fashion has been accomplished in order to mediate drug effects in a site-specific or time-controlled manner (Table 9.1), specifically emphasizing their application or potential utility in immunotherapy.

9.3.1 Intracellular

Intracellular compartments represent appealing immunotherapeutic targets. Notable examples include the cytoplasm and endosome, which serve as protein reservoirs for the pathways of peptide antigen presentation by the major histocompatibility complexes [30]. The vesicular membranes of the endosomes are also enriched in TLRs [31]. Release mechanisms that facilitate drug release after cellular uptake are thus desirable in a variety of immunotherapeutic scenarios.

9.3.1.1 pH

Within cells, pH-triggered release can be very effective for endosomal drug targeting since the cellular endosome, the cytoplasmic vesicle involved in large molecule uptake, possesses an acidic pH of approximately 6.0 [32]. Once a delivery vehicle has entered the endosome, the ionic functional groups within constituent polymers are ionized due to the acidic pH, causing localized repulsions among them. The polymer swells, decreasing interactions with the payload, which is subsequently released from the delivery vehicle [32]. Materials with this endosomal pH-triggered response may be naturally derived, such as the polysaccharide chitosan, or synthetic polycationic polymers such as methacrylates and poly(ethylenimine).

An illustrative example is a recent study by Yoshizaki and colleagues, in which egg yolk-based liposomes containing OVA were mated to the pH-sensitive polymer 3-methylglutarylated hyperbranched poly(glycidol) [33]. In vitro characterization of cultured dendritic cells treated with the OVA-loaded fluorescent liposomes revealed that after trafficking to the endosome, the polymer capsule was disrupted, allowing the OVA protein access to the cytosol. To increase susceptibility to the acidic environment, liposomes were also modified with the cationic lipid 3,5-didodecy-loxybenzamidine (TRX) [33]. After subcutaneous immunization with these OVA-loaded liposomal formulations, the development of serum anti-OVA antibody titers and splenocyte production of IFN-γ in response to OVA restimulation was found to be enhanced with increased amounts of TRX [33]. This effect was hypothesized to result from TRX increasing the magnitude of the electrostatic attraction between the outer liposomal membrane and the pH-responsive polymer. As a result, potential interference between the pH-responsive polymer and the endosomal membrane was avoided, such that OVA was most concentrated within the endosome [33].

9.3.1.2   Reduction/Oxidation

In addition to pH, oxidation/reduction potentials are known to vary between extra- and intracellular environments as well as within the cells themselves. The primary intracellular oxidizing source is from reactive oxygen species, which are elevated in chronic inflammatory conditions such as atherosclerosis, cancer, and diabetes mel-litus. These oxidizing molecules accept electrons from other molecules or chemical bonds and when in excess may induce cellular damage [34]. Drug delivery nanofor-mulations can exploit redox sensitivity by employing bioactive polymers that may change hydrophilicity in oxidative environments (e.g., poly(propylene) sulfide) or are biologically derived (e.g., sodium alginate) and integrated with particular cross-linkers to contain disulfide bond triggers. These disulfide bonds are then cleaved by endogenous reductant species, which donate an electron to the linkage [35].

Enhancing redox responsiveness has been shown to be effective in enhancing the cellular and humoral immune response to protein antigen, and thus has val-ue in boosting vaccine efficiency. As one example, Li et al. developed nanogels composed of alginate bound to the cationic polymer polyethylenimine for the en-capsulation of OVA [36]. 3,3′-Dithiobis(sulfosuccinimidylpropionate) (DTSSP), a cross-linker containing a disulfide bond, was then covalently attached to the formed nanogels such that the mass ratio of nanogel to cross-linker was 10:1. This ratio was identified to confer maximum protein antigen (OVA) loading and particle sta-bility. The rationale for this design was that the nanogel would be reduced after dendritic cell uptake since the cross-linker would be compromised in the abundance of intracellular molecules containing oxidizable thiols, such as the amino acid cys-teine and the antioxidant glutathione. After intraperitoneal immunization, efficacy of immune stimulation in mice was compared among nanogels with and without the cross-linker as well as OVA adsorbed to aluminum hydroxide gel (Alum), an adjuvant in longstanding clinical use. The reducible nanogels provided a more pronounced stimulatory effect than the clinical mainstay as well as similar non-reducible nanosystems [36]. Mice immunized with OVA-loaded reducible nanogels also developed much greater anti-OVA IgG$_1$ isotype antibody titers and detectable titers of anti-OVA IgG$_2$ isotype antibody after immunization in contrast to Alum and nanogels without DTSSP [36]. Additionally, only immunization with nanogels composed of the reducible DTSSP increased the number of OVA-specific CD8+ cytotoxic splenocytes [36].

9.3.2   Extracellular

The extracellular environment presents additional challenges to achieving targeted release since differences in pH and oxidation/reduction potential do not vary widely among tissues. Exceptions to this can include the hypoxic and acidic tumor microenvironment, which will be lower in both oxygen concentration and pH and contain higher levels of redox and enzymatic species [37]. For the purposes of this discussion, we instead briefly discuss enzymatic-triggered release as this is a widely explored area of investigation and development. The extraphyisological mechanism of release via light triggering will also be discussed.

9.3.2.1   Enzymatic

Certain pathologies are associated with the upregulation of specific enzymes, such as matrix metalloproteinases, which are responsible for much of the tissue remodeling in cancer [37]. The selective expression of such disease-associated enzymes can be exploited to achieve target site-restricted degradation of a carrier and release of loaded drug. A particularly innovative approach designed to facilitate increased intratumoral drug penetration is the degradation of tumor-accumulating nanoparticles ~100 nm in diameter to a smaller, more diffusive size regime via the proteolytic activity of matrix metalloproteinases within the tumor microenvironment [38]. The advantage of such a multistage system is the ability to overcome perivascular entrapment by the dense collagenous extracellular matrix within solid tumors that stymies drug bioactivity [39]. Such a scheme has the significant potential to improve the intratumoral penetration of immunotherapeutic drugs in solid tumors.

9.3.2.2   Light

Chemical linkages or other mechanisms of drug entrapment that are degraded upon exposure to light of specific wavelengths offer the opportunity for spatiotemporally controlled drug targeting and release. Near infrared (NIR) light is primarily used due to its superior depth of tissue penetration as well as minimal scattering and absorbance by hemoglobin and water.

Gold at the nanoscale significantly absorbs in the NIR regime and is therefore extensively used in NIR light-activated systems [40]. Tao et al. used a multifunctional gold nanorod platform to trigger in a site-specific manner the release of two co-delivered therapeutics to achieve simultaneous cytotoxicity against the tumor and immune stimulation [41]. To accomplish this, the nanorods were PEGylated and coated with TLR9 ligand CpG. Doxorubicin was then readily integrated into the CpG nanorods, as it intercalates into DNA. NIR irradiation of the CpG-doxo-rubicin nanorods caused burst drug release that led to >80% of total payload to be released within 20 min [41]. A single, direct injection of CpG-doxorubicin nanorods into murine hepatocellular carcinomas markedly reduced tumor growth when the tumor was irradiated 2 h post injection, presumably due to tumor-localized drug release [41]. Light-triggered targeting and release systems thus have great potential to achieve localized and efficient drug release that could benefit tumor immuno-therapy and immunosuppressive treatment regimens for transplantation, where lo-coregional rather than systemic drug effects are desirable.

9.4   Conclusion and Future Perspectives

Advances in engineered nanoformulations for immunotherapeutic drug delivery have accompanied the rapidly growing interest in clinical immunotherapy. Several material systems have been developed that improve efficacy by overcoming deliv-ery barriers detrimental to drug bioactivity. Notable examples include prolonged drug circulation times and lymph node-targeted delivery. The utility of material nanoformulations to facilitate co-delivery of multiple drugs with synergistic activ-ity has also been demonstrated, as has the potential for drug delivery and immuno-therapeutic activity to be enhanced via active targeting. Important innovations in biomaterial and bioconjugate chemistry have furthermore led to the development of triggered drug release mechanisms that increase the precision and control of drug bioactivity within targeted subcellular compartments and/or tissues. As the diversity and versatility of biomaterial-based immunotherapeutic technologies improve, the contributions of these advances in nanoformulation to clinical problems in arthritis, atherosclerosis, cancer, as well as other pathologies is only expected to increase.

References

1. Allen TM, Cullis PR. Drug delivery systems: entering the mainstream. Science. 2004;303(5665):1818–22.
2. Levey AS, Bosch JP, Lewis JB, Greene T, Rogers N, Roth D. A more accurate method to es-timate glomerular filtration rate from serum creatinine: a new prediction equation. Ann Intern Med. 1999;130(6):461–70.
3. Liao W, Lin J-X, Leonard WJ. IL-2 family cytokines: new insights into the complex roles of IL-2 as a broad regulator of T helper cell differentiation. Curr Opin Immunol. 2011;23(5):598–604.
4. Miki K, Nagaoka K, Harada M, et al. Combination therapy with dendritic cell vaccine and IL-2 encapsulating polymeric micelles enhances intra-tumoral accumulation of antigen-specific CTLs. Int Immunopharmacol. 2014;23(2):499–504.
5. Acquavella N, Kluger H, Rhee J, et al. Toxicity and activity of a twice daily high-dose bolus interleukin 2 regimen in patients with metastatic melanoma and metastatic renal cell cancer. J Immunother. 2008;31(6):569–76.

6. Donohue JH, Rosenberg SA. The fate of interleukin-2 after in vivo administration. J Immunol. 1983;130(5):2203–8.
7. Van Furth R, Cohn Z, Hirsch J, Humphrey J, Spector W, Langevoort H. The mononuclear phagocyte system: a new classification of macrophages, monocytes, and their precursor cells. Bull World Health Org. 1972;46(6):845.
8. Karakoti AS, Das S, Thevuthasan S, Seal S. PEGylated inorganic nanoparticles. Angew Chem Int Ed. 2011;50(9):1980–94.
9. Kaminskas LM, Boyd BJ, Karellas P, et al. The impact of molecular weight and PEG chain length on the systemic pharmacokinetics of PEGylated poly l-lysine dendrimers. Mol Pharm. 2008;5(3):449–63.
10. Ryan GM, Kaminskas LM, Bulitta JB, McIntosh MP, Owen DJ, Porter CJ. PEGylated poly-lysine dendrimers increase lymphatic exposure to doxorubicin when compared to PEGylated liposomal and solution formulations of doxorubicin. J Control Release. 2013;172(1):128–36.
11. Randolph GJ, Angeli V, Swartz MA. Dendritic-cell trafficking to lymph nodes through lymphatic vessels. Nat Rev Immunol. 2005;5(8):617–28.
12. Thomas SN, Schudel A. Overcoming transport barriers for interstitial-, lymphatic-, and lymph node-targeted drug delivery. Curr Opin Chem Eng. 2015;7:65–74.
13. Roozendaal R, Mempel TR, Pitcher LA, et al. Conduits mediate transport of low-molecular-weight antigen to lymph node follicles. Immunity. 2009;30(2):264–76.
14. Thomas SN, Rutkowski JM, Pasquier M, et al. Impaired humoral immunity and tolerance in K14-VEGFR-3-Ig mice that lack dermal lymphatic drainage. J Immunol. 2012;189(5):2181–90.
15. Reddy ST, Rehor A, Schmoekel HG, Hubbell JA, Swartz MA. In vivo targeting of dendritic cells in lymph nodes with poly(propylene sulfide) nanoparticles. J Control Release. 2006;112(1):26–34.
16. Porter CJ. Drug delivery to the lymphatic system. Crit Rev Ther Drug Carrier Syst. 1997;14(4):333–93.
17. Dane KY, Nembrini C, Tomei AA, et al. Nano-sized drug-loaded micelles deliver payload to lymph node immune cells and prolong allograft survival. J Control Release. 2011;156(2):154–60.
18. Rehor A, Hubbell JA, Tirelli N. Oxidation-sensitive polymeric nanoparticles. Langmuir. 2004;21(1):411–7.
19. de Titta A, Ballester M, Julier Z, et al. Nanoparticle conjugation of CpG enhances adjuvancy for cellular immunity and memory recall at low dose. Proc Natl Acad Sci. 2013;110(49):19902–7.
20. Swartz MA, Lund AW. Lymphatic and interstitial flow in the tumour microenvironment: linking mechanobiology with immunity. Nat Rev Cancer. 2012;12(3):210–9.
21. Thomas SN, Vokali E, Lund AW, Hubbell JA, Swartz MA. Targeting the tumor-draining lymph node with adjuvanted nanoparticles reshapes the anti-tumor immune response. Biomaterials. 2014;35(2):814–24.
22. Kwong B, Liu H, Irvine DJ. Induction of potent anti-tumor responses while eliminating systemic side effects via liposome-anchored combinatorial immunotherapy. Biomaterials. 2011;32(22):5134–47.
23. Takeda K, Akira S. TLR signaling pathways. Semin Immunol. 2004 Feb;16(1):3–9.
24. Laing P, Bacon A, McCormack B, Gregoriadis G, Frisch B, Schuber F. The 'co-delivery' approach to liposomal vaccines: application to the development of influenza-A and hepatitis-B vaccine candidates. J Liposome Res. 2006;16(3):229–35.
25. Bal SM, Hortensius S, Ding Z, Jiskoot W, Bouwstra JA. Co-encapsulation of antigen and toll-like receptor ligand in cationic liposomes affects the quality of the immune response in mice after intradermal vaccination. Vaccine. 2011;29(5):1045–52.
26. Gorelik L, Flavell RA. Immune-mediated eradication of tumors through the blockade of transforming growth factor-beta signaling in T cells. Nat Med. 2001;7(10):1118–22.

27. Park J, Wrzesinski SH, Stern E, et al. Combination delivery of TGF-β inhibitor and IL-2 by nanoscale liposomal polymeric gels enhances tumour immunotherapy. Nat Mater. 2012;11(10):895–905.

28. Cruz LJ, Rosalia RA, Kleinovink JW, Rueda F, Löwik CWGM, Ossendorp F. Targeting nanoparticles to CD40, DEC-205 or CD11c molecules on dendritic cells for efficient CD8+ T cell response: a comparative study. J Control Release. 2014;192(0):209–18.

29. Bandyopadhyay A, Fine RL, Demento S, Bockenstedt LK, Fahmy TM. The impact of nanoparticle ligand density on dendritic-cell targeted vaccines. Biomaterials. 2011;32(11):3094–105.

30. Hubbell JA, Thomas SN, Swartz MA. Materials engineering for immunomodulation. Nature. 2009;462(7272):449–60.

31. Gleeson PA. The role of endosomes in innate and adaptive immunity. Semin Cell Dev Biol. 2014;31:64–72.

32. Schmaljohann D. Thermo-and pH-responsive polymers in drug delivery. Adv Drug Del Rev. 2006;58(15):1655–70.

33. Yoshizaki Y, Yuba E, Sakaguchi N, Koiwai K, Harada A, Kono K. Potentiation of pH-sensitive polymer-modified liposomes with cationic lipid inclusion as antigen delivery carriers for cancer immunotherapy. Biomaterials. 2014;35(28):8186–96.

34. Matsuda M, Shimomura I. Increased oxidative stress in obesity: implications for metabolic syndrome, diabetes, hypertension, dyslipidemia, atherosclerosis, and cancer. Obes Res Clin Pract. 2013;7(5):e330–41.

35. Huo M, Yuan J, Tao L, Wei Y. Redox-responsive polymers for drug delivery: from molecular design to applications. Polym Chem. 2014;5(5):1519–28.

36. Li P, Luo Z, Liu P, et al. Bioreducible alginate-poly(ethylenimine) nanogels as an antigen-delivery system robustly enhance vaccine-elicited humoral and cellular immune responses. J Control Release. 2013;168(3):271–9.

37. Schmalfeldt B, Prechtel D, Harting K, et al. Increased expression of matrix metalloproteinases (MMP)-2, MMP-9, and the urokinase-type plasminogen activator is associated with progression from benign to advanced ovarian cancer. Clin Cancer Res. 2001;7(8):2396–404.

38. Wong C, Stylianopoulos T, Cui J, et al. Multistage nanoparticle delivery system for deep penetration into tumor tissue. Proc Natl Acad Sci. 2011;108(6):2426–31.

39. Chauhan VP, Stylianopoulos T, Boucher Y, Jain RK. Delivery of molecular and nanoscale medicine to tumors: transport barriers and strategies. Annu Rev Chem Biomol Eng. 2011;2(1):281–98.

40. Tomina N, Sankaranarayanan J, Almutairi A. Photochemical mechanisms of light triggered release from nanocarriers. Adv Drug Del Rev. 2012;64(11):1005–20.

41. Tao Y, Ju E, Liu Z, Dong K, Ren J, Qu X. Engineered, self-assembled near-infrared photothermal agents for combined tumor immunotherapy and chemo-photothermal therapy. Biomaterials. 2014;35(24):6646–56.

Susan N. Thomas is an assistant professor at the Georgia Institute of Technology who trained at the University of California Los Angeles, The Johns Hopkins University, and École Polytechnique Fédérale de Lausanne. As a bioengineer, she has expertise in lymphatic–immune physiology as well as drug formulation for immunotherapy applications.

Chapter 10
Biomaterial-Based Modulation of Cancer

Fnu Apoorva and Ankur Singh

10.1 Introduction

Cancer is the second most common cause of death in the USA, accounting for one out of every four deaths. The American Cancer Society estimated that during 2014 there were around 1.65 million new cases of cancer diagnosed in the USA and close to half a million cancer-related deaths [1]. While complete treatment of cancer still remains a formidable challenge, the twentieth century has brought about significant research and development in cancer therapy, such as surgical resection [2], radiotherapy and chemotherapy [3], and more recently, immunotherapy, [4]. Such advancements have successfully prolonged the lives of cancer patients (Fig. 10.1).

Research in oncology has enhanced our understanding of tumor microenvironments, and the molecular mechanism behind cancer. Hanahan et al. [5, 6] proposed six characteristic hallmarks of cancer: (a) sustaining proliferative signaling, (b) evading growth suppressors, (c) resisting cell death, (d) enabling replicative immortality, (e) inciting angiogenesis, and (f) activating invasion and metastasis. In the process of transforming from normal to a neoplastic state, cells acquire these traits and turn tumorigenic, and ultimately malignant. Several of the above characteristics render tumors resistant to drug response. It is also evident that cancer could be prevented and treated with tumor-specific immunotherapy; however, it is the complexity of the tumor microenvironment, and its robust immune-evading mechanisms, such as high antigenic mutation, secretion of immunosuppressive cytokines, and down regulation of major histocompatibility complex (MHC) expressions, which make it difficult for the immune system to recognize and kill cancer cells [7]. Studies suggest that the immune system is capable of recognizing and

A. Singh (✉) · F. Apoorva
Sibley School of Mechanical and Aerospace Engineering, Cornell University, Ithaca, NY 14853, USA
e-mail: as2833@cornell.edu
F. Apoorva
e-mail: fa235@cornell.edu

© Springer International Publishing Switzerland 2015
L. Santambrogio (ed.), *Biomaterials in Regenerative Medicine and the Immune System*,
DOI 10.1007/978-3-319-18045-8_10

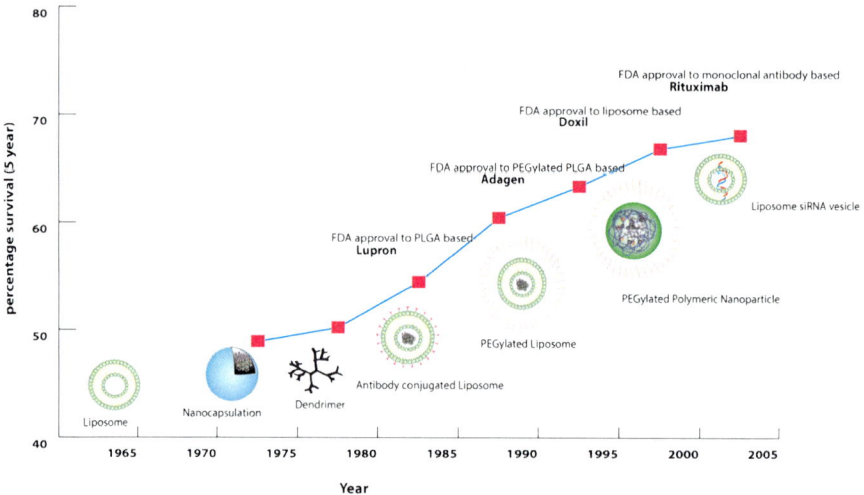

Fig. 10.1 Percentage of cancer patients surviving for more than 5 years since the date of diagnosis has remarkably improved in the last three decades. The timeline shows the invention of key immune modulating biomaterials and FDA approval of these biomaterial-based cancer drugs. *PLGA* poly(lactide-co-glycolide), *FDA* the Food and Drug Administration

regulating cancerous tumors, but tumors modulate certain immune checkpoints to evade the system [8, 9]. The theoretical framework within which anticancer therapies are being engineered can be based on the two-signal theory of T cell activation. Although a two-signal theory of B cell activation was proposed by Brestcher and Cohn [10], it was expanded to T cell activation by Lafferty and Cunningham [11]. According to the two-signal theory, T cell activation requires two factors: (a) recognition of antigen peptides and (b) production of co-stimulatory and inhibitory signals by accessory surface molecules on T cells. The activation of T cells is a cumulative effect of positive and negative signals from signals from regulatory T cell. Cytotoxic T-lymphocyte antigen-4 (CTLA-4) and programmed death-1 receptor (PD-1) are two negative regulators with significant roles in immune response [9, 12–14]. In 2011, the Food and Drug Administration (FDA) approved anti-CTLA-4-based monoclonal antibody ipilimumab for the treatment in of advanced-stage melanoma [15] and anti-PD-1-based therapeutics were in the clinical trial stage [16]. While CTLA-4 modulates the early activation of naïve and memory T cells (Fig. 10.2), PD-1 is responsible for controlling T cell activities in peripheral tissue via its ligands PD-L1 and PD-L2. In clinical trials, PD-1 has shown promising results for the treatment of melanoma, kidney cancer, and non-small cell lung cancer (NSLM) [17, 18]. The blockade of cytokines within tumors is another strategy that has been explored by few laboratories. Some of the agents employed to successfully achieve this blockade are interleukin (IL)-10, IL-13, and transforming growth factor beta (TGF-β) [19].

The progress in understanding the molecular mechanism of cancer, its microenviroment, and the need for targeted drug delivery has inspired the development of immune modulating biomaterials and immune therapies for cancer [20–22]. The

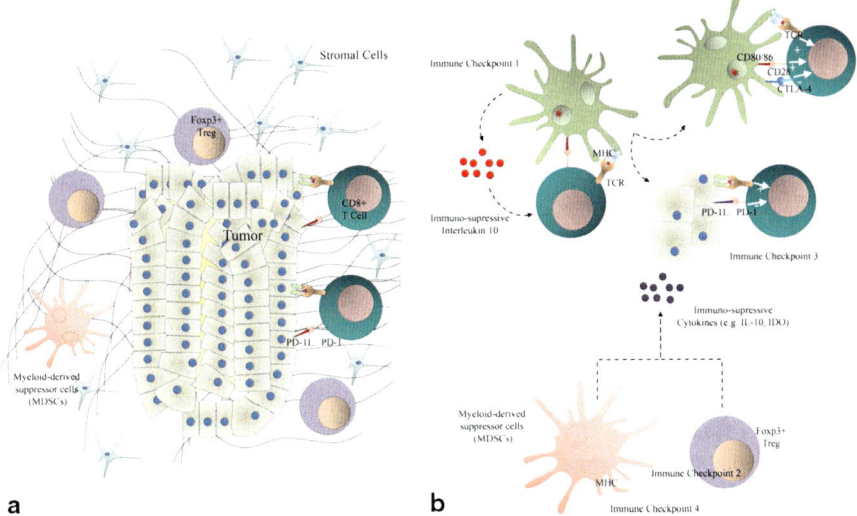

Fig. 10.2 Immunosuppressive tumor microenvironment. **a** Tumors consist of a wide array of immune and stromal cells along with malignancy-supportive extracellular matrix and cytokines secreted by infiltrating cells such as myeloid-derived suppressor cells (*MDSCs*), dendritic cells (*DCs*), neutrophils, natural killer cells, and lymphocytes. **b** Immune checkpoints 1–4 regulate key components involved in the generation of immune response. The cytotoxic T-lymphocyte-associated antigen-4 (*CTLA4*)-mediated immune checkpoint is induced in T cells when an antigen is first presented by DCs to the T cells. The programmed cell death protein 1 (*PD1*) pathway-mediated immune checkpoint is induced when activated T cells with high expression levels of PD1 encounter PD1 ligands on tumor tissues, which suppress the T cell response. PD1 ligands are upregulated in tumor tissues in response to interferon-γ produced by activated T cells. Cytokines such as IL-10 and Indoleamine 2,3-dioxygenase *(IDO)* produced by DCs, Tregs, and other cells, function as immunosuppressive checkpoints against cancer. (Adapted with permission from Purwada et al. [14]). *CD* cluster of differentiation

rational design approaches—based on protein engineering and immune-modulating biomaterials—for biologics and materials have improved the efficacy of conventional cancer therapeutics; first by the virtue of targeted drug delivery, and second by directing sustained tumor antigen-specific T cell response [23–25]. The advances in immune modulating biomaterials, combined with nano–micro-technology and tissue engineering, promises new opportunities in cancer therapy. This chapter presents an overview of biomaterials for cancer cell modulation by manipulating the functioning of the immune system and/or the tumor microenvironment. We further present a brief review of functionalized biomaterial-based tissue engineering and their significance in cancer research. We discuss emerging immunomodulatory therapies for cancer and the significance of biomaterials in immunomodulatory therapies, followed by a description of biomaterials in targeted drug delivery. We also discuss immune modulation by adjuvants, and current challenges in adjuvant-based cancer therapies, and eventually conclude with the future challenges in oncology and role of immune modulating biomaterials in meeting these challenges.

10.2   Modulation of 3D Tumor Microenvironment

Within a tissue, cells are surrounded by the extracellular matrix (ECM), which consists of a hierarchy of micro- and nanoscale fibers. This hierarchical arrangement
provides cells with a support structure and modulates cell cell interaction by stor-
ing, releasing, and activating a wide range of biological factors [26]. One of the
applications of tissue engineering is to synthesize a scaffold/matrix that mimics
biological, physical, and biochemical environments of a natural ECM [27, 28]. Mi-
croarchitecture and surface features of the biomaterial plays an important role in
cell morphology, signaling, adhesion, migration, proliferation, and differentiation
[26, 29]. By applying the concepts of micro- and nanotechnology to tissue engineering, researchers have been able to engineer biomaterials with 3D topography and
multiscale features that closely emulate the hierarchical architecture and functionality of tumor-specific ECM [27, 28].

The tumor microenvironment is characterized by altered physiological properties
such as composition, stiffness, growth factors, etc. Often these alterations contribute to the germination of cancer and its subsequent progression. In the past, tissue
engineering has been applied to study the physiology of tumor microenvironments
and to determine the various mechanisms to modulate them to develop novel cancer
therapies. Since cells in a 3D microenvironment are closer to reality, emphasis has
been on engineering 3D tumor cultures. Scaffold-based 3D culture of tumor cells
develop necrotic areas which closely resemble in vivo cancer [30]. Hutmacher et
al. demonstrated that 3D fibrin matrices are favorable to the growth of endothelial
progenitor cells, which leads to the development of blood vessels resembling lu-
men structures [31], a feature that cannot be replicated in 2D cultures. These blood
vessel-like structures enable 3D cultures to be more suitable for the study of angio-
genesis-like, environment-sensitive processes. Poly(lactide-*co*-glycolide) (PLGA)
polymer-based scaffolds are biologically passive, porous, and rigid, and one of the
first 3D culture systems employed to study human oral squamous carcinoma cells
where cultured cancer cells formed tumors with proliferation, growth characteristics, and marker expression similar to animal models [32]. The growing need for
a rational culture environment design has inspired the development of functionalized semi-synthetic hydrogels representing 3D culture models [33]. For instance,
Chaudhary et al. studied the effect of malignant phenotypes on the stiffness of ECM
and found that increasing the stiffness of normal mammary epithelial cells results in
malignant phenotype, while this effect is missing when accompanied by an increase
in basement-membrane ligands [29]. It was concluded that stiffness and structural
composition are sensed via the β4 integrin, Rac1, and PI3K pathway. Thus, hydrogels and scaffolds in which ECM stiffness and cell–matrix adhesion can be indepen-
dently controlled can be apt models for studying the tumor environment and various
mechanisms to modulate its physiological properties. We recently reported a facile
approach to micro-manufacture composite bio-adhesive microgels of maleimide-
functionalized polyethylene glycol (PEG-MAL) with an interpenetrating network
of ionically cross-linked gelatin and silicate nanoparticles (gelatin-NP) based on
Michael-type addition polymerization and ionic gelation processes [34]. In these
studies (Fig. 10.3), Patel et al. reported that the viscoelastic properties of these hy-

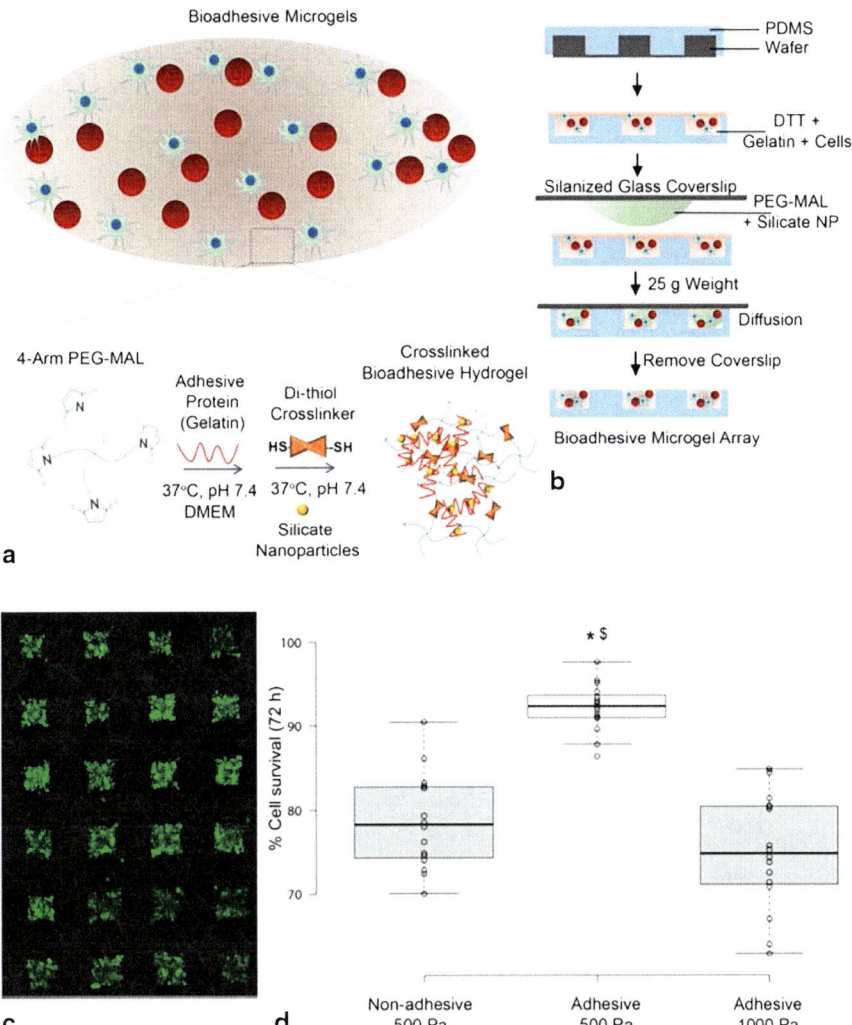

Fig. 10.3 Bioadhesive, cell-encapsulated interpenetrating polymer networks (*IPN*) of PEG-MAL and gelatin-silicate nanoparticles (NP). **a** Schematic of bioadhesive cell-supportive microenvironment consisting of 4-arm PEG-MAL crosslinked with dithiothreitol (*DTT*) and coated with a stable IPN of gelatin with silicate NP. The 4-Arm PEG-MAL undergoes a Michael-type addition reaction with thiol groups on DTT and gelatin forming an ionic gelation complex with NPs at 37 °C and pH 7.4. The PEG component provides structural support for cells, while the gelatin-NP component provides adhesive ligands for cell spreading and signaling. *Red spheres* represent suspension cells and green cells are anchorage dependent. **b** Schematic representing microfabrication of bioadhesive microgels. Component B consists of 4-arm PEG-MAL precursors mixed with silicate NPs and media with or without cells and poured onto a polydimethylsiloxane (*PDMS*) microwell mold. Component A consisted of a well-mixed solution of gelatin with DTT and media, placed on a sigmacote-coated glass slide and aligned with component B on each micromold, allowing the polymers to diffuse and mix. After 1 min, glass slides were removed leaving behind an array of cell-encapsulated microgels. **c** Calcien-stained fluorescent micrograph of a representative microgel array with cervical cancer cells (each well represents a 300-μm-wide gel). **d** Effect of hydrogel stiffness on cell survival. *Box plot* represents percent cell survival of HeLa cells after

drogels are independent of their composition of adhesive additives—which has a unique effect on cell spreading, viability, and proliferation of cervical cancer and leukemia cell line models. Such engineered systems provide a platform to modulate the behavior of cancer cells and decouple some of the underlying microenvironment factors that induce tumorigenesis, affect tumor growth, and further tumor progression [35].

10.3   Modulation of Tumor Vascularization and Anti-angiogenesis for Cancer Therapeutics

Another application of biomaterials-based tumor modulation is in modeling the microenvironment with tumor-mimicking vasculature. Zheng et al. reported a microfluidic environment for 3D cell culture that develops a network of confluent and perfusable microvessels in type-I collagen for in vitro study of angiogenesis and thrombosis. The development of desired vasculature in a scaffold is key to regenerative medicine and the study of physiochemical phenomena, such as pharmacotoxicology, vascular and blood biology, tumor angiogenesis/thrombosis, etc. The microfluidic vasculature demonstrated desired characteristics, such as morphology, barrier function, angiogenic remodeling, etc., enabled the study of interaction between blood components and the endothelium, and indicated recapitulation of characteristics of healthy microvessels, as well as pathological tissues. It was found that geometry and luminal flow determine important biophysical mechanisms via which the aforementioned characteristics are modulated. When vasculogenic medium is applied exogenously through a vascular network with perfusion, it mimics the proangiogenic environment of a solid tumor and shows characteristics such as relatively unorganized CD31 staining at cell–cell contact and sprouting of endothelial cells into the collagen matrix with lumens. Interestingly, in the presence of proangiogenic signals, the region of endothelial sprouting was marked by solute extravasations (Fig. 10.4), a characteristic of tumor neovasculature.

Pro-angiogenic factors such as vascular endothelial growth factor (VEGF), basic fibroblast growth factor (bFGF), and IL-8 play a key role in regulation of tumor angiogenesis. Fischbach et al. demonstrated that secretion of these factors could be upregulated in 3D poly(lactide-*co*-glycolide) (PLG) cell culture; the secretion of VEGF and bFGF were moderately higher, and IL-8 was two orders of magnitudes higher than its 2D counterpart. The increased secretion of VEGF and bFGF were related to the reduction in partial O_2 pressure, whereas enhancement of IL-8 was at-

---

72 h of culture in microgels with approximately 550 and 1100 Pa storage modulus, as determined from bulk gels analysis at 10 kHz frequency. $n = 16$. $*p < 0.05$ compared to nonadhesive hydrogels with same 500 Pa modulus; $^{\$}p < 0.05$ compared to adhesive hydrogels with 1000 Pa modulus. All adhesive hydrogels were composed of equivalent amount of adhesive ligand. (Adapted with permission from Patel et al. [34])

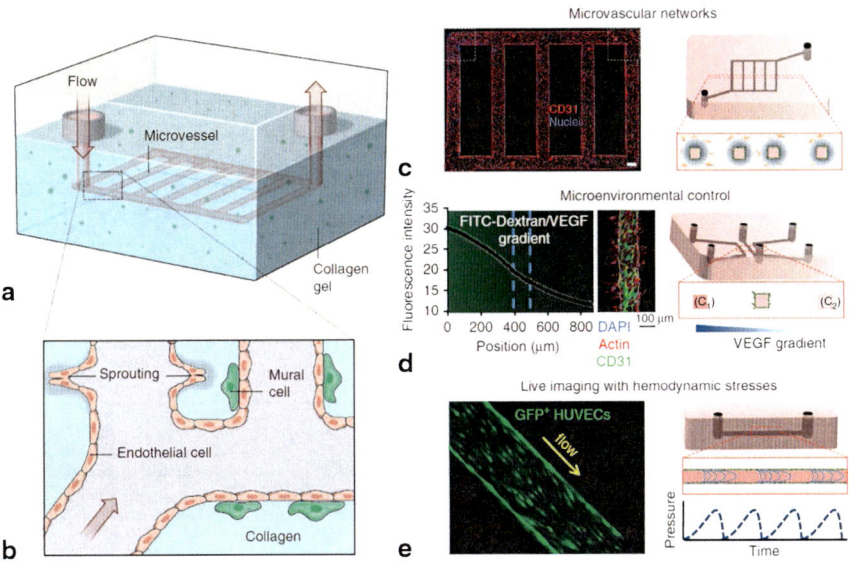

Fig. 10.4 Development of a microvascular network in vitro using a microfluidic 3D cell culture environment. **a, b** Schematic showing morphology of endothelium, its sprouting, perivascular association, and subsequent blood perfusion, **c** confocal microscopic endothelialized microfluidic vessels, **d** mimicking endothelial cell sprouting for study of angiogenesis, and **e** alignment of endothelial cells in the direction of the flow when subjected to physiological shear stress and flow. (Adapted with permission from Morgan et al. [36]) *FITC* fluorescein isothiocyanate, *VEGF* vascular endothelial growth factor, *GFP* green florescent protein

tributed to the 3D nature of scaffold. Fischbach et al. predict that composition of the ECM, integrin expression, and the cell–ECM interaction may play an important role in modulation of upregulation of IL-8. It is believed that the effectiveness of cancer therapeutics, whose working principle is to inhibit angiogenesis via blockading the VEGF signal, can be considerably improved by using a combination regimen that targets inhibition of VEGF as well as IL-8.

Anti-angiogenesis is the process of inhibiting tumor cells from recruiting new blood vessels, which results in limiting the nutrients and food supplies, eventually leading to the elimination of tumor cells [37–39]. Proposed by Judah Folkman in 1971 [40], the first anti-angiogenesis cancer therapeutic was approved by the FDA in 2004, in the form of Avastin (also called bevacizumab), to be used in combination with standard chemotherapy for metastatic colon cancer. Avastin blocks angiogenesis by inhibiting vascular endothelial growth factor A (VEGF-A), a protein signal often associated with angiogenesis in various diseases, including cancer. Since its first approval in 2004, it has been used as therapy for lung cancer, renal cancer, ovarian cancer, breast cancer, and cancerous brain tumors. While it is an elegant technique for cancer therapy, the inhibition of blood vessel growth could adversely interfere with other physiological processes, for example healing and general maintenance of the body. Other adverse effects of anti-angiogenesis therapy-induced-

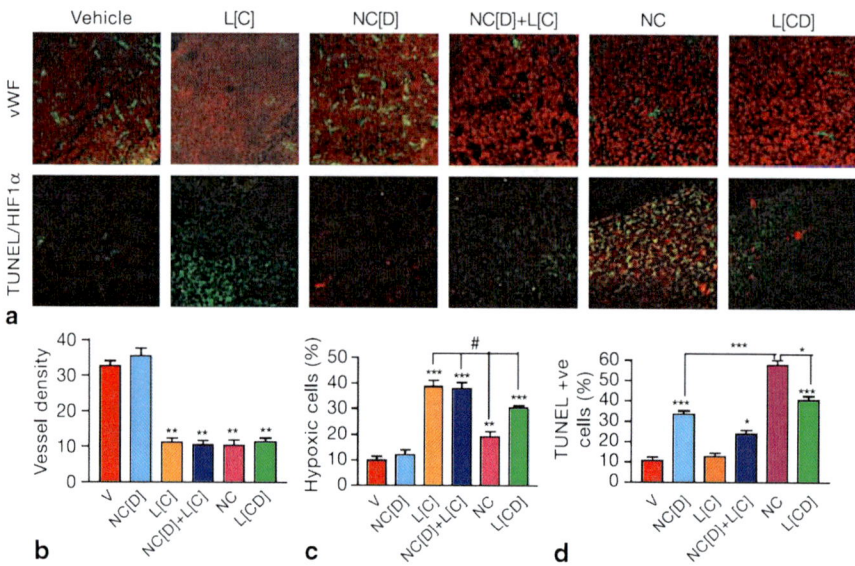

Fig. 10.5 Angiogenesis: targeted drug delivery: **a** the effect of nanocell treatment on tumor vasculature and apoptosis, **b–d** tumor vessel density, **e–g** effect of hypoxia-inducible factor 1 (*HIF1*) and VEGF level. (Adapted with permission from Sengupta et al. [41]). *TUNEL* terminal deoxynucleotidyl transferase-mediated dUTP nick end-labeling, *vWF* von Willebrand factor, *VEGF* vascular endothelial growth factor

tumoral-hypoxia include: (a) genes capable of reducing the pro-apoptotic effects of chemotherapy are upregulated resulting in reactive resistance of tumors and (b) it enhances metastatic and invasive potential of tumor cells. In order to overcome these adverse effects, Sengupta et al. developed a delivery vehicle for controlled sequential exposure of tumors to cytotoxic agents post anti-angiogenesis-induced vascular shutdown [41]. The delivery system, called nanocell, consists of nuclear PLGA nanoparticles (NP) enveloped in a PEGylated-phospholipid block-copolymer. The NPs conjugate chemotherapeutic agents, namely Doxorubicin, which is known to induce apoptosis via DNA intercalation, and a lipid part, loaded with an anti-angiogenesis agent, combretastatin-A4. After delivery of the drug inside the tumor, the disruption of the lipid envelope results in rapid release of combretastatin-A4. This drug disrupts the cytoskeleton structure leading to a vascular collapse and trapping of NPs inside the tumor. The NPs release cytotoxic agent Doxorubicin slowly over an extended period of time to kill the tumor cells. Sengupta et al. noticed that combrestastatin-A4 reaches significant levels in 12 h, whereas the doxorubicin-PLGA system degrades over an extended period of 15 days, as shown in Fig. 10.5.

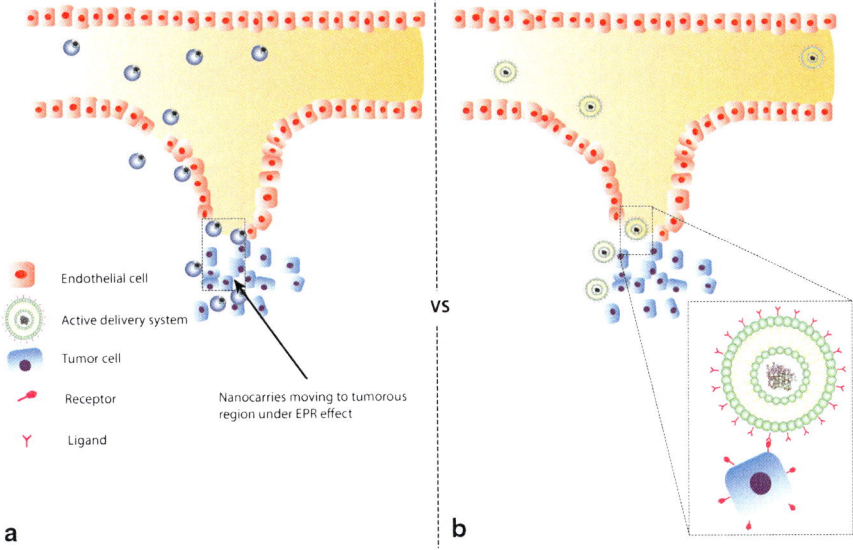

Fig. 10.6 Modes of biomaterials drug delivery to tumors **a** EPR-mediated passive drug delivery in which drug-loaded nanocarriers extravasate and accumulate in pathogenic region (e.g., tumor) because of enhanced vascularity and lower pH, increasing the local drug concentration **b** active drug delivery where antibody-conjugated drug cargo (in this case, drug-encapsulated liposome) interacts with tumor receptors and discharges the therapeutic load selectively inside tumor

10.4 Biomaterials as Drug Delivery Vehicles

In the 1960s, it was discovered that liposomes could be used as protein and drug carriers for disease treatment, and since then, a variety of biomaterials have been developed for drug delivery systems [42]. Currently, there are more than 247 nano-medicines approved by the USFDA or in the clinical trial stage [43]. Nanotechnology has been successfully employed to synthesize modular therapeutic/delivery devices (such as polymeric particles carrying drug cargo) with desired structure and molecular composition. This has promoted the stability, specificity, and controlled release of drugs to achieve safer, more effective therapeutics. In the past, polymeric micro- and nanoparticles (polyplexes), lipid NPs, and emulsions have been employed for controlled-release targeted delivery [23, 24, 44]. For instance, DNA vaccine conjugated with cationic polymer complex results in significant improvement in cellular uptake of DNA, while maintaining the integrity of the DNA and immunogenicity of the antigen [45].

The tumor microenvironment has different physiochemical properties than their healthy counterparts. For example, tumor microenvironments are characterized by a high density of proliferating endothelial cells, pericyte deficiency, formation of anomalous basement membrane, and destruction of lymphatic vessels. This enhances permeation and accumulation of NPs (20–200 nm) in the tumor region, an

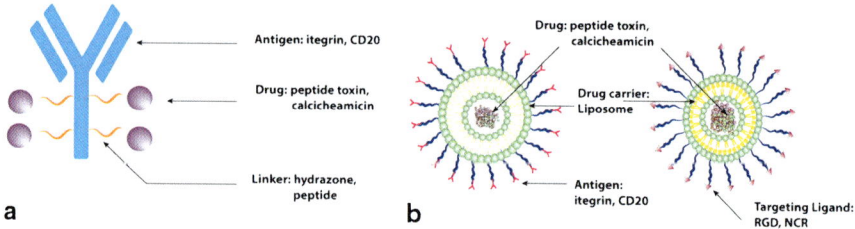

Fig. 10.7 Targeted delivery of cancer therapeutics: **a** antibody immunoconstructions consisting of a monoclonal antibody conjugated with therapeutic molecules **b** targeted nanocarriers in which targeting ligands are grafted on surface of therapeutic encapsulated nanocarriers. The ligands can be monoclonal antibodies or non-antibody ligands (e.g., peptidic). *RGD* integrin receptor-binding Arg-Gly-Asp

effect called the enhanced permeability and retention (EPR) effect [46], as shown in (Fig. 10.6). Another difference between tumor cells and healthy cells is their relative glycolysis rate, which results in a difference of pH. While healthy tissue and blood maintain a pH of 7.4, most tumorous regions have a pH in the range of 6.0–7.0, thus making it easier for neutral or anionic charges to diffuse across the membrane [47]. Passive delivery systems often exploit EPR and low pH inside the tumor for targeted drug delivery (Fig. 10.6). Key characteristics of passive systems are: (a) nanocarriers in the size range of 10–100 nm, (b) neutral or anionic surface charge, and (c) nanocarriers must evade reticuloendothelial cells that phagocytose foreign material. For instance, Joshi et al. reported endogenous lung surfactant-inspired, pH-responsive, nanovesicle aerosols for lung cancer [48]. Joshi et al. proposed that pulmonary surfactant mimetic NPs that offer pH responsive release, specifically in tumors, may overcome the concerns related to pulmonary toxicity and nonspecific targeting of NPs. The researchers developed developed lung surfactant mimetic and pH-responsive lipid nanovesicles for aerosol delivery of paclitaxel in metastatic lung cancer. In these studies, 100–200–nm-sized nanovesicles showed improved fusogenicity and cytosolic drug release, specifically with cancer cells, thereby resulting in improved cytotoxicity of paclitaxel in B16F10 murine melanoma cells and cytocompatibility with normal lung fibroblasts (MRC 5). The nanovesicles showed airway patency similar to that of endogenous pulmonary surfactant and did not elicit inflammatory response in alveolar macrophages. Aerosol administration, while significantly improving the biodistribution of paclitaxel in comparison to Taxol (intravenous, i.v.), also showed significantly higher metastases inhibition (~75%) in comparison to that of i.v. Taxol and i.v. Abraxane. No signs of interstitial pulmonary fiborisis, chronic inflammation, or any other pulmonary toxicity were observed with nanovesicle formulation. Overall, these nanovesicles may be a potential platform to efficiently deliver hydrophobic drugs as aerosols in metastatic lung cancer and other lung diseases, without causing pulmonary toxicity.

In active targeting, tumor-specific ligands are grafted on nanocarriers (Fig. 10.7), which bind to specific receptors on the target cells. Three kinds of active drug delivery vehicles have been explored in extensive detail [49, 50]: (a) monoclonal antibody-based systems, (b) immunoconstruction-based systems, and (c) targeted nanocarriers. Monoclonal antibodies target-specific antigens that regulate proto-oncogenes responsible for cancer cell proliferation. These monoclonal antibodies act as drugs as well as targeting ligands. Many therapies employ antibodies just for targeting; the main goal is to deliver another therapeutic molecule grafted on top of the antibody. Such combinations are called immunoconstructions; yttrium–ibri-tumomab tiuxetan (Zevalin®), denileukin diftitox (Ontak®) and gemtuzumab ozo-gamicin (Mylotarg®) are some examples approved by the FDA [51]. In targeted nanocarriers, the targeting agents are molecules (mostly peptides) which bind to specific receptors [52].

Dhar et al. studied targeted delivery of cisplatin, a chemotherapeutic agent used for treating a range of cancers [53]. Cisplatin's therapeutic efficacy is restricted due to its toxicity and intrinsic and acquired resistances. Dhar et al. designed Pt(IV)-encapsu-lated NPs of poly(D,L-lactic-*co*-glycolic acid)-poly(ethylene glycol) functionalized with a prostate-specific membrane antigen A10 (PSMA Apt). This NP's lethal dose of platinum(IV) compound c,t,c-[Pt(NH3)2(O2CCH2CH2CH2CH2CH3)2Cl2] was targeted to prostate cancer cells LNCaP, and it was observed that the effective-ness of PSMA targeted Pt-NP-Apt-based therapy was an order of magnitude more effective than its free counterpart, cisplatin. Fluorescence microscopy suggested that PSMA aptamers conjugated to the surface of NPs, and these particles were uptaken by PSMA-expressing prostate cancers via receptor-mediated endocytosis. Cytotoxicity studies on LnCaP (PSMA+) cells revealed that under normal condi-tions, the IC50 of targeted Pt-NP-Apt was 0.03 µM, whereas non-targeted (Pt-NP) had an IC50 value of 0.13 µM; cisplatin, on the other hand, had an IC50 value of 2.4 µM. In the case of PC3 (PSMA-) cells, the IC50 of non-targeted particles (Pt-NP) and targeted Pt-NP-Apt were comparable (~0.1 µM), whereas the IC50 for cisplatin was 0.18 µM. These results demonstrate that Pt-NP-Apt are 80 times more effective than free cisplantin.

An effective drug delivery system may have one or more of the following char-acteristics: (a) longevity (b) targetibility, and (c) stimuli-sensitivity. Most of the conventional drug delivery systems lack one or more of these desired qualities, mo-tivating study and design of multifunctional drug delivery systems [54, 55]. In one study, Negussie et al. synthesized cyclic NGR peptide-targeted, temperature-sensi-tive liposomes with doxorubicin, which preferentially bind to CD13/aminopeptidi-dase N (often overexpressed on tumor cells) [56]. This multifunctional nanocarrier was designed for targeted delivery to tumor cells and subsequently temperature induced a triggering of chemotherapeutics (doxorubicin) release. A pH-responsive targeted nanocarrier is another multifunctional drug delivery system. Sawant et al. developed a nanocarrier with PEGylated liposomes (modified with TAT peptide) and PEG-phosphatidylethanolamine (PEG-PE)-based micelles [57]. For targeting, these carriers were grafted with monoclonal antimyosin antibody 2G4 via pNP-PEG-PE moieties; in order to make the carrier pH sensitive, a hydrazine bond was inserted between PEG and PE (PEG-Hz-PE). Combination cancer therapy, on the other hand, may prove to be more effective against multidrug resistant cancers than their conventional counterparts. Generally, in order to prevent cancer-cell resistance

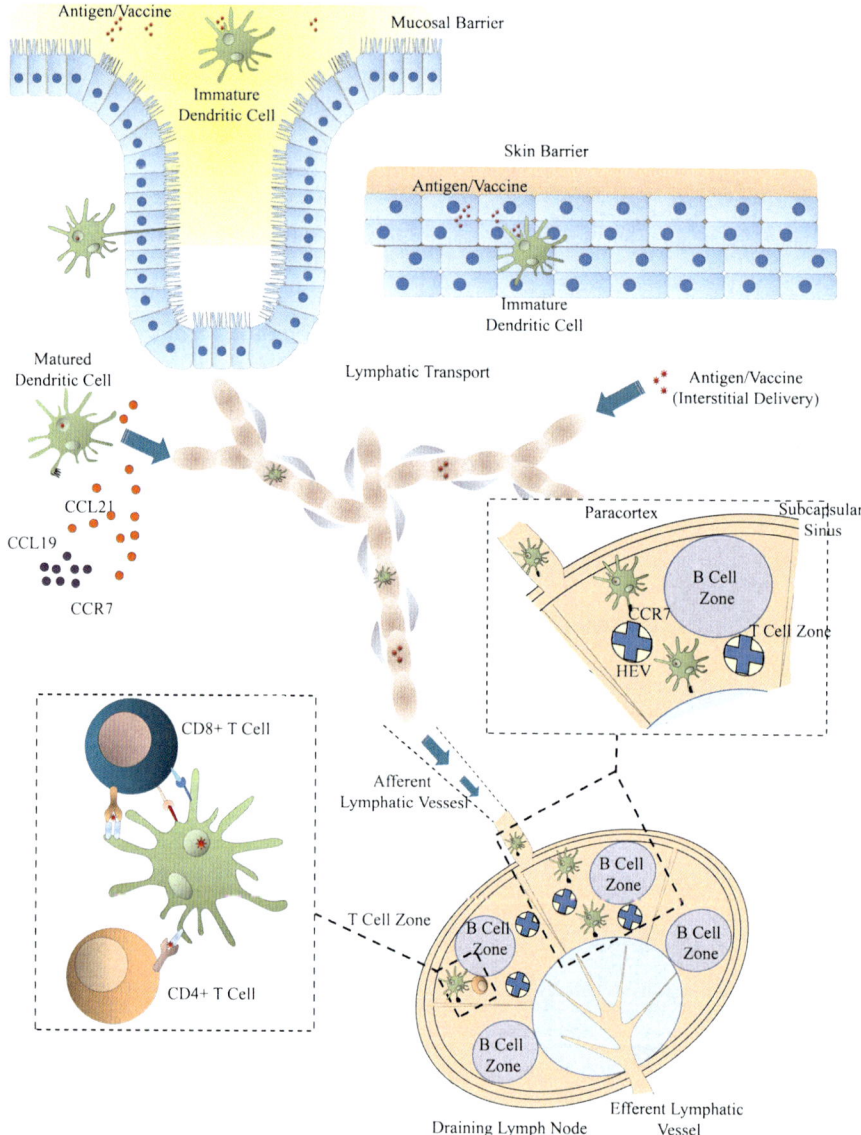

Fig. 10.8 Mechanism of interaction between immune system and nanovaccines for targeting of T cells and dendritic cells. (Adapted with permission from Purwada et al. [14]). *HEV* hepatitis E virus

anticancer drug are combined with suppressors of cellular resistance within a single multifunctional nanocarrier. Saad et al. studied a cationic liposome nanocarrier-based delivery system (NDS), loaded with anticancer drug doxorubicin, a cellular-resistance suppressant, and siRNA targeted to multidrug resistance-associated protein gene (MRP1) and B cell lymphoma 2 (BCL2) mRNA for multidrug-resistant cancer therapy [58]. They found that NDS remarkably improved the efficiency of chemotherapy, significantly higher than what is achieved by applying these components separately.

10.5   Biomaterials for Immune Therapy Against Cancer

Inside a draining lymph node or other secondary lymphoid tissue, antigen-presenting cells localize in the vicinity of T cells for activation. DCs are the most important antigen-presenting cells as they initiate and modulate the immune response. Immature DCs are found in peripheral tissues and blood, where they look out for pathogens like bacteria, virus, toxins, etc., and kill them via nonspecific or receptor-mediated phagocytosis. Post phagocytosis, immature DC cells turn into mature DCs and present fragments of phagocytized pathogen in the form of antigen complexes on their major histocompatibility complexes (MHC). Mature DC cells possess high levels of C-C chemokine receptor type 7 (CCR7), which guide them to lymph nodes under the gradient of C-C motif chemokine (CCL)-19 and CCL-21 chemokines, secreted by lymph node cells (Fig. 10.8). Lymph node T cells can recognize MHC-antigen complexes, which trigger activation of CD8+ T cells, responsible for killing pathogens, as well as infected and/or cancerous cells. DCs fragment exogenous molecules in the endosome region with proteases and present them on MHC class II receptors. Antigen complexes on MHC class II receptors bind with CD4+ helper cells to activate them into CD8+ cytotoxic cells; subsequently, these complexes help CD8+ cells multiply in the presence of cytokines secreted by DCs and other T cells. Eventually, these cells migrate to infectious regions in response to inflammatory cytokines produced by infectious cells. In the case of cancer, tumor cells not only have relatively low expression of MHC on their surface, but they also secrete a protein signal which suppresses immune reaction of DCs and cytotoxic T lymphocyte (CTLs). These abilities help tumor cells evade the immune system.

In order to mimic this process, artificial high-antigen-density platforms (for example, magnetic beads, liposomes, exosomes, or polymer particles) have been studied [23, 44, 59]. These platforms occasionally exploit cytokines and IL-2, since these regulatory immune molecules play an important role in T cell expansion. Fadel et al. developed a carbon-nanotube-polymer composite with PLGA NPs and a magnetite artificial APC platform for expanding T cells for cancer immunotherapy [60]. This method first exploits nanotubes' high aspect ratio and nanotopography to enhance cell–cell attractions and long-term cultures, and second applies magnetite for enrichment of carbon nanotube–polymer composite (CNP)-expanded T cells by magnetic separation. These result in a clustered presentation of T cell stimuli, peptide-loaded MHC-I and co-stimulatory ligand, anti-CD28. IL-2 is encapsulated in bioinylated PLGA NPs together with magnetite that serves for Il-20's paracrine delivery. Fadel et al. demonstrated that, in a murine melanoma model, this method expands cytotoxic T cells in vitro to clinical standard and significantly delays tumor growth. In addition, when tested on human cells they found that, in comparison

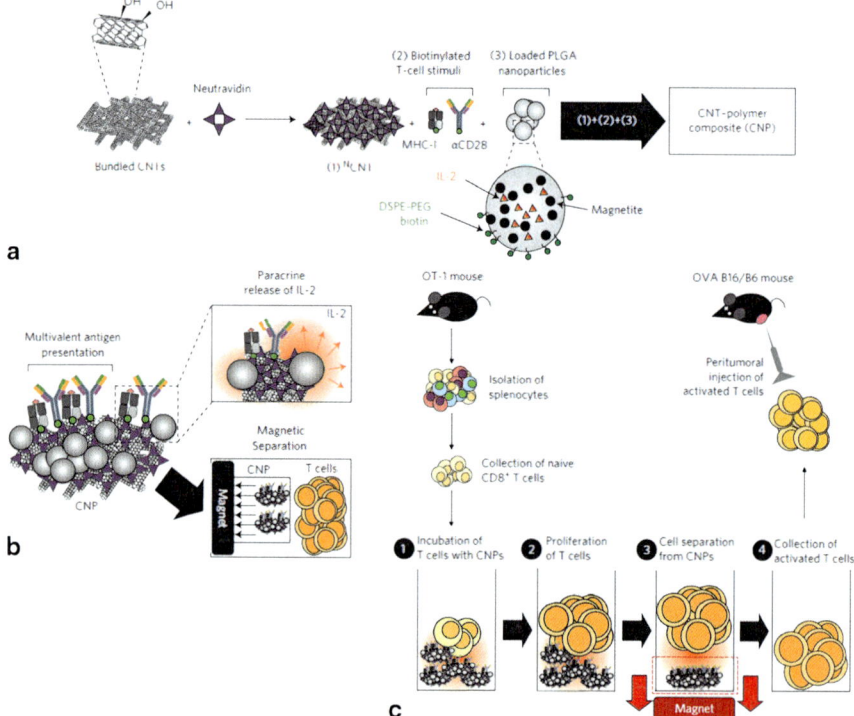

Fig. 10.9 T cell therapy: **a** first carbon nanotubes (*CNTs*) bind with neutravidin to present biotinylated T cell stimuli. PLGA particles encapsulate magnetite and IL-2, **b** multivalent antigen presentation, paracrine delivery of cytokine IL-2, and magnetic separation of CNPs from T cell, and **c** T cell stimulation process, and subsequent separation of T cells using CNPs; First OT-1 CD8+ T cells were purified from splenocytes, followed by a 3-day incubation with CNPs. Activated cells are injected in mice model. (Adapted with permission from Fadel et al. [60]) *DSPE-PEG* distearoylphosphatidylethanolamine-PEG

with human DCs, CNPs had a remarkable expansion of human Epstein-Barr Virus (EBV)-specific CD8+ T cells (Fig. 10.9). Currently, T cell therapy is restricted to clinical trials, though preliminary results of these trials were remarkable. For example, ACT treatment on patients with acute lymphoblastic leukemia (ALL) demonstrated complete disappearance of cancer, and these patients remained tumor free for an extended period of time.

In 2010, the FDA-approved Provenge (Sipuleucel-T), the first DC-based therapeutic cancer vaccine that exploits patients' own immune system to eliminate cancer [61]. This therapy is prescribed for prostate cancer patients, and has been found to prolong patients' life by 4 months. DC-targeting-based preventive and therapeutic vaccines are a personalized medication technique in which DCs are isolated from the blood of the patient and exposed to antigens and various other maturation stimuli before reinjecting them into the patient [62–66]. The challenges in these treatments are primarily due to medical and logistical complexities. Biomaterial-based antigen delivery offers a simpler alternative to cell-based therapy owing to its in vivo activation capability and simple synthesis technique. These biomaterial-based antigen delivery systems can also be surface-functionalized with antigens, homing molecules, etc. for more effective targeted delivery. In order to meet these challenges, biomaterials with tailored immunogens and transport properties have been synthesized to achieve efficient antigen delivery to DCs. For instance, liposome-encapsulated ovalbumin enhances cross-presentation of exogenous antigens and leads to efficient antigen specific CD8+ T cell response. Kim et al. have published an excellent review on biomaterials for modulation of DCs and their therapeutic significance in the context of cancer [67].

Biomaterials-based fusion of DCs and tumor cells to create a hybrid vaccine is another efficient strategy in cancer therapy. Liu et al. studied a cancer vaccine based on a hybrid fusion between J558-IL-4 myeloma cells secreting IL-4 and immature/relatively mature DCs [68]; DCs derived from bone marrow (BM) culture were fused with tumor cells using PEG. This hybrid vaccine was found to induce enhanced immune protection in mice models, which was attributed to high expression of immune molecules on relatively mature DCs, and their superior ability to stimulate antitumor T cell response. Liu et al. developed an injectable thermoresponsive hydrogel-based vaccine that can modulate DCs and studied its therapeutic values using a mouse model [68]. This is a two-step hybrid therapy in which DCs are first recruited to mPEG-PLGA hydrogel through sustained release of a chemoattractant, followed by loading of these resident DCs with cancer antigens through viral or nonviral vectors. This results in activation of resident cells, which subsequently initiates antigen presentation and produces the desired immune response. Liu et al. employed granulocyte-macrophage colony-stimulating factor (GM-CSF) as the chemoattractant, and DC-targeted Lentiviral vector (DC-LV-OVA) and OVA protein as viral and nonviral vectors respectively [68]. GM-CSF mostly released within five days of administration, and DC count peaked on the seventh day of vaccination.

Gene therapy, developed for treating hereditary diseases, is now widely recognized as a potential candidate for therapy of acquired diseases, such as cancer or infections [20, 69–74]. Gene silencing by RNA interference (RNAi) for drug target validation is currently in the clinical trial stage. Gene therapeutics have two important components: first, RNAi-based drugs that target the molecular mechanism of cancer, and second, the appropriate drug delivery system [75]. If present in the cytoplasmic or endosomal region, dsRNA can be recognized by RNA-dependent kinases to induce an interferon response that leads to the release of pro-inflammatory cytokines via the NF-kappa-B pathway. A similar response is found in the case of melanoma differentiation gene 5 (MDA-5) [76]. Likewise, non-self RNA on the cell surface can be sensed via toll-like receptors (TLRs), which leads to NF-kB activation. There are numerous ways of reducing unfavorable immune responses of siRNA, which include, but are not limited to, optimizing the length of siRNA, reducing the dosing amount, and chemical modification of RNAi payloads. The delivery of RNAi is challenging because of its high molecular weight (~13 kDa), poly-anionic nature (~40 negative phosphate groups), and its poor stability in blood and serum [74, 77]. Both active and passive delivery systems have been studied in the context of RNAi delivery: passive systems relying on the increased permeability of the cancerous region and the propensity of liposomes to accumulate there, and active systems, which consist of the antibodies and ligands that assist nanocarriers in delivery to targeted cells and tissues. Cationic lipids/polymer-based nonviral vectors have shown promising results in targeted delivery of RNAi [72–74].

Another area of key importance is in situ lymph node engineering through immunomodulation. We have previously shown that simultaneous delivery of DC cytokine silencing RNAi and lymphoma tumor antigens using injectable biomaterials lead to significant modulation of tumor response in mouse models of lymphoma by creating an artificial immune priming center at the site of injection and then regulating the response in lymph nodes [73]. More recently, Liu et al. developed a cancer molecular vaccine strategy that exploits structure-based programming for targeting lymph nodes [78]. This idea holds significance because it lays down design rules for amphiphile conjugates to modulate the interaction of lymph node-active compounds with the immune system that can be applicable to a wide range of immunomodulatory therapeutics. This vaccine strategy (Fig. 10.10) is inspired by the *sentinel lymph node mapping* procedure, which employs radioactive or colored tracers that selectively bind with serum albumin, a common carrier of fatty acids. The underlying hypothesis behind structural programming is that modification of antigens and adjuvants with lipophilic albumin-binding domains promote accumulation in the lymph node. Liu et al. constructed a range of amphiphilic 20-base phosphorothioate-stabilized cytosine-phosphate-guanosine (CpG) oligonucleotides 5′-linked to lipophilic tails, as shown in Fig. 10.10. Most of these conjugated oligonucleotides lacked interaction within an aqueous solution, but after incubating in serum they could migrate with albumin. In C57BL/6 mice, injection of amph-CpGs promoted marginally higher uptake of C18-and Cho-CpG in lymph nodes than CpGs.

Interestingly, lipo-CpG accumulated eightfold more than CpG, though accumulation was independent of TLR9 recognized CpG motifs and was unrelated to increased nuclease resistance of lipid-modified phosphorothioate (PS)-backbone CpGs. The lipid modification of amphiphiles increased the internalization by DCs and significantly improved the frequency of antigen-specific, cytokine-producing CD8+ T cells compared with CpGs. Besides, lipo-CpGs increase antibody response and CD8+ T cell response. Altogether, amph-CpGs enhance the potency of molecular adjuvant significantly, and at the same time lower side effects. The potency of this programmed antigen and adjuvant-based targeting was tested with amph-peptide (DSPE-PEG-peptide) conjugates of a model HIV antigen (AL11 epitope from SIV Gag), melanoma antigen Trp2, and a peptide derived from human papillomavirus (HPV) antigen E7. Amph-peptides accumulation in lymph node improved and further demonstrated enhanced expansion of antigen-specific, cytokine-producing CD8+ T cells. In animal models with TC-1 tumors, amph-vaccines triggered sustained regression of large TC-1 tumors, and in B16F10 melanoma amph-vaccines slowed down the tumor growth rate, whereas the soluble vaccine exhibited little to no effect in either case.

Pradhan et al. suggest that in order for immunotherapy to succeed, there should be a fine balance between the T helper cell (Th)1 and Th2 response induced by DCs

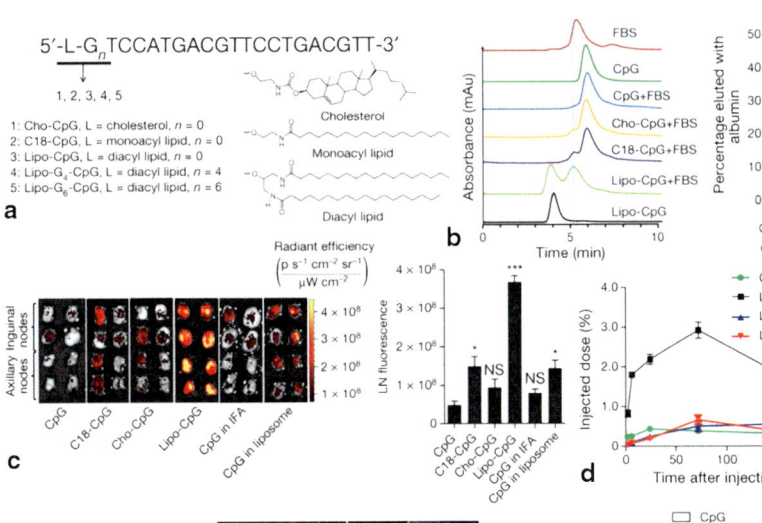

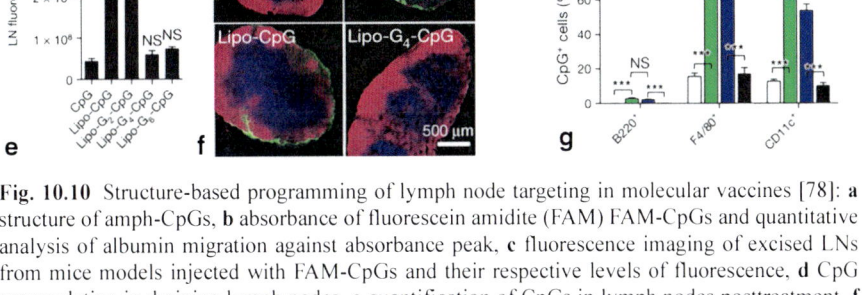

Fig. 10.10 Structure-based programming of lymph node targeting in molecular vaccines [78]: **a** structure of amph-CpGs, **b** absorbance of fluorescein amidite (FAM) FAM-CpGs and quantitative analysis of albumin migration against absorbance peak, **c** fluorescence imaging of excised LNs from mice models injected with FAM-CpGs and their respective levels of fluorescence, **d** CpG accumulation in draining lymph nodes, **e** quantification of CpGs in lymph nodes posttreatment, **f** immunohistochemistry of inguinal lymph nodes, and **g** flowcytometry analysis of CpG+ cells from lymph node. (Adapted with permission from Liu et al. [78])

and various other antigen-presenting cells [79]. It is a general practice in cancer immune therapy to employ TLRs and their specific agonist to regulate innate and adaptive immunity. For instance, TLR9 agonist cytosine-phosphate-guanosine oligodeoxynucleotides (CpG ODN) and synthetic TLR3 agonist Poly (I:C) have been widely exploited for synergistic Th1 polarizing effects in cancer immune therapy. While these TLR agonists can modulate DCs' activation, maturation, and their presentation to naïve T cells, they also lead to secretion of immunosuppressive IL10 cytokines and Th1 specific IL12 cytokines, which diminishes the ability of DCs to stimulate a strong Th1 response and polarizes immunity towards Th2. Pradhan et al. employed surface-functionalized cationic PLGA microparticles, which can deliver antigens and various TLR ligands simultaneously to immature DCs [79]. Thus, these microparticles mimic pathogens and help to regulate immune stimulation and cytokine activity. The microparticles, when included with CpG ODN and IL-10 siRNA, enhance the immune protection of an idiotype DNA vaccine in prophylactic murine models of B cell lymphoma. We have extensively reviewed the effect of biomaterials, its shape and size, and combinatorial delivery approaches such as hydrogels and scaffolds in Purwada et al. [14] and Singh et al. [20].

10.6   Challenges and Future Direction in Biomaterials

Although oncology research and drug development has significantly improved the survival rate (5 years by ~50% since 1970) of cancer patients on average, some cancers still pose a formidable challenge; for instance in pancreatic, liver, and lung cancer, improvement in survival rate over the last 30 years has been dismal [1]. In these "incurable" cancers, there are frequent cases of reoccurrence (relapse of cancer), and often patients develop drug resistance. The lack of insight into the mechanism by which cancer relapses or develops a drug resistance needs imminent attention to be able to improve existing therapies. This knowledge would assist design of biomaterials with tailored physiochemical properties to be able to arrest these incurable tumors. The key focus will be the synthesis of appropriate biomaterials interfaces that can suitably modulate the immune system to disrupt the mechanisms responsible for cancer relapse and resistance to therapy. In addition, spontaneous regression of a tumor is a rare phenomenon, and a relatively less researched area. Previously, it has been attributed to multiple mechanisms such as differentiation, apoptosis, and genetic crisis. The understanding of various mechanisms that lead to spontaneous regression will provide important understanding of tumor modulation and help us engineer suitable immune-modulating biomaterials that can emulate the microenvironments and signals responsible for spontaneous regression. It is also important to understand the structure–activity relationship of immunomodulating biomaterials, since it will give us an insight into the physiochemical properties that are key to the rational design of drug and drug delivery vehicles. Currently, success of a therapy strongly depends on the stage of cancer at the time of diagnosis and the type of malignancy. A key challenge is the design of therapies and biomaterials independent of malignancy and stage, without compromising the efficacy of therapy. Also, some promising therapies, such as immune therapy, are highly personalized; in these cases, the key challenge is to engineer biomaterials specific to the need of patients. Apart from improving the efficacy of cancer therapy, there is a need to reduce the side effects by synthesizing safer biomaterials and controlling their accumulation in nonspecific organs.

References

1. Society AC. http://www.cancer.org/research/cancerfactsstatistics/cancerfactsfigures2014/index. 2014. Accessed: Dec 15, 2014
2. Simmonds PC, Primrose JN, Colquitt JL, Garden OJ, Poston GJ, Rees M. Surgical resection of hepatic metastases from colorectal cancer: a systematic review of published studies. Br J Cancer. 2006;94(7):982–99.
3. Wagner AD, Grothe W, Haerting J, Kleber G, Grothey A, Fleig WE. Chemotherapy in advanced gastric cancer: a systematic review and meta-analysis based on aggregate data. J Clin Oncol. 2006;24(18):2903–9.
4. Gattinoni L, Powell DJ, Jr., Rosenberg SA, Restifo NP. Adoptive immunotherapy for cancer: building on success. Nat Rev Immunol. 2006;6(5):383–93.
5. Hanahan D, Weinberg RA. Hallmarks of cancer: the next generation. Cell 2011;144(5):646–74.
6. Hanahan D, Weinberg RA. The hallmarks of cancer. Cell 2000;100(1):57–70.
7. Zimmer J. Immunotherapy: natural killers take on cancer. Nature. 2014;505(7484):483.
8. Mittendorf EA, Sharma P. Mechanisms of T-cell inhibition: implications for cancer immunotherapy. Expert Rev Vaccines. 2010;9(1):89–105.
9. Pardoll DM. The blockade of immune checkpoints in cancer immunotherapy. Nat Rev Cancer. 2012;12(4):252–64.
10. Bretscher P, Cohn M. A theory of self-nonself discrimination. Science. 1970;169(3950):1042–9.
11. Lafferty KJ, Cunningham AJ. A new analysis of allogeneic interactions. Aust J Exp Biol Med Sci. 1975;53(1):27–42.
12. Callahan MK, Wolchok JD. At the bedside: CTLA-4- and PD-1-blocking antibodies in cancer immunotherapy. J Leukoc Biol. 2013;94(1):41–53.
13. Korman AJ, Peggs KS, Allison JP. Checkpoint blockade in cancer immunotherapy. Adv Immunol. 2006;90:297–339.
14. Purwada A, Roy K, Singh A. Engineering vaccines and niches for immune modulation. Acta Biomaterialia. 2014;10(4):1728–40.
15. Lipson EJ, Drake CG. Ipilimumab: an anti-CTLA-4 antibody for metastatic melanoma. Clin Cancer Res. 2011;17(22):6958–62.
16. Topalian SL, Hodi FS, Brahmer JR, Gettinger SN, Smith DC, McDermott DF, et al. Safety, activity, and immune correlates of anti-PD-1 antibody in cancer. N Engl J Med. 2012;366(26):2443–54.
17. Brahmer JR. Immune checkpoint blockade: the hope for immunotherapy as a treatment of lung cancer? Semin Oncol. 2014;41(1):126–32.
18. Brahmer JR, Drake CG, Wollner I, Powderly JD, Picus J, Sharfman WH, et al. Phase I study of single-agent anti-programmed death-1 (MDX-1106) in refractory solid tumors: safety, clinical activity, pharmacodynamics, and immunologic correlates. J Clin Oncol. 2010;28(19):3167–75.
19. Young DA, Lowe LD, Booth SS, Whitters MJ, Nicholson L, Kuchroo VK, et al. IL-4, IL-10, IL-13, and TGF-beta from an altered peptide ligand-specific Th2 cell clone down-regulate adoptive transfer of experimental autoimmune encephalomyelitis. J Immunol. 2000;164(7):3563–72.
20. Singh A, Peppas NA. Hydrogels and scaffolds for immunomodulation. Adv Mater. 2014;26(38):6530–41.
21. Orive G, Ali OA, Anitua E, Pedraz JL, Emerich DF. Biomaterial-based technologies for brain anti-cancer therapeutics and imaging. Biochim Biophys Acta. 2010;1806(1):96–107.
22. Devaud C, John LB, Westwood JA, Darcy PK, Kershaw MH. Immune modulation of the tumor microenvironment for enhancing cancer immunotherapy. Oncoimmunology 2013;2(8):e25961.

23. Peer D, Karp JM, Hong S, Farokhzad OC, Margalit R, Langer R. Nanocarriers as an emerging platform for cancer therapy. Nat Nanotechnol. 2007;2(12):751–60.

24. Zhang L, Gu FX, Chan JM, Wang AZ, Langer RS, Farokhzad OC. Nanoparticles in medicine: therapeutic applications and developments. Clin Pharmacol Ther. 2008;83(5):761–9.

25. Levy-Nissenbaum E, Radovic-Moreno AF, Wang AZ, Langer R, Farokhzad OC. Nanotechnology and aptamers: applications in drug delivery. Trends Biotechnol. 2008;26(8):442–9.

26. Fischbach C, Chen R, Matsumoto T, Schmelzle T, Brugge JS, Polverini PJ, et al. Engineering tumors with 3D scaffolds. Nat Methods. 2007;4(10):855–60.

27. Choi NW, Cabodi M, Held B, Gleghorn JP, Bonassar LJ, Stroock AD. Microfluidic scaffolds for tissue engineering. Nat Mater. 2007;6(11):908–15.

28. Zheng Y, Chen J, Craven M, Choi NW, Totorica S, Diaz-Santana A, et al. In vitro microvessels for the study of angiogenesis and thrombosis. Proc Natl Acad Sci U S A. 2012;109(24):9342–7.

29. Chaudhuri O, Koshy ST, Branco da Cunha C, Shin JW, Verbeke CS, Allison KH, et al. Extracellular matrix stiffness and composition jointly regulate the induction of malignant phenotypes in mammary epithelium. Nat Mater. 2014;13(10):970–8.

30. Yamada KM, Cukierman E. Modeling tissue morphogenesis and cancer in 3D. Cell 2007;130(4):601–10.

31. Hutmacher DW, Horch RE, Loessner D, Rizzi S, Sieh S, Reichert JC, et al. Translating tissue engineering technology platforms into cancer research. J Cell Mol Med. 2009;13(8A): 1417–27.

32. Holy CE, Shoichet MS, Davies JE. Engineering three-dimensional bone tissue in vitro using biodegradable scaffolds: investigating initial cell-seeding density and culture period. J Biomed Mater Res. 2000;51(3):376–82.

33. Zhang S, Gelain F, Zhao X. Designer self-assembling peptide nanofiber scaffolds for 3D tissue cell cultures. Semin Cancer Biol. 2005;15(5):413–20.

34. Patel RG, Purwada A, Cerchietti L, Inghirami G, Melnick A, Gaharwar AK, et al. Microscale bioadhesive hydrogel arrays for cell engineering applications. Cell Mol Bioeng. 2014;7(3):394–408.

35. Cayrol F, Diaz Flaque MC, Fernando T, Yang SN, Sterle HA, Bolontrade M, et al. Integrin alphavbeta3 acting as membrane receptor for thyroid hormones mediates angiogenesis in malignant T cells. Blood 2015;125(5):841–51.

36. Morgan JP, Delnero PF, Zheng Y, Verbridge SS, Chen J, Craven M, et al. Formation of microvascular networks in vitro. Nat Protoc. 2013;8(9):1820–36.

37. Folkman J. Angiogenesis in cancer, vascular, rheumatoid and other disease. Nat Med. 1995;1(1):27–31.

38. Carmeliet P, Jain RK. Angiogenesis in cancer and other diseases. Nature. 2000;407(6801): 249–57.

39. Risau W. Mechanisms of angiogenesis. Nature. 1997;386(6626):671–4.

40. Folkman J, Merler E, Abernathy C, Williams G. Isolation of a tumor factor responsible for angiogenesis. J Exp Med. 1971;133(2):275–88.

41. Sengupta S, Eavarone D, Capila I, Zhao G, Watson N, Kiziltepe T, et al. Temporal targeting of tumour cells and neovasculature with a nanoscale delivery system. Nature. 2005;436(7050):568–72.

42. Alexis F, Pridgen EM, Langer R, Farokhzad OC. Nanoparticle technologies for cancer therapy. Handb Exp Pharmacol. 2010;(197):55–86.

43. Etheridge ML, Campbell SA, Erdman AG, Haynes CL, Wolf SM, McCullough J. The big picture on nanomedicine: the state of investigational and approved nanomedicine products. Nanomedicine. 2013;9(1):1–14.

44. Brigger I, Dubernet C, Couvreur P. Nanoparticles in cancer therapy and diagnosis. Adv Drug Deliv Rev. 2002;54(5):631–51.

45. Pannier AK, Shea LD. Controlled release systems for DNA delivery. Mol Ther. 2004;10(1): 19–26.

46. Maeda H, Wu J, Sawa T, Matsumura Y, Hori K. Tumor vascular permeability and the EPR effect in macromolecular therapeutics: a review. J Control Release. 2000;65(1–2):271–84.

47. Maeda H, Sawa T, Konno T. Mechanism of tumor-targeted delivery of macromolecular drugs, including the EPR effect in solid tumor and clinical overview of the prototype polymeric drug SMANCS. J Control Release. 2001;74(1–3):47–61.

48. Joshi N, Shirsath N, Singh A, Joshi KS, Banerjee R. Endogenous lung surfactant inspired pH responsive nanovesicle aerosols: pulmonary compatible and site-specific drug delivery in lung metastases. Sci Rep. 2014;4:7085.

49. Danhier F, Feron O, Preat V. To exploit the tumor microenvironment: passive and active tumor targeting of nanocarriers for anti-cancer drug delivery. J Control Release. 2010;148(2):135–46.

50. Reichert JM, Valge-Archer VE. Development trends for monoclonal antibody cancer therapeutics. Nat Rev Drug Discov. 2007;6(5):349–56.

51. Allen TM. Ligand-targeted therapeutics in anticancer therapy. Nat Rev Cancer. 2002;2(10):750–63.

52. Akinc A, Querbes W, De S, Qin J, Frank-Kamenetsky M, Jayaprakash KN, et al. Targeted delivery of RNAi therapeutics with endogenous and exogenous ligand-based mechanisms. Mol Ther. 2010;18(7):1357–64.

53. Dhar S, Gu FX, Langer R, Farokhzad OC, Lippard SJ. Targeted delivery of cisplatin to prostate cancer cells by aptamer functionalized Pt(IV) prodrug-PLGA-PEG nanoparticles. Proc Natl Acad Sci U S A. 2008;105(45):17356–61.

54. Torchilin VP. Multifunctional nanocarriers. Adv Drug Deliv Rev. 2006;58(14):1532–55.

55. Cheng Z, Al Zaki A, Hui JZ, Muzykantov VR, Tsourkas A. Multifunctional nanoparticles: cost versus benefit of adding targeting and imaging capabilities. Science. 2012;338(6109):903–10.

56. Negussie AH, Miller JL, Reddy G, Drake SK, Wood BJ, Dreher MR. Synthesis and in vitro evaluation of cyclic NGR peptide targeted thermally sensitive liposome. J Control Release. 2010;143(2):265–73.

57. Sawant RM, Hurley JP, Salmaso S, Kale A, Tolcheva E, Levchenko TS, et al. "SMART" drug delivery systems: double-targeted pH-responsive pharmaceutical nanocarriers. Bioconjug Chem. 2006;17(4):943–9.

58. Saad M, Garbuzenko OB, Minko T. Co-delivery of siRNA and an anticancer drug for treatment of multidrug-resistant cancer. Nanomedicine (Lond). 2008;3(6):761–76.

59. Han S, Mahato RI, Sung YK, Kim SW. Development of biomaterials for gene therapy. Mol Ther. 2000;2(4):302–17.

60. Fadel TR, Sharp FA, Vudattu N, Ragheb R, Garyu J, Kim D, et al. A carbon nanotube-polymer composite for T-cell therapy. Nat Nanotechnol. 2014;9(8):639–47.

61. Kantoff PW, Higano CS, Shore ND, Berger ER, Small EJ, Penson DF, et al. Sipuleucel-T immunotherapy for castration-resistant prostate cancer. N Engl J Med. 2010;363(5):411–22.

62. Boudreau JE, Bonehill A, Thielemans K, Wan Y. Engineering dendritic cells to enhance cancer immunotherapy. Mol Ther. 2011;19(5):841–53.

63. Brody JD, Engleman EG. DC-based cancer vaccines: lessons from clinical trials. Cytotherapy. 2004;6(2):122–7.

64. Gilboa E. DC-based cancer vaccines. J Clin Invest. 2007;117(5):1195–203.

65. Jahnisch H, Fussel S, Kiessling A, Wehner R, Zastrow S, Bachmann M, et al. Dendritic cell-based immunotherapy for prostate cancer. Clin Dev Immunol. 2010;2010:517493.

66. Radford KJ, Tullett KM, Lahoud MH. Dendritic cells and cancer immunotherapy. Curr Opin Immunol. 2014;27:26–32.

67. Kim J, Mooney DJ. In vivo modulation of dendritic cells by engineered materials: towards new cancer vaccines. Nano Today. 2011;6(5):466–77.

68. Liu Y, Zhang W, Chan T, Saxena A, Xiang J. Engineered fusion hybrid vaccine of IL-4 gene-modified myeloma and relative mature dendritic cells enhances antitumor immunity. Leuk Res. 2002;26(8):757–63.

69. Fang B, Roth JA. The role of gene therapy in combined modality treatment strategies for cancer. Curr Opin Mol Ther. 2003;5(5):475–82.

70. Fujiwara T, Grimm EA, Roth JA. Gene therapeutics and gene therapy for cancer. Curr Opin Oncol. 1994;6(1):96–105.

71. Wadhwa PD, Zielske SP, Roth JC, Ballas CB, Bowman JE, Gerson SL. Cancer gene therapy: scientific basis. Annu Rev Med. 2002;53:437–52.

72. Singh A, Suri S, Roy K. In-situ crosslinking hydrogels for combinatorial delivery of chemokines and siRNA-DNA carrying microparticles to dendritic cells. Biomaterials. 2009;30(28):5187–200.

73. Singh A, Qin H, Fernandez I, Wei J, Lin J, Kwak LW, et al. An injectable synthetic immune-priming center mediates efficient T-cell class switching and T-helper 1 response against B cell lymphoma. J Control Release. 2011;155(2):184–92.

74. Singh A, Nie H, Ghosn B, Qin H, Kwak LW, Roy K. Efficient modulation of T-cell response by dual-mode, single-carrier delivery of cytokine-targeted siRNA and DNA vaccine to antigen-presenting cells. Mol Ther. 2008;16(12):2011–21.

75. Blau HM, Springer ML. Gene therapy–a novel form of drug delivery. N Engl J Med. 1995;333(18):1204–7.

76. Kang DC, Gopalkrishnan RV, Wu Q, Jankowsky E, Pyle AM, Fisher PB. mda-5: an interferon-inducible putative RNA helicase with double-stranded RNA-dependent ATPase activity and melanoma growth-suppressive properties. Proc Natl Acad Sci U S A. 2002;99(2): 637–42.

77. Rudzinski WE, Aminabhavi TM. Chitosan as a carrier for targeted delivery of small interfering RNA. Int J Pharm. 2010;399(1–2):1–11.

78. Liu H, Moynihan KD, Zheng Y, Szeto GL, Li AV, Huang B, et al. Structure-based programming of lymph-node targeting in molecular vaccines. Nature 2014;507(7493):519–22.

79. Pradhan P, Qin H, Leleux JA, Gwak D, Sakamaki I, Kwak LW, et al. The effect of combined IL10 siRNA and CpG ODN as pathogen-mimicking microparticles on Th1/Th2 cytokine balance in dendritic cells and protective immunity against B cell lymphoma. Biomaterials 2014;35(21):5491–504.

Prof. Ankur Singh is an assistant professor in the Sibley School of Mechanical & Aerospace Engineering at Cornell University, where he directs the Immunotherapy and Cell Engineering Laboratory. Dr. Singh has expertise in the engineering of biomaterials-based platforms for immune cell modulation, cell–biomaterial interactions, tumor microenvironment, nanoengineered materials, and stem and somatic cell mechanobiology. He is a recipient of the 2014 Cellular and Molecular Bioengineering Young Innovators award.

Chapter 11
Targeting Liposomes to Immune Cells

Matthew Levy and Deborah Palliser

11.1   Introduction

Immune response modifiers are powerful tools for potentiating multiple therapies. Reagents that modify the activation of antigen-presenting cells (APCs), in particular dendritic cells (DCs), can affect downstream T cell responses, which would be useful for treating diverse diseases including cancers and autoimmunity. Robust immune responses are needed for effective tumor therapies, and their induction requires multiple components. Strong tumor-specific cluster of differentiation (CD4)+ and CD8+ T cell responses are critical for tumor clearance, and these are elicited by DCs that present tumor antigen on major histocompatibility complex-1 (MHC I) and MHC II, express maturation markers (CD80, CD86) and secrete inflammatory cytokines (type I interferon, IL-12). Similarly, effective vaccines elicit CD4+ T cell responses, with a requirement for CD8+ T cell activation for diseases including human immunodeficiency virus (HIV)-1 and herpes simplex virus (HSV)-2. Conversely, inhibition of immune responses would be a desirable outcome in conditions including autoimmunity and atopy. Modification of DC responses, resulting in a reduction of antigen-specific CD4+ and CD8+ T cell responses, could alleviate these diseases.

Activation of DCs occurs following an encounter with the pathogen. Two elements are required for induction of effective immune responses: uptake of pathogen, resulting in ligation of receptors (e.g., toll-like receptors, TLRs; retinoic acid inducible gene, RIGs) and presentation of antigenic peptides via MHC I and MHC II. The only clinically approved agent that incorporates these components is the prostate cancer vaccine Provenge, an individualized protein-based

D. Palliser (✉)
Department of Microbiology and Immunology,
Albert Einstein College of Medicine, 1300 Morris Park Ave, Bronx, NY 10461, USA
e-mail: deborah.palliser@einstein.yu.edu

M. Levy
Department of Biochemistry
Albert Einstein College of Medicine, 1300 Morris Park Ave, Bronx, NY 10461, USA
e-mail: matthew.levy@einstein.yu.edu

© Springer International Publishing Switzerland 2015
L. Santambrogio (ed.), *Biomaterials in Regenerative Medicine and the Immune System,*
DOI 10.1007/978-3-319-18045-8_11

vaccine that treats leukapheresed peripheral blood mononuclear cells (PBMCs) with a prostate-specific antigen (prostatic acid phosphatase) fused to the immune modifier Granulocyte–macrophage colony-stimulating factor (GMCSF). A major disadvantage of this therapy is the high cost. This is due to the complex proto- col required for harvesting, modulating, and reinfusing a patient's PBMCs. Fur- thermore, survival is not significantly extended. The ex-vivo manipulation of the PBMCs may limit its efficacy: the cultured DCs are reported to be inefficient an- tigen presenters, and they exhibit decreased migratory and stimulatory functions. An approach that targets antigen to DCs in situ may circumvent these problems. Antibodies, specific for receptors that are highly expressed on DCs (e.g., the man- nose receptor or DEC205) linked to tumor- and HIV-derived antigens are being evaluated in clinical trials for their ability to elicit humoral and T cell responses [1] (NCT01127464; NCT02166905).

Therefore, manipulation of DC function represents a potentially powerful ap- proach for modifying immune responses. However, factors including DC hetero- geneity, the choice of delivery agent for antigen and adjuvant, as well as the type of adjuvant used will significantly influence the immune responses elicited. One prom- ising approach for modulating DC responses is the formulation of antigens (plus or minus adjuvants) in nanoparticles (NPs). The wide array of NPs that differ in size, composition, and structure should enable delivery of appropriate cargoes to various DC subtypes, at various locations, resulting in inhibition or activation of DCs.

In this chapter, we outline recent advances in NP formulations used for delivery of cargoes. We summarize how NP targeting to specific DC subsets can elicit dis- tinct immune responses that are required to control various infections. Recent stud- ies making use of novel NP formulations have been designed to activate immunity, and the evolution of these platforms into clinical reagents are discussed. However, because the field of NP and NP delivery is rather vast, here, we focus largely on de- livery using liposomes, which provide not only a means to deliver protein antigens but can also be designed to efficiently encapsulate nucleic acids including mes- senger RNA (mRNA), small interfering RNA (siRNA), and microRNA (miRNA) (Fig. 11.1).

11.2   The Role of DCs in Directing Immune Responses

11.2.1   Overview

Responses to pathogens are initiated by recognition of conserved molecular patterns (termed pathogen-associated molecular patterns, PAMPs) by pathogen recognition receptors (PRRs) expressed on DCs. Following internalization and processing of the pathogen, antigenic epitopes are presented by MHC II and MHC I, usually re- sulting in activation of T cells. However, DCs are a heterogeneous population of cells, and their function is dependent on their origin and location. For example, sys- temic administration, compared with local injection (e.g., subcutaneous) of cargo

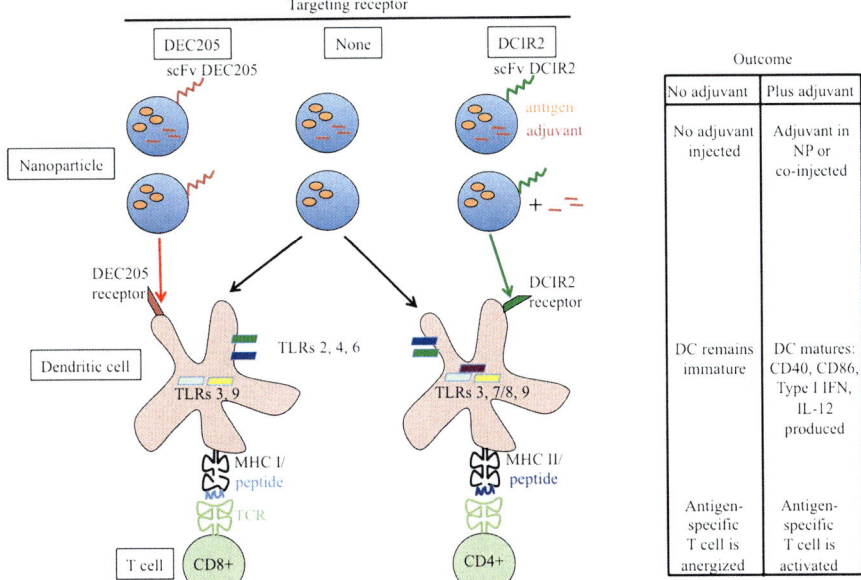

Fig. 11.1 Activation of distinct T cell subsets following injection of targeted or nontargeted NPs in the presence or absence of adjuvant. Directing antigen to DEC205+ DCs results in antigen presentation via MHC I and ligation of antigen-specific TCR on CD8+ T cells. Similarly, directing antigen to DCIR2+ DCs results in antigen presentation via MHC II and ligation of antigen-specific TCR on CD4+ T cells. In the absence of a targeting moiety, any DC subset may be able to take up the NP (depending on factors including NP size and site of injection), which can result in antigen presentation via MHC I and MHC II. Adjuvant can be encapsulated or co-injected. In the presence of adjuvant, DCs will mature due to ligation of PRRs including TLRs by PAMPs expressed by the pathogen (or adjuvant). TLRs expressed by DEC205+ and DCIR2+ DCs are shown: DEC205+ DCs do not express TLR7. DC maturation and antigen presentation will result in T cell activation. In the absence of adjuvant, DCs will present antigen, but they will not undergo maturation, resulting in energy of cognate T cells. *DSPE* 1,2-distearoryl-sn-glycero-3-phosphoethanolamine, *DMPE* Dimyristoylphosphatidylethanolamine, *DOTAP* 1,2-dioleoyl-3-trimethylammonium-propane, *DLinDMA* 1,2-dilinoleyloxy-3-dimethylaminopropane These acronyms are part of the other figure legend (you have inverted the legends)

will result in uptake by disparate DC populations which can result in induction of markedly different immune responses. Therefore, for DC uptake, in addition to NP size and composition, the design of an NP platform must consider the DC population to be targeted, and the route of administration. Strategies such as attachment of targeting ligands, the type of cargo encapsulated, and the inclusion of any adjuvant will also determine the nature of the immune responses elicited.

11.2.2   Heterogeneity of DCs

The two major subsets of DCs in mice are classical DCs (cDCs) and plasmacytoid DCs (pDCs). pDCs secrete large amounts of type I interferon in response to viral infections, and they are present in the thymus and secondary lymphoid organs [2]. cDCs derive from a common DC progenitor (pre-cDC), and they share functional features including expression of CD11c and MHC II, ability to process and present antigens to T cells, and common molecular signatures [3, 4]. One subset of cDCs is present in lymphoid organs, while a second set is migratory and circulates from tissues to draining lymph nodes. cDCs present in the lymphoid organs are categorized as CD8α+DEC205+ or CD8α−DCIR2+, while certain subsets of migratory cDCs express markers including CD103 [5].

As cDCs are present at the pathogen–host interface, they are well positioned to initiate host responses. Expression of innate immune receptors such as the TLRs and NOD-like receptors (NLRs) act not only as maturation stimuli following antigenic encounter but can also be utilized as uptake receptors by NPs coated with specific ligands (Fig. 11.2). A variation of this approach has been used to determine the function of specific cDC subsets: antibodies specific for DC-expressed receptors conjugated to various antigens have revealed differential functions. For example, a DEC205 antibody linked to antigen identified the ability of this DC population to present antigenic epitopes via MHC II and MHC I following injection into mice [6, 7]. When an antigen was coupled to a dendritic cell inhibitory receptor (DCIR2)-specific antibody, uptake by the DCIR2+ DC population resulted in antigen presentation predominantly via MHC II [8]. Inherent differences in CD8α+

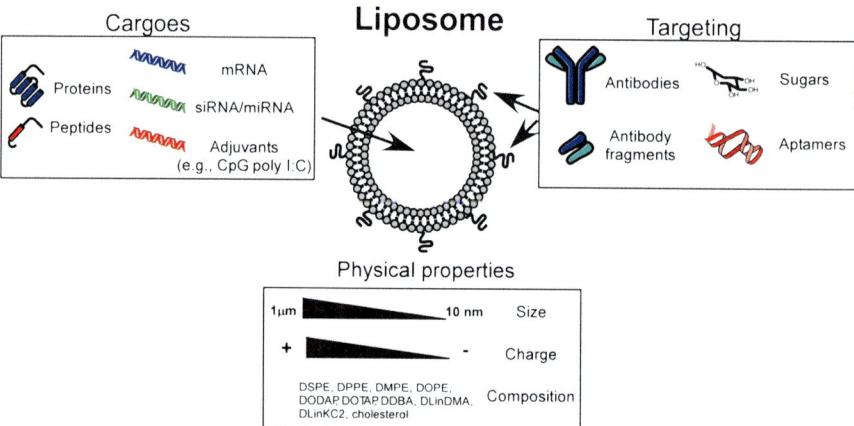

Fig. 11.2 Summary of strategies which can be used to develop liposome-based vaccines. The physical properties of liposomes can be tuned by altering their size and surface chemistry. They can be used to encapsulate a variety of different cargos including proteins, peptides, mRNA, siRNA, and miRNA or adjuvants. Importantly, the cargoes can be encapsulated individually or in combination with one another to more finely tune and control immune responses. Targeting agents including antibodies, antibody fragments (e.g., single-chain antibodies) small molecules (sugars), or even aptamers can be grafted to the liposome surface and have the potential to enhance uptake by specific cell types or subsets. *mRNA* messenger RNA, *siRNA* small interfering RNA, *miRNA* microRNA, *DEC205* CD205, *scFv* single-chain variable fragment, *DC* dendritic cell, *DCIR2* dendritic cell inhibitory receptor 2, *TLRs* toll-like receptors, *MHC* major histocompatibility complex, *CD* cluster of differentiation, *NP* nanoparticle, *IFN* interferon, *IL* interleukin

versus CD8α− DCs, which include maintaining an alkaline pH in phagosomes to limit the destruction of antigens and antigenic access to the cytosol, partially explain their altered functions [9, 10].

In humans, the counterparts to CD8α+ and CD8α− DCs are respectively, blood dendritic cell antigen (BDCA)3+ (CD141) and BDCA1+ (CD1c) myeloid DCs. Although initial studies reported enhanced cross-presentation by the BDCA3+ population, subsequent work shows that both populations, as well as pDCs can cross-present antigen [10, 11]. Antigen targeting to various DC receptors (e.g., DCIR, CD205, and CD40) resulted in cross-presentation. However, the efficiency of antigen presentation has been shown to differ depending on the intracellular compartment accessed following receptor-mediated uptake. For example, uptake via CD205 delivers antigen to late endosomes, whereas CD40- or CD11c-mediated uptake targets early endosomes [12]. Both routes result in cross-presentation in BDCA3+ DCs, whereas antigen delivery to early endosomes is required for BDCA1+ DCs [13]. Therefore, unlike murine DCs, multiple human DC subsets can cross-present antigen.

Overall, the cross-presentation pathways utilized by DC subsets following antigen uptake are generally conserved between mice and humans. However, under defined conditions, multiple human DC subsets have the ability to cross-present antigen. Clearly, these differences in human and mouse DC function need to be considered when designing NPs for delivery to DCs.

11.3   Cargoes and Adjuvants

Other considerations for DC targeting include the type of cargo to be delivered (e.g., antigen, siRNA) as well as any requirement for adjuvant. When antigen is delivered to steady-state DCs, the cognate T cells are tolerized [6, 7]. Therefore, depending on the desired outcome, co-injection of an adjuvant together with antigen may be needed. Studies have demonstrated the ability of DC targeting to downregulate T cell responses, resulting in deletion of diabetogenic effector T cells or protection from autoimmune diseases such as experimental autoimmune encephalomyelitis (EAE) [14, 15]. This outcome is mediated, at least partially, via the induction of IL-10-producing Foxp3+ regulatory T cells and production of transforming growth factor-β (TGF-β) by DCs [16, 17].

DCs can be modulated by overexpression or knockdown of genes. siRNAs have been delivered to DCs using various platforms. Human and murine DCs have been targeted using CpG conjugated to siRNA specific for the oncogene signal transducer and activator of transcription 3 (STAT3 [18, 19]. DNA oligonucleotides bind the DEC205 receptor, making this a useful strategy for DC targeting [20, 21]. Combining an adjuvant with knockdown of an oncogene may be a useful approach for overcoming the strongly immunosuppressive tumor environment. Conversely, delivery of siRNAs that target co-stimulatory receptors has also been reported and could be useful for treatment of autoimmune diseases. siRNAs targeting CD40, packaged into NPs, were reported to decrease CD40 expression in vivo, and reduce mixed lymphocyte responses in vitro [22].

11.3.1 Cargo Format

The molecular makeup of the encapsulated cargo will be a major determinant of NP composition. A wide variety of cargoes including proteins, DNA, or mRNA can be encapsulated. However, the efficiency of encapsulation is directly determined by components that make up the NP. Proteins, often consisting of inactivated pathogens or subunits, have been used extensively as vaccines and elicit effective humoral immunity. Incorporation of proteins or peptides into NPs, when co-delivered with adjuvant, can activate both cellular and humoral immunity. The efficiency of incorporation of proteins, or subunits thereof, into NPs is correlated with factors such as size and charge. Similar considerations apply for nucleic acids, and the use of cationic lipids facilitates the incorporation of DNA and RNA moieties. However, as outlined below, therapeutic use of many conventional cationic lipids is precluded due to issues of toxicity. Newer formulations are being synthesized that minimize toxic effects (see "Advancements in lipid formulation"), and these platforms may be useful for delivery to DC subsets. One example, discussed in more detail below, describes the use of 1,2-dilinoleyloxy-3-dimethylaminopropane (DLinDMA) NPs [23]. These NPs, encapsulating siRNAs, are in clinical trials for cargo delivery to hepatocytes. A follow-up study has shown the ability of these NPs for targeting DCs in vivo. However, a relatively high concentration of NPs was required, and it is likely that significant improvement in cargo delivery can be achieved by targeting these NPs via an appropriate cell surface receptor.

11.3.2 Inclusion of Adjuvants

Whether antigen and adjuvant should be included in NPs is another factor that will need to be experimentally determined for any NP platform and will be disease dependent. For example, for vaccines that require strong antibody responses, a recent study found that packaging of adjuvants into separate NPs was required to elicit optimal responses [24]. However, improved T cell responses, in particular activation of CD8+ T cells were observed when antigen and adjuvant were present in the same NP [25–27]. One explanation for induction of enhanced CD8+ T cell responses comes from studies that show a requirement for antigen and adjuvant uptake by the same APC. In particular, the adjuvant and antigen may need to be present within the same intracellular compartment to elicit antigen-specific T cell responses [28]. Recent evidence suggests that direct interaction of APCs via their PRRs with pathogen, or pathogenic fragments, may be critical for supporting survival and differentiation of CD8+ T cells into effector cells [29]. Therefore, uptake of NPs by APCs would provide a mechanism for such directed delivery of antigen.

There are several advantages to encapsulating both components: the antigen and adjuvant are stabilized and will be protected from degradation by serum proteases and nucleases. By restricting uptake of adjuvant to specific DC subsets, off-target effects can be limited. For example, adjuvant recognition by TLRs will occur on multiple cell type s-DCs, other APCs (e.g., macrophages, monocytes), as well as somatic cells such as epithelial cells, fibroblasts, and keratinocytes. This can result in induction of nonspecific immune responses including high levels of inflammatory cytokines in the serum, activation of multiple immune cells, and splenomegaly. Exacerbation of autoimmune disease has also been reported [30, 31].

11.3.3   Should DCs Be Targeted via Defined Receptors?

This section reviews some recent literature that uses unmodified NPs ("Passive uptake") as well as NPs that are modified by conjugation of targeting agents including antibodies (Abs), antibody fragments (e.g., single-chain antibodies; scFv), and carbohydrates (e.g., C-type lectins ("Active targeting")) .

There are multiple advantages to targeting defined DC (or any APC) subsets. As outlined previously, any targeting strategy will limit off-target toxicities associated with systemic administration of adjuvants. This can be achieved merely by encapsulating an adjuvant, as factors including NP size, composition, and route of injection will determine which cells are encountered and whether the NPs are preferentially taken up. However, passive targeting may limit induction of optimal antigen-specific immune responses due to nonspecific uptake of NPs by multiple types of APCs. For example, a major requirement for many vaccines is induction of effective and durable CD8+ T cell responses. This requires uptake of antigen and adjuvant by specific DCs (in particular CD8α+DEC205+ DCs), resulting in antigen presentation to CD8+ T cells. Therefore, if NPs are taken up by multiple APCs, uptake by CD8α+ DCs may not be optimal, resulting in ineffective CD8+ T cell responses. Targeting specific APC subsets may also allow for reduced dosing, which should further reduce toxic responses.

A major drawback to using a targeted approach is that by restricting delivery to specific cell types, specific immune responses can be limited. For example, as mentioned in the "Heterogeneity of DCs" section, it could be envisaged that if antigenic components are encapsulated in DCIR2 antibody-coated NPs, CD8+ T cell responses may not be effectively elicited. Therefore, careful consideration is required regarding which immune response components must be elicited for targeting a specific disease, when designing an NP platform.

11.4   Types of Nanoparticles

A variety of different NP formulations have been utilized for both the passive (non-targeted) and active (targeted) delivery of cargoes to immune cells including those composed of synthetic polymers such as poly(lactic-*co*-glycolic acid) (PLGA) [32], natural polymers such as polyglutamic acid (a polypeptide) [33–35] or chitosan (an oligosaccharide) [36, 37] as well as lipids, typically in the form of liposomes) [38]. NP-based platforms have also been employed which make use of the ability of cationic proteins, typically protamine, to condense mRNA into nanoscale aggregates which serve to protect this fragile cargo from serum nucleases as well as to enhance immune cell uptake [39].

11.4.1   Passive Uptake

Particles in and of themselves are easily recognized, endocytosed, and processed by APCs, in particular monocytes, macrophages, and DCs. Particle uptake is affected by many parameters including route of delivery. To some degree, uptake can also be altered by altering the physicochemical properties (size, shape, charge, etc.) of the particles and has been best characterized using polymeric (polystyrene) particles. For example, in a recent study, Blank et al. used defined polystyrene particles ranging from 20 nm to 1 μm in diameter and observed that while alveolar macrophages took up the majority of inhaled particles independent of size, DCs in the trachea displayed increased uptake of 20 nm particles, and those in the lung parenchyma displayed increased uptake of both 20 and 50 nm particles. These smaller particles also displayed greater uptake by migrating DCs in the lung-draining lymph nodes [40]. In a similar study, when compared with 500 nm polystyrene particles, 50 nm particles displayed increased uptake by DCs in the lung [41]. Enhanced localization of 40–50 nm particles to DEC205+CD40+CD86+ DCs was observed when similar polystyrene particles were delivered intradermally [42]. As a whole, the data support the notion that DCs prefer smaller, "virus-sized" particles.

Charge and composition also can affect uptake by liposomes and other NPs. For example, the addition of a positive surface charge has been shown to enhance the uptake of polystyrene NPs [43] as well as liposomes [44] by DCs in vitro. However, the uptake enhancements due to charge are likely not DC specific; cationic liposomes have been used for years as delivery agents for nucleic acids to a variety of different cell types (reviewed in [45]). The net positive surface charge on these particles has been shown to enhance association with the cell surface leading to enhanced uptake and oftenenhanced endosomal escape. Unfortunately, most cationic liposomes display varied levels of cytotoxicity [46]. The use of these lipids has also been reported to result in association with negatively charged serum proteins leading to opsonization and increased clearance [47]. While this, in fact, can be used to one's advantage in vaccine development as it leads to enhanced uptake by cells in

the mononuclear phagocyte system (MPS) including DCs, it likely precludes targeting of specific cell subsets.

11.4.2 Active Targeting

Particle uptake can further be enhanced using targeting agents specific for surface
receptors on the target cell. In the case of DCs, a number of molecular targets including Fc-receptors, C-type lectins (DEC205, Clec9a, CD207, CD209), and even
cell-surface integrins (CD11c) have been identified which can enhance delivery to
these cells. Of particular interest are the C-type lectins which display DC specificity
and can also be used to target specific DC subsets, for example, to target $CD8\alpha+$
DCs [48]. Many of these receptors (DEC205 and Clec9a) have also been shown to
be effective at enhancing cross-presentation when ligands targeting these receptors
are fused to protein antigens. These cellular targets are thus not only key players in
mobilizing cytotoxic T lymphocyte (CTL) responses but may also prove unique for
the targeted delivery of nucleic acid cargoes such as mRNA and/or siRNA which
require access to the cytosol to function, a requisite for cross-priming.

To this end, a variety of C-type lectin ligands including carbohydrates (e.g., mannose for targeting CD206) receptor-specific antibodies or antibody fragments [22,
48, 49], and aptamers [50] selected to target these receptors have been described.
Perhaps the most widely used are small-molecule carbohydrates, such as mannose,
which has been previously used to generate mannose-functionalized liposomes. Numerous studies have now demonstrated that such liposomes preferentially target
and enhance uptake by macrophages and DCs both in vitro and in vivo [51]. In a recent example, Unger et al. utilized lipids bearing the CD209-binding glycans Lewis
B or Lewis X to enhance uptake by DCs, increasing the efficiency of presentation
100-fold when compared to nontargeted liposomes or soluble antigen alone [52]. In
vivo efficacy was subsequently demonstrated using transgenic mice engineered to
express human CD209 under the control of the murine CD11c promoter.

However, targeting with carbohydrates likely does not afford the same level of
specificity which can be achieved through the use of antibodies. In light of this,
Altin and coworkers were among the first to demonstrate the ability to target liposomes to DCs in vivo using scFv targeting both DEC205 as well as CD11c [49]. In
both cases, following systemic injection, antibody targeting was shown to enhance
liposome uptake by splenic DCs as determined by flow cytometry and microscopy.
Importantly, when loaded with ovalbumin (OVA) and coadministered with the adjuvant lipopolysaccharide (LPS), both the DEC205 and CD11c targeted liposomes
produced stronger OVA-specific CTL responses than a nontargeted control. In
many respects, these experiments mimic those initially reported by Steinman and
coworkers in which antibodies fused to antigens lead to enhanced uptake, antigen
presentation, and subsequent immune responses. However, here, the use of an NP
provides a means to co-encapsulate and target both antigen and adjuvant, a feat
which is considerably more difficult if one is seeking to combine a targeting agent,

and antigen and an adjuvant into a single particle. In addition, the NP can serve as a shield, restricting the interaction of the adjuvant and its cognate receptor to target cells, a problem encountered by Diebold and colleagues when they tried to make tripartite fusions of DEC205 antibodies, antigen, and CpG [53].

More recently, antibodies targeting DEC205 have been employed for the targeted delivery of siRNA to DCs. In this work, antibodies targeting DEC205 were grafted on to the surface of PEGylated cationic liposomes containing an anti-CD40 siRNA. DC-specific targeting was observed both in vitro and in vivo leading to knockdown in the expression of CD40 and, more importantly, inhibition of in vitro allotypic responses [22]. The ability to specifically target and deliver siRNA to DCs provides a potentially important advance in vaccine development as it provides a means to specifically control gene expression levels in these cells which may allow for better tuning of immunogenic responses.

11.5   Advancements in Liposome Formulations

In recent years, largely from efforts focused on targeting the liver, significant advances have been achieved in the formulation of liposomes for the encapsulation of nucleic acids (reviewed in [54]) In general, commonly employed, commercially available cationic lipids are inefficient at delivery and, in many cases, are cytotoxic and/or immunostimulatory [46]. "Next-generation" ionizable lipids including DLinDMA and DLinKC2–DMA are neutral under physiological conditions and thus avoid this complication [55, 56]. These become positively charged and fusogenic following endocytosis and acidification of endosomes and thus, not only offer significant enhancements in potency (doses as low as 10 µg/kg in animals [55, 57]) but also display minimal toxicity and innate immunogenicity [57]. Thus, while the initial thrust on using these materials for delivery has focused on delivering siRNA to the liver, more recent efforts have begun to explore targeting immune cells, in particular APCs, including DCs. Most relevant, a recent study by Basha et al. utilized four different ionizable lipids (DLinDMA, DLinDAP, DLinK-DMA, and DLinKC2-DMA) to assess siRNA delivery and knockdown gene expression in APCs following systemic injection [23]. The best-performing lipid, DLinKC2–DMA, demonstrated both improved knockdown efficiency, but more importantly, less toxicity than the other lipids tested. The more common and commercially available lipids were not utilized in this study, so it is not possible to make direct comparisons to these lipids. Interestingly, in contrast to the passive uptake studies described above using polystyrene particles which suggest smaller particle size is favored for DC uptake, here, the use of larger particle sizes (240 and 360 nm) not smaller particles (80 nm) resulted in better specificity for APCs. However, in this case, the improved specificity appears to be a consequence of the fact that the larger particles are taken up less efficiently by the liver. Thus, improved targeting may be realized, provided that liver uptake can be avoided or if more active mechanisms for targeting DCs are utilized.

siRNAs are not the only nucleic acids which have been delivered using these new lipid formulations. For example, Geall and coworkers recently used DLinD-MA liposomes to generate a vaccine against the respiratory syncytial virus (RSV) by encapsulating and delivering an mRNA encoding the RSV fusion glycoprotein, RSV-F. To enhance protein production, the mRNA was flanked by alpha viral non-structural proteins which allow the mRNA to replicate following access to the cell cytosol [58]. For these experiments, the vaccine was delivered into the muscle with the bulk of protein expressed by muscle cells. This approach could be used for the delivery of mRNA to DCs using the strategies previously described: systemically relying on passive uptake by splenic APCs [23], or a targeted method using antibodies [22, 49] [22] or carbohydrates.

Finally, there have been recent significant advances in using liposomes to encapsulate both protein antigens and adjuvants. Most notably, by utilizing a thiol-reactive maleimide functionalized lipid and the dithiol, dithiothreitol (DTT), Moon and coworkers developed a system for cross-linking lipid headgroups with the liposomes to create interbilayer-cross-linked multilamellar vesicles (ICMVs). The resulting liposomes display significantly enhanced stability compared to non-cross-linked liposomes, yet remained biodegradable by lipases. The cross-linked nature of the particles also resulted in significantly improved loading capacities, >25% for three different protein cargoes tested, and much slower release kinetics. Taking advantage of the ability to utilize particles to co-encapsulate cargos, a combination of OVA and the adjuvant, monophosphoryl lipid A, were found to elicit antibody titers ~1000-fold greater than that observed following treatment with liposome [59]. The approach has since been adapted to enhance humoral responses to a malaria antigen [60]. Here again, it is facile to envision that a more targeted approach that results in antigen and adjuvant delivery directly to DCs or other APCs may enhance and fine-tune immune responses.

11.6 Conclusion

We have outlined how NPs, in particular liposomes, can be utilized to deliver antigen and adjuvants to DCs as a method to modulate immune responses. Due to the ability of DCs to both activate immunity, when appropriate adjuvants are provided and diminish responses, in the absence of adjuvants, delivery of cargoes to DCs represents a powerful method for immune modulation. Targeting antigen to distinct DC subsets should allow for either the activation (when combined with adjuvant) or diminution (in absence of adjuvant) of defined antigen-specific T cell responses. In contrast with treatments that globally turn T cell responses off, for example, steroids, or on, for example, anti-cytotoxic T lymphocyte antigen 4 (CTLA4) or anti-PD1, this approach should provide a major advancement in safety and efficacy of multiple therapies.

With the continued improvement of NP formulation, resulting in synthesis of components that display low levels of toxicity, NPs encapsulating antigens, potentially together with anti-inflammatory compounds or siRNAs targeting co-stimulatory molecules, could provide useful tools for alleviating autoimmune diseases through targeting specific T cells. Similarly, targeting cancer antigens plus adjuvants to DCs, such as DEC205+ DCs, may enhance antitumor responses, for example, clinical trials #NCT02129075.

Overall, it is becoming evident that antigen/adjuvant targeting to APCs, in particular DCs, using nontoxic NPs will provide safe and effective therapies for multiple diseases.

References

1. Morse MA, Chapman R, Powderly J, et al. Phase I study utilizing a novel antigen-presenting cell-targeted vaccine with Toll-like receptor stimulation to induce immunity to self-antigens in cancer patients. Clin Cancer Res. 2011;17(14):4844–53.
2. Villadangos JA, Young L. Antigen-presentation properties of plasmacytoid dendritic cells. Immunity. 2008;29(3):352–61.
3. Miller JC, Brown BD, Shay T, et al. Deciphering the transcriptional network of the dendritic cell lineage. Nat Immunol. 2012;13(9):888–99.
4. Steinman RM, Idoyaga J. Features of the dendritic cell lineage. Immunol Rev. 2010;234(1): 5–17.
5. Hashimoto D, Miller J, Merad M. Dendritic cell and macrophage heterogeneity in vivo. Immunity. 2011;35(3):323–35.
6. Bonifaz L, Bonnyay D, Mahnke K, Rivera M, Nussenzweig MC, Steinman RM. Efficient targeting of protein antigen to the dendritic cell receptor DEC-205 in the steady state leads to antigen presentation on major histocompatibility complex class I products and peripheral CD8+ T cell tolerance. J Exp Med. 2002;196(12):1627–38.
7. Hawiger D, Inaba K, Dorsett Y, et al. Dendritic cells induce peripheral T cell unresponsiveness under steady state conditions in vivo. J Exp Med. 2001;194(6):769–79.
8. Dudziak D, Kamphorst AO, Heidkamp GF, et al. Differential antigen processing by dendritic cell subsets in vivo. Science. 2007;315(5808):107–11.
9. Savina A, Peres A, Cebrian I, et al. The small GTPase Rac2 controls phagosomal alkalinization and antigen crosspresentation selectively in CD8(+) dendritic cells. Immunity. 2009;30(4):544–55.
10. Segura E, Amigorena S. Cross-presentation by human dendritic cell subsets. Immunol Lett. 2014;158(1–2):73–8.
11. Segura E, Durand M, Amigorena S. Similar antigen cross-presentation capacity and phagocytic functions in all freshly isolated human lymphoid organ-resident dendritic cells. J Exp Med. 2013;210(5):1035–47.
12. Chatterjee B, Smed-Sorensen A, Cohn L, et al. Internalization and endosomal degradation of receptor-bound antigens regulate the efficiency of cross presentation by human dendritic cells. Blood. 2012;120(10):2011–20.
13. Cohn L, Chatterjee B, Esselborn F, et al. Antigen delivery to early endosomes eliminates the superiority of human blood BDCA3+ dendritic cells at cross presentation. J Exp Med. 2013;210(5):1049–63.
14. Mukhopadhaya A, Hanafusa T, Jarchum I, et al. Selective delivery of beta cell antigen to dendritic cells in vivo leads to deletion and tolerance of autoreactive CD8+ T cells in NOD mice. Proc Natl Acad Sci U S A. 2008;105(17):6374–9.

15. Ring S, Maas M, Nettelbeck DM, Enk AH, Mahnke K. Targeting of autoantigens to DEC205(+) dendritic cells in vivo suppresses experimental allergic encephalomyelitis in mice. J Immunol. 2013;191(6):2938–47.

16. Yamazaki S, Iyoda T, Tarbell K, et al. Direct expansion of functional CD25+ CD4+ regulatory T cells by antigen-processing dendritic cells. J Exp Med. 2003;198(2):235–47.

17. Yamazaki S, Dudziak D, Heidkamp GF, et al. CD8+ CD205+ splenic dendritic cells are specialized to induce Foxp3+ regulatory T cells. J Immunol. 2008;181(10):6923–33.

18. Kortylewski M, Swiderski P, Herrmann A, et al. In vivo delivery of siRNA to immune cells by conjugation to a TLR9 agonist enhances antitumor immune responses. Nat Biotechnol. 2009;27(10):925–32.

19. Zhang Q, Hossain DM, Nechaev S, et al. TLR9-mediated siRNA delivery for targeting of normal and malignant human hematopoietic cells in vivo. Blood. 2013;121(8):1304–15.

20. Caminschi I, Meuter S, Heath WR. DEC-205 is a cell surface receptor for CpG oligonucleotides. Oncoimmunology. 2013;2(3):e23128.

21. Lahoud MH, Ahmet F, Zhang JG, et al. DEC-205 is a cell surface receptor for CpG oligonucleotides. Proc Natl Acad Sci U S A. 2012;109(40):16270–5.

22. Zheng X, Vladau C, Zhang X, et al. A novel in vivo siRNA delivery system specifically targeting dendritic cells and silencing CD40 genes for immunomodulation. Blood. 2009;113(12):2646–54.

23. Basha G, Novobrantseva TI, Rosin N, et al. Influence of cationic lipid composition on gene silencing properties of lipid nanoparticle formulations of siRNA in antigen-presenting cells. Mol Ther. 2011;19(12):2186–200.

24. Kasturi SP, Skountzou I, Albrecht RA, et al. Programming the magnitude and persistence of antibody responses with innate immunity. Nature. 2011;470(7335):543–7.

25. Speiser DE, Schwarz K, Baumgaertner P, et al. Memory and effector CD8 T-cell responses after nanoparticle vaccination of melanoma patients. J Immunother. 2010;33(8):848–58.

26. Ali OA, Emerich D, Dranoff G, Mooney DJ. In situ regulation of DC subsets and T cells mediates tumor regression in mice. Sci Transl Med. 2009;1(8):8ra19.

27. Ali OA, Verbeke C, Johnson C, et al. Identification of immune factors regulating antitumor immunity using polymeric vaccines with multiple adjuvants. Cancer Res. 2014;74(6):1670–81.

28. Blander JM, Medzhitov R. Toll-dependent selection of microbial antigens for presentation by dendritic cells. Nature. 2006;440(7085):808–12.

29. Kratky W, Reis e Sousa C, Oxenius A, Sporri R. Direct activation of antigen-presenting cells is required for CD8+ T-cell priming and tumor vaccination. Proc Natl Acad Sci U S A. 2011;108(42):17414–9.

30. Bourquin C, Anz D, Zwiorek K, et al. Targeting CpG oligonucleotides to the lymph node by nanoparticles elicits efficient antitumoral immunity. J Immunol. 2008;181(5):2990–8.

31. Ronaghy A, Prakken BJ, Takabayashi K, et al. Immunostimulatory DNA sequences influence the course of adjuvant arthritis. J Immunol. 2002;168(1):51–6.

32. Hamdy S, Haddadi A, Hung RW, Lavasanifar A. Targeting dendritic cells with nano-particulate PLGA cancer vaccine formulations. Adv Drug Deliv Rev. 2011;63(10–11):943–55.

33. Uto T, Wang X, Sato K, et al. Targeting of antigen to dendritic cells with poly(gamma-glutamic acid) nanoparticles induces antigen-specific humoral and cellular immunity. J Immunol. 2007;178(5):2979–86.

34. Akagi T, Wang X, Uto T, Baba M, Akashi M. Protein direct delivery to dendritic cells using nanoparticles based on amphiphilic poly(amino acid) derivatives. Biomaterials. 2007;28(23):3427–36.

35. Uto T, Akagi T, Yoshinaga K, Toyama M, Akashi M, Baba M. The induction of innate and adaptive immunity by biodegradable poly(gamma-glutamic acid) nanoparticles via a TLR4 and MyD88 signaling pathway. Biomaterials. 2011;32(22):5206–12.

36. Prego C, Paolicelli P, Diaz B, et al. Chitosan-based nanoparticles for improving immunization against hepatitis B infection. Vaccine. 2010;28(14):2607–14.

37. Arca HC, Gunbeyaz M, Senel S. Chitosan-based systems for the delivery of vaccine antigens. Expert Rev Vaccines. 2009;8(7):937–53.
38. Schwendener RA. Liposomes as vaccine delivery systems: a review of the recent advances. Ther Adv Vaccines. 2014;2(6):159–82.
39. Kallen KJ, Heidenreich R, Schnee M, et al. A novel, disruptive vaccination technology: self-adjuvanted RNActive((R)) vaccines. Hum Vacc Immunother. 2013;9(10):2263–76.
40. Blank F, Stumbles PA, Seydoux E, et al. Size-dependent uptake of particles by pulmonary antigen-presenting cell populations and trafficking to regional lymph nodes. Am J Resp Cell Mol Biol. 2013;49(1):67–77.
41. Hardy CL, Lemasurier JS, Mohamud R, et al. Differential uptake of nanoparticles and microparticles by pulmonary APC subsets induces discrete immunological imprints. J Immunol. 2013;191(10):5278–90.
42. Fifis T, Gamvrellis A, Crimeen-Irwin B, et al. Size-dependent immunogenicity: therapeutic and protective properties of nano-vaccines against tumors. J Immunol. 2004;173(5): 3148–54.
43. Foged C, Brodin B, Frokjaer S, Sundblad A. Particle size and surface charge affect particle uptake by human dendritic cells in an in vitro model. Int J Pharm. 2005;298(2):315–22.
44. Foged C, Arigita C, Sundblad A, Jiskoot W, Storm G, Frokjaer S. Interaction of dendritic cells with antigen-containing liposomes: effect of bilayer composition. Vaccine. 2004; 22(15–16):1903–13.
45. Zuhorn IS, Engberts JBFN, Hoekstra D. Gene delivery by cationic lipid vectors: overcoming cellular barriers. Eur Biophys J Biophy. 2007;36(4–5):349–62.
46. Lv H, Zhang S, Wang B, Cui S, Yan J. Toxicity of cationic lipids and cationic polymers in gene delivery. J Control Release. 2006;114(1):100–9.
47. Ishida T, Harashima H, Kiwada H. Liposome clearance. Biosci Rep. 2002;22(2):197–224.
48. Idoyaga J, Lubkin A, Fiorese C, et al. Comparable T helper 1 (Th1) and CD8 T-cell immunity by targeting HIV gag p24 to CD8 dendritic cells within antibodies to Langerin, DEC205, and Clec9A. Proc Natl Acad Sci U S A. 2011;108(6):2384–9.
49. van Broekhoven CL, Parish CR, Demangel C, Britton WJ, Altin JG. Targeting dendritic cells with antigen-containing liposomes: a highly effective procedure for induction of antitumor immunity and for tumor immunotherapy. Cancer Res. 2004;64(12):4357–65.
50. Wengerter BC, Katakowski JA, Rosenberg JM, et al. Aptamer-targeted antigen delivery. Mol Ther. 2014;22(7):1375–87.
51. Kelly C, Jefferies C, Cryan SA. Targeted liposomal drug delivery to monocytes and macrophages. J Drug Deliv. 2011;2011:727241.
52. Unger WW, van Beelen AJ, Bruijns SC, et al. Glycan-modified liposomes boost CD4+ and CD8+ T-cell responses by targeting DC-SIGN on dendritic cells. J Control Release. 2012;160(1):88–95.
53. Kreutz M, Giquel B, Hu Q, et al. Antibody-antigen-adjuvant conjugates enable co-delivery of antigen and adjuvant to dendritic cells in cis but only have partial targeting specificity. PLoS One. 2012;7(7):e40208.
54. Tam YY, Chen S, Cullis PR. Advances in lipid nanoparticles for siRNA delivery. Pharmaceutics. 2013;5(3):498–507.
55. Jayaraman M, Ansell SM, Mui BL, et al. Maximizing the potency of siRNA lipid nanoparticles for hepatic gene silencing in vivo. Angew Chem Int Ed Engl. 2012;51(34):8529–33.
56. Semple SC, Akinc A, Chen J, et al. Rational design of cationic lipids for siRNA delivery. Nat Biotechnol. 2010;28(2):172–6.
57. Novobrantseva TI, Borodovsky A, Wong J, et al. Systemic RNAi-mediated gene silencing in nonhuman primate and rodent myeloid cells. Mol Ther Nucleic Acids. 2012;1:e4.
58. Geall AJ, Verma A, Otten GR, et al. Nonviral delivery of self-amplifying RNA vaccines. Proc Natl Acad Sci U S A. 2012;109(36):14604–9.

59. Moon JJ, Suh H, Bershteyn A, et al. Interbilayer-crosslinked multilamellar vesicles as synthetic vaccines for potent humoral and cellular immune responses. Nat Mater. 2011;10(3): 243–51.
60. Moon JJ, Suh H, Li AV, Ockenhouse CF, Yadava A, Irvine DJ. Enhancing humoral responses to a malaria antigen with nanoparticle vaccines that expand Tfh cells and promote germinal center induction. Proc Natl Acad Sci U S A. 2012;109(4):1080–5.

Dr. Matthew Levy is associate professor of Biochemistry. He trained in the laboratories of Stanley Miller (UCSD) and Andrew Ellington (UT, Austin) prior to joining the faculty at Albert Einstein College of Medicine. He is expert in developing, synthesizing and engineering aptamers, ribozymes and deoxyribozymes and engineering proteins. His current lab is focused on developing new methods as well as new aptamers which target cells for the development of novel vaccines and cancer therapies.

Dr. Deborah Palliser is associate professor in the Department of Microbiology and Immunology. She trained in the laboratories of Dr Herman Eisen (MIT) and Dr Judy Lieberman (Harvard Medical School) prior to joining the faculty at Albert Einstein College of Medicine. She is an expert in the fields of dendritic cell biology and T cell biology. She has extensive expertise in the targeted delivery of various cargoes to cells of the immune system, including antigens (e.g. heat shock fusion proteins) and siRNAs.

Chapter 12
Integrated Biomaterial Composites for Accelerated Wound Healing

Viness Pillay, Pradeep Kumar and Yahya E. Choonara

12.1 Introduction

Wound healing is a complex process requiring translational therapeutic approaches capable of accelerating the overlapping inflammatory, migratory, proliferative, and remodeling phases, thereby allowing rapid tissue filling and near-to-normal skin barrier restoration. Advanced biomaterial-based strategies in the form of wound dressings or fillings form an inherent part of such translational approaches wherein these specialized materials (1) provide and maintain balanced hydration levels through the absorption and control of the wound exudates, (2) facilitate the exchange of gases across the wound site, (3) prevent secondary infections, (4) control and localize the release of incorporated bioactives, and (5) demonstrate extracellular matrix (ECM) equivalent physicomechanical characteristics in terms of elasticity, biocompatibility, and biodegradability. A single biomaterial encompassing all these biological, physicochemical, and physicomechanical characteristics is yet to be discovered (natural) or designed (synthetic or semisynthetic). However, combining two or more biomaterials with contrasting and/or complimenting properties may offer additive benefits with a possibility of acquiring novel features and unique advantages over individual components. This chapter provides an insight into the most recent advances in integrated biomaterial composite systems being tried and tested

V. Pillay (✉) · P. Kumar · Y. E. Choonara
Wits Advanced Drug Delivery Platform, Department of Pharmacy and Pharmacology,
University of the Witwatersrand, Johannesburg, 7 York Road, Parktown 2193, South Africa
e-mail: viness.pillay@wits.ac.za

P. Kumar
e-mail: pradeep.kumar@wits.ac.za

Y. E. Choonara
e-mail: yahya.choonara@wits.ac.za

© Springer International Publishing Switzerland 2015
L. Santambrogio (ed.), *Biomaterials in Regenerative Medicine and the Immune System*,
DOI 10.1007/978-3-319-18045-8_12

for accelerated and efficient wound healing applications. In addition, a brief account of nano-reinforced and nano-manipulated biomaterial archetypes is provided. Furthermore, the inherent bioactive incorporation into and their subsequent release from the devices is discussed with an emphasis on the effect of chemical makeup of the biomaterials on the biomedical performance of the composites.

12.2   Multi-biomaterial Archetypes

The most simplistic approach of introducing multifunctionality into a wound dressing is to amalgamate two or more biomaterials employing simple blending leading to the formation of, but not limited to, interpenetrating polymer networks (IPNs), polyelectrolyte complexes, stimuli responsive systems, multilayered architectures, and nanoconstructs. These specialized systems, alone as well as in combination with wound healing bioactives and even cells, have shown enhanced and accelerated re-epithilialization, connective tissue formation, morphogenesis, and even blood vessels formation. Although these simplistic archetypes provide immense wound healing benefits, the biomaterial selection and design is a crucial aspect of their performance. In a recent study, Jhong et al. (2014) developed a pseudo-zwitterionic biomaterial dressing consisting of two mixed charged polymers, namely 3-sulfo-propyl methacrylate (SA; negatively charged component) and [2-(methacryloyloxy)ethyl]trimethylammonium (TMA; positively charged component) deposited on polytetrafluoroethylene (ePTFE) membranes. The balanced charged approach, in comparison to single charge bias approach, provided better hydration properties, decreased the adsorption and adhesion of fibrinogen and cells such as erythrocytes, leucocytes, thrombocytes, and fibroblasts, showed enhanced hemocompatibility and blood clotting, and prevented bacteria colonization to a greater extent (Fig. 12.1) [1].

In an interesting study, Ngadaonye et al. (2014) developed a chitosan-poly (N, N-diethylacrylamide) (PDEAAm)-based thermoresponsive wound dressing which demonstrated reduced swelling at lower temperatures and proved to be an important strategy for traumatic and easy dressing removal after the bioactive and/or antibacterial agents have been eluted. Additionally, the specialized wound dressing provided efficient water vapor transmission rate—breathing membrane—with no bacterial permeability, thereby allowing better control of exudate absorption and evaporation [2].

Chitosan (CS), by far, is the most researched biomaterial for wound dressing applications owing to its bioactive properties and cationic nature. Hsu and Hsieh (2014) demonstrated that polymer membranes consisting of CS and hyaluronan provided a perfect platform for the self-assembly of adult adipose-derived stem cells as evident from the enhanced expression of cytokine (fibroblast growth factor, FGF1; vascular endothelial growth factor, VEGF; and C-C chemokine ligand 2, CCL2) and migration-associated (C-X-C chemokine receptor type 4, CXCR4; matrix metalloproteinase-1, MMP1) genes with the localization of spheroids into

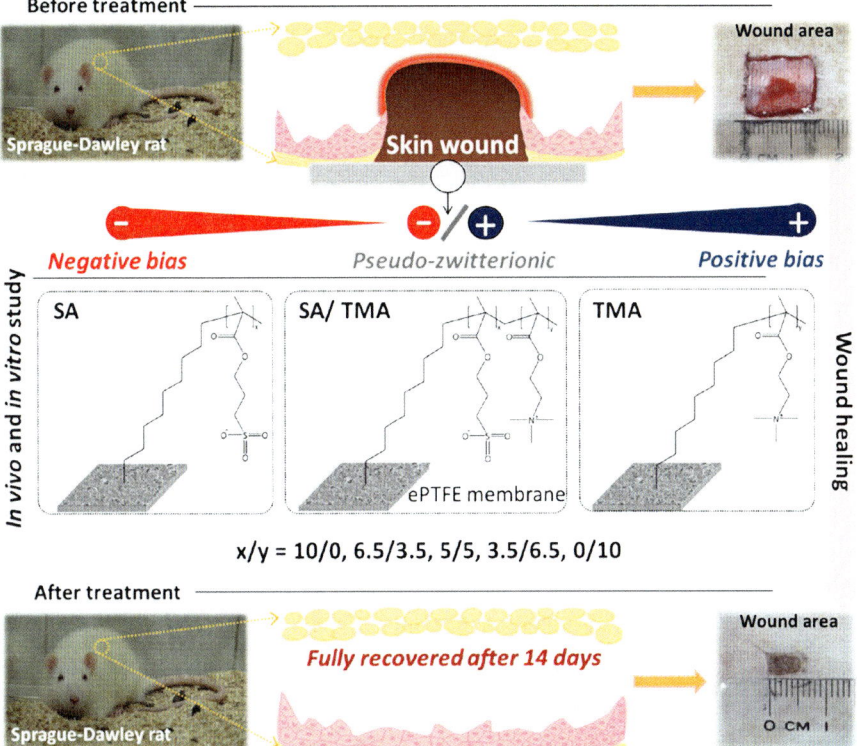

Fig. 12.1 Schematic illustration of the use of SA and TMA monomers in controlled ratios for surface modification of ePTFE membranes, in order to yield ePTFE-g-pSA/pTMA intended to be used as wound dressings [1]. *SA* 3-sulfopropyl methacrylate, *TMA* [2-(methacryloyloxy)ethyl] trimethylammonium

regenerated wound microvessels proving the role of paracrine effects in the enhanced angiogenesis and faster wound closure [3]. Exploring the polyelectrolyte forming properties of CS-alginate (Alg) composites, Han et al. designed layered bi- and biopolymeric composites loaded with ciprofloxacin hydrochloride to provide antimicrobial therapy to exposed wounds. Preliminary results confirmed sustained release of the antimicrobial for prolonged period with effective inhibition of microbial growth for 7 days. The wound dressings further provided the much needed "moist microenvironment" as compared to the control gauge (Fig. 12.2) [4].

In line with the above study, Yang et al. designed polycation/polyanion layer-by-layer assemblies (LBL) of CS/Alg, CS/poly(γ-glutamic acid) (CS/PGA), and CS/poly(aspartic acid) (CS/PAsp) over a styrene–butadiene–styrene (SBS) block copolymer membrane. The LBL-modified SBS membranes displayed semipermeable properties allowing the passage of water vapors with no bacterial transport. The extent of wettability and fibronectin adsorption were directly and inversely proportion, respectively, to the availability of COO− and −O=C−N− functional groups on the CS/polyanion surface. The researchers further proposed that addition

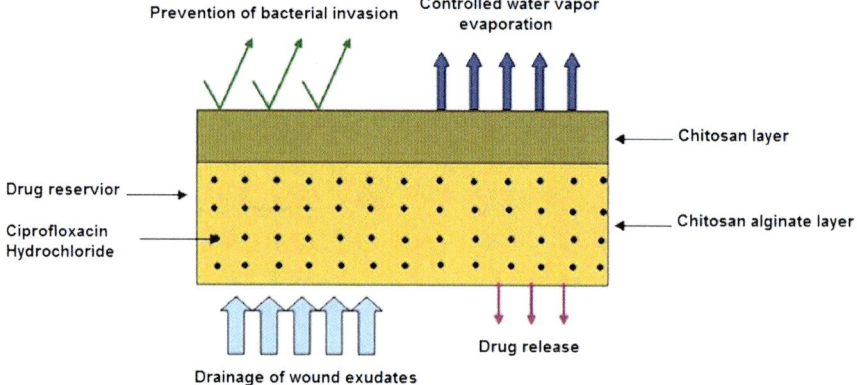

Fig. 12.2 Design of the Alg/Chs bilayer composite membrane [4]

of an ECM (gelation-hyaluronan/chondronitin-6-sulfate) to the above system to give an asymmetrically structured bilayer dressing allowing normalized dermal/epidermal recovery [5]. Harkins et al. investigated CS–cellulose (CEL) composites (CEL+CS) for their capability to provide antimicrobial along with antiinflammatory activities. The wound healing property of CEL+CS composites was further attributed to their superior whole blood absorption activity as well as reduction and inhibition of microphages-mediated tumor necrosis factor-α (TNF-a) and interleukin-6 (IL-6) levels and production, respectively [6]. The functional properties of electrospun poly(ε-caprolactone) (PCL) non-woven mats were further enhanced by coating the nanofibers with CS and encapsulating tea tree oil (TTO) as the active ingredient. The addition of CS imparted bio- and cyto-compatibility and platelet aggregation properties to the wound dressing while the TTO provided the much needed antiinfective properties resulting into accelerated wound closure through collagen generation and tissue regeneration and integration [7].

In addition to CS, several researchers have and are employing the excellent biofunctionality, biocompatibility, and manufacturability of silk fibroin (SF) to develop hybrid biopolymer composites for wound healing applications. The following are the most recent studies employing SF as one of the biopolymeric components in the design and performance of wound dressing:

1. Aramwit et al. developed non-crosslinked salt-leached sericin/poly(vinyl alcohol) (PVA)/glycerin scaffolds and compared them with their freeze-dried counterparts. The economical and time-efficient salt-leaching technique provided better poro-mechanical properties with lowered bioadhesion and faster biodegradation (and hence faster sericin release) leading to improved L929 mouse fibroblast cells survival [8].

2. Shahverdi et al. tested SF and poly(lactide-*co*-glycolic acid) (PLGA) scaffolds (1:2) for chronic wound dressing applications and the in vitro and in vivo studies concluded that adding SF to PLGA improved the surface wettability of the

scaffolds resulting into enhanced attachment, proliferation, migration, and viability of L929 mouse fibroblast cells [9].

3. Nayak and Kundu prepared dually crosslinked (glutaraldehyde and aluminium chloride), transforming growth factor β1 loaded and freeze-dried 3D scaffolds composed of sericin (silk protein from non-mulberry silkworm) and carboxymethyl cellulose (1:1) with tuneable strength, swelling, and stability. The scaffolds demonstrated superior cell carrying capability and minimal immunogenicity making them a perfect candidate for "cellular wound dressing material" [10].

4. Sukumar et al. developed SF/dextrin-based scaffolds loaded with bioactive cactus extract and recombinant human epidermal growth factor (rhEGF). The researchers concluded that the superior compressive strength and modulus of the scaffold was directly proportional to the dextrin content, the antibacterial activity of the scaffold was attributed to dextrin and the rhEGF, and the antidiabetic potential of cactus extract could provide interesting interventional advantages in case of diabetic wound dressing applications [11].

Anionic polymers and biomaterials such as hyaluronic acid, Algs, pectins, and certain cellulose derivatives provide bio-chemico-physico-compatible wound dressings with the dermal tissues. Li et al. explored the highly porous character of bacterial nanocellulose (BC) to obtain hyaluronan (HA)-reinforced cellulose nanocomposites (BCHA-NC) employing a solution impregnation method. The BCHA-NC provided two unique advantages in terms of wound dressing applications: (1) enhancement of hydrophilicity by incorporation of HA which could be beneficial for protein adsorption and cellular adhesion and (2) decreased surface roughness as compared to BC-alone membranes leading to better cellular proliferation. In addition to abovementioned adhesion–proliferation balance, the lowered water loss and increased elongation at break properties makes BCHA-NC a perfect candidate for wound dressing applications [12]. In a multipolymeric composite study, Zhou et al. prepared lyophilized Gelatin-Alg-HA (GAH) hydrogel scaffolds crosslinked via carbodiimide chemistry. The scaffolds demonstrated unique morphological characteristics with stacked sheets forming irregular interconnected pores (50–500 μm) uniformly distributed across the scaffold cross-section. Interestingly, the poro-morphology of the GAH scaffolds was independent of the ration of the components. However, an increase in Alg concentration significantly enhanced the wettability and water transmission rate of the scaffolds enhancing the tuneability of the polymeric architecture for wound dressing applications [13].

In a mechanistic study, da Cunha et al. investigated the influence of stiffness on the biology of dermal fibroblasts and on the wound healing genetic programs. The substrate for fibroblast consisted of a 3D IPNs of Alg and collagen-I in situ crosslinked with divalent Ca^{++} ions. The crosslinking levels were varied to produce IPNs with storage modulus ranging from 50 to 1200 Pa. Interestingly, the fibroblasts acquired elongated, spindle-shaped morphology at lower stiffness values while at higher stiffness values, spherical-shaped fibroblasts were observed. Additionally, higher stiffness matrices down regulated the expression of chemokine ligand 2, transgelin, colony stimulating factor 2, connective tissue growth factor and MMP-9, whereas the expression of inflammatory cytokines (IL-10, IL-1b, and COX2), genes expressing cell adhesion and ECM molecules (integrin a4, MMP-1, and vitronectin), and hepatocyte growth factor was upregulated. The study proved that scaffold stiffness alone can effectively regulate the wound healing process independent of the scaffold porosity, ratio of polymers employed or the ligand density [14]. In an innovative approach, De Cicco et al. employed supercritical-assisted atomization (SAA) technology to produce Alg (high-mannuronic content) and pectin (amidated) blend microparticles loaded with an antibiotic (gentamicin sulfate). This specialized free-flowing dry powder proved to be a site-filling wound dressing material capable of limiting bacterial proliferation. The high mannuronic acid content of the Alg contributed to cytokine production while the amidation of pectin enhanced the in situ gelling of the microparticle network hydrogel [15].

3D printing is the future of tissue engineering and regenerative medicine to obtain customized polymeric architectures based on computational interfaces. However, only few studies have been conducted to generate robust wound dressings. Rees et al. produced advanced tailor-made wound dressing material via 3D-printing carboxymethylated-periodated nanocellulose nanofibrils (width:length::20 nm:200 nm), as the bioink, over (2,2,6,6-tetramethylpiperidin-1-yl)oxidanyl-oxidized nanocellulose film sustrates. The nanofibrils were fixed with the film substrates through Ca^{++}-mediated ionic crosslinking followed by freeze-drying, thereby forming a porous patterned 3D architecture capable of acting as a bioresponsive structure with controlled antimicrobial/bioactive release properties—an added advantage for wound dressing applications [16].

12.3   Nanofibrous Composites as Wound Dressings

Chemically burned corneas present a challenging and complicated wound scenario needing immediate attention—human amniotic membrane (HAM) being the most widely employed therapeutic intervention. Ye et al. designed specialized collagen-HA nanofibrous membranes surface modified with CS. The optically transparent membrane showed HAM-mimicking mechanical properties with selective adhesion of corneal and conjunctival epithelial cells (Fig. 12.3). When tested in vivo in alkali-damaged cornea model (rats), the nanofibrous membranes demonstrated significantly enhanced re-epithelialization of the corneal tissue as early as 1 week [17].

Electrospinning provides a simplistic approach for the efficient and reproducible fabrication of wound dressings at a nanoscale with almost 100% incorporation efficiency for the bioactives. Additionally, even a small modification in the polymer composition can be evidently observed in the final formulations. In a recent study, Fu et al. incorporated curcumin into poly(ε-caprolactone)-poly(ethylene glycol)-poly(ε-caprolactone) (PCEC) nanofibrous mats and observed that addition of curcumin increased the conglutination between the fibrous architecture along with the formation of toruloid structures which was further attributed to hydrogen-

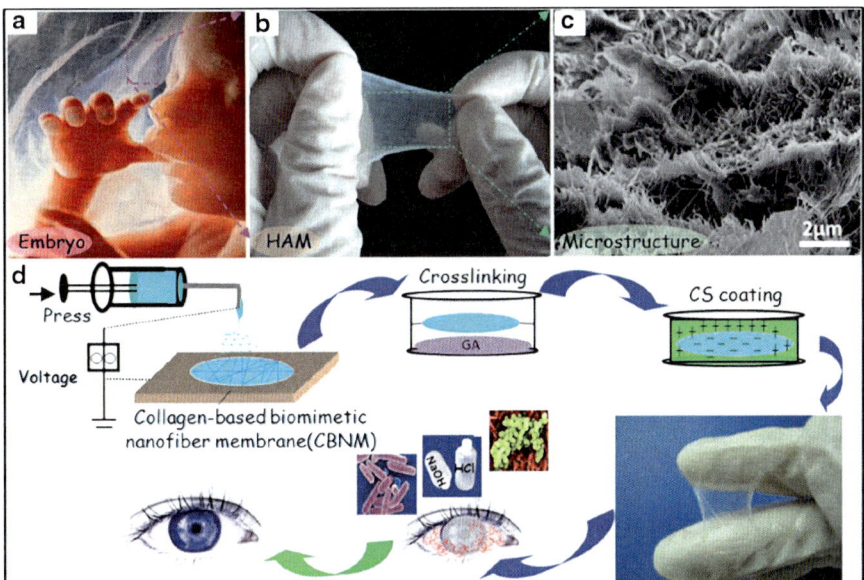

Fig. 12.3 Schematic procedure for fabricating CBNMs that mimic the microstructure and architecture of devitalized tissue membranes (e.g., HAM) via electrospinning and chitosan (CS) modification for the treatment of damaged corneas. **a–c** Macroscopic morphology and microstructure of the base side of HAM, composed of nanoscale collagen fibers randomly arranged. **d** Schematic procedure for CBNM fabrication [17]. *HAM* human amniotic membrane, *CBNMs* carbon nanomaterials

bonding interactions between curcumin and PCEC. The incorporation of curcumin and hence the unique morphology of the fibers accelerated the re-epithelialization and blood vessel formation in a full-thickness dermal defect model [18].

The choice of adhesion- and nonadhesion-based wound dressings is based on whether the dressing is intended to be resorbable (scaffolds for full tissue regeneration) or non-resorbable (removable wound dressings), respectively. Vatankhah et al. fabricated nanofibrous cellulose acetate/gelatin membranes mimicking the ECM as skin substitutes. The researchers confirmed that varying the ratio of cellulose acetate:gelatin can assist in the design of adherent (25:75) and non-adherent dressings (75:25) which in turn was reliant on the surface hydrophilicity and hydrophobicity of these membranes, respectively [19].

Given the severity of certain wounds (full-thickness wounds), the release of bioctives and antimicrobial agents from the wound dressings need to be extended for prolonged period, occasionally to several weeks. However, such polymeric architectures have financial implications because of higher polymer content and specialized fabrication requirements. Zhao et al. proposed the use of low-cost bioactive biomaterials—a CS/sericin (2.5:1) composite—for the preparation of electrospun nanofibrous membranes. The nanofibers displayed antimicrobial activities against gram-negative *(Escherichia coli)* and gram-positive *(Bacillus subtilis)* bacteria at a minimal nanofiber concentration of 0.4 mg/ml [20]. In line with the above study, Çalamak et al. combined nano- and bionanotextile technologies to fabricate electrospun antibacterial polyethylenimine containing SF scaffolds. The bionanotextile so developed demonstrated strong antibacterial activities against *Staphylococcus aureus* (gram positive) and *Pseudomonas aeruginosa* (gram negative) and further prevented bacterial adhesion [21].

Electrospinning of high-viscosity polymers such as CS poses several challenges in terms of electrospinnablity as well as the finally obtained macro-range size of the fibers. To obtained fine-sized nanofibers, additives such as polyethylene oxide (PEO) can be incorporated into the CS solution. However, this addition deteriorates the mechanical strength of the CS membrane. Naseri et al. incorporated chitin crystals within CS-PEO membranes to produce reinforced nanofibrous membranes with a tensile strength of 64.9 MPa and storage modulus of 10.2 GPa along with high surface area (35 $m^2\ g^{-1}$) and water vapor transmission rate (1290 and 1548 g $m^{-2}day^{-1}$) in corroboration with that of an injured wound [22]. The wound healing potential of chitin was further confirmed by Huang et al. wherein they successfully fabricated pure chitin nanofibers (for the first time) employing a NaOH-urea aqueous system. Both woven and non-woven fibers were obtained, respectively, through freezing and hot-pressing (after solvent exchange) techniques. According to the researchers, the hot-pressing technique produced self-aggregation leading to packing and bonding of the chitin fibers. When tested in a full-thickness would model, the non-woven membranes showed "excellent ability to accelerate healing, owing to the retainment of the intrinsic α-chitin structure" [23].

Due to the inability of electrospinning technique to form nanofibers with nonconductive polymers and to avoid the use of toxic solvents, Xu et al. employed Forcespinning® technology to fabricate binary (CS/PVA) and ternary (tannic acid/CS/PVA; TA/CS/PVA) composite nanofibrous membranes with CS dissolved in 4%w/w/ citric acid solution. The membranes so formed were further cured at 140 °C to formed water-stable wound dressings. The ternary composite demonstrated enhanced antibacterial activities against *E. coli* which was attributed to the synergistic effect of CS (interaction of protonated amino groups of CS with negatively charged surfaces of bacteria) and tannic acid (microbial enzymes' inhibition) [24]. The researchers further developed crosslinked, water-stable TA/CS/pullulan (TA/CS/PL) ternary membranes (employing Forcespinning® technology) with a rapid water-uptake ability. The ECM-mimicking structure of the membranes allowed extensive cell proliferation and attached forming interlayered cellular membranes and effectively provided antibacterial activity [25].

The wound healing activity and efficiency of polymeric architectures can be further be enhanced by the incorporation chemical entities responsive to external stimuli, thereby enhancing the release of bioactives such as epidermal growth factors (EGF). Jin et al. incorporated a photosensitive polymer (poly(3-hexylthiophene) (P3HT)) into EGF-loaded, core-shelled, gelatin/poly(L-lactic acid)-*co*-poly-(e-caprolactone) nanofibers (Gel/PLLCL/P3GF(cs)) prepared by coaxial spinning. In an in vitro wound healing model, the Gel/PLLCL/P3GF(cs) fibers, after stimulation with light, initiated the differentiation of adipose-derived stem cells (ASCs) into keratinocyte morphology which confirmed their potential as possible photosensitive skin grafts capable of reestablishing the dermal architecture [26].

Polyurethane (PU) is commonly employed for wound dressing applications due to its oxygen permeability and barrier properties. However, PU is too soft and hydrophobic to provide clinical handling and exudate absorption properties to the dressing. Unnithan et al. blended PU with two electrospinnable polymers—cellulose acetate and zein—to impart hydrophilicity, cytocompatibility, and blood clotting ability to the PU nanofibrous membranes. Physicomechanically, PU-cellulose acetate (CA)-zein membranes demonstrated balanced elasticity and hardness with increased hydrophilicity as compared to pristine PU membranes. The cationic functionality of zein interacted with the negatively charged blood cells leading to accelerated blood clotting along with platelet adhesion and activation making them a perfect candidate for diabetic wound healing [27].

SF is a versatile polymer which has extensively been employed in nanofibrous wound dressing applications due to its easy fabricability via electrospinning and effective modifiability with both hydrophobic and hydrophilic polymers. Chutipakdeevong et al. hybridized SF with polycaprolactone and developed a SF-coated PCL-based biomimetic electrospun nanofibrous surface functionalized with fibronectin. The SF-PCL fibrous mats provided a perfect platform with balanced mechanical, chemical, biocompatible, and functional properties as concluded from the dermal fibroblast support and proliferation analysis [28]. Kim et al. conjugated the amine functional groups of prefabricated SF nanofibers with the dialdehydic dextran (obtained via sodium-periodate-mediated oxidation) to obtain enhanced water binding and biocompatibility. The moisture holding property of the surface modified SF membranes demonstrated accelerated collagen regeneration and elastic tissue recovery in an in vivo full-thickness wound model in rats [29].

In a nonconventional approach, Ong et al. attempted electrospinning of a polymer bipolymeric nanofibrous composite (PCL/gelation) onto a monopolymeric film (polyurathene; Tegaderm (TG)) and tested the efficacy of fibroblast-seeded TG-NF(+) and -unseeded TG-NF( ) scaffolds in a partial thickness wound model in pigs. The TG-NF(+) wound dressing displayed complete re-epithelialization, caused no inflammatory response, prevented wound deterioration, and showed no signs of infection. Additionally, the fibroblast-loaded dressings led to "higher collagen coverage" in the post-healing granular tissue—further justifying the potential of electrospinning and cell-based interventions in wound healing [30].

12.4   Inorganic Materials Incorporated into Wound Dressings

The physicomechanical properties of hydrogel scaffolds and membranes fabricated using synthetic or natural polymers can further be enhanced by the addition of inorganic materials such as, but not limited to, silica particles, clay components, metal ions, and bioactive glass. Below is a list of some very recent research studies demonstrating the contribution of inorganic components towards the wound healing performance of these bio-inorgano-composites:

1. Ag–graphene-incorporated hydrogels: Fan et al. synthesized Ag–graphene hydrogel composites by crosslinking acrylic acid with N, N'-methylene bisacrylamide (BIS) in the presence of glucose reduced Ag–graphene composites. The addition of Ag–graphene (a) enhanced the mechanical properties of the hydrogel such as tensile strength and elongation at break by acting as a matrix filler; (b) increased the conductivity of the hydrogel towards cell adhesion and growth due to its porous nature and biocompatibility; (c) maintained moist environment at the wound site via excellent water-carrying capacity; (d) provided antibacterial action due to the presence of silver ions; and (e) provided time-efficient and higher wound healing ratio than the controls [31].

2. Magnetite incorporated anionic hydrogels: Grumezescu et al. developed an anionic polymer hydrogel composed of sodium Alg and carboxymethylcellulose (SA-CMC) followed by addition of usnic acid loaded magnetite nanoparticles (Fe_3O_4@UA; 10 nm) to the hydrogels forming CMC/Fe_3O_4@UA and $AlgNa/Fe_3O_4$@UA, respectively. Human foetal progenitor cells maintained their spindle-shaped structure when cultured on the hydrogels and migrated to deeper layers of the scaffold confirming the porous geometry and balanced adhesion–proliferation characteristics of the composites. The scaffolds, however, delayed the progression of cell cycle ($G1{\rightarrow}S{\rightarrow}G2$) which further needs to be investigated. The scaffolds effectively prevented *S. aureus* colonization and provided prefect mechanical support for eukaryotic cells [32].

3. Montmorillonite incorporated gelatin hydrogels: Kevadia et al. intercalated ciprofloxacin (CPX) into the layered montmorillonite (MMT) architecture and reacted the CPX-MMT complex with gelatin polypeptide to form a multifunctional 3D composite hydrogel via electrostatic ion-exchange interactions. The hydrogel composites displayed accelerated healing when tested in an in vitro scratch wound healing method (Fig. 12.4) [33].

4. Bioactive glass introduced nanofibrous membranes: Bioactive glass (SiO_2–CaO–B_2O_3–P_2O_5–CuO–ZnO–K_2O–Na_2O) when added to electrospun gelatin/CS nanofibrous membranes rendered an increase in water contact angle and a decrease in hydrophilicity, thereby providing tuneable water responsiveness to the multifunctional polymer membranes. Additionally, the presence of bioactive glass enhanced the release of the bioactive ions (better protection against infection) with an enhanced biodegradation profile (complete degradation within 4 weeks post operation) [34].

5. Copper (II) Alg hydrogels: Klinkajon and Supaphol synthesized copper (II) crosslinked Alg hydrogels employing a two-step crosslinking process—preparation of soft gels lightly crosslinked with Cu^{++} ions followed by dip-crosslinking to obtain dimensional stability. The antibacterial (against *E coli, S aureus,* methicillin-resistant *S aureus, Staphylococcus epidermidis,* and *S pyogenes*) and blood-coagulating activities (platelet activation with fibrin and prothrombotic

coagulation) of the hydrogels were directly proportional to the concentration of Cu^{++} ions—a perfect example of customized antibacterial wound dressing easily controllable via modification of Alg and/or Cu^{++} concentration [35].

6. Silica–collagen biocomposite: Perumal et al. explored the synergistic wound healing potential of uniformly distributed mupirocin-loaded silica microsphere (Mu-SM) embedded into a collagen scaffold. The Mu-SM/collagen scaffolds revealed complete epithelialization within ≈2 weeks as compared to ≈2.5 and ≈3 weeks in the case of collagen alone and control groups, respectively. The wound closure analysis revealed that the complete epithelialization was observed at 14.2 ± 0.44 days for Mu-SM-loaded collagen, whereas this was 17.4 ± 0.44 days and 20.6 ± 0.54 days for collagen and control groups, respectively. The addition of silica particles imparted better bioadhesion and wound contraction efficiency to the collagen scaffold leading to increased "fibroblast proliferation" and enhanced "dermal collagenisation" [36].

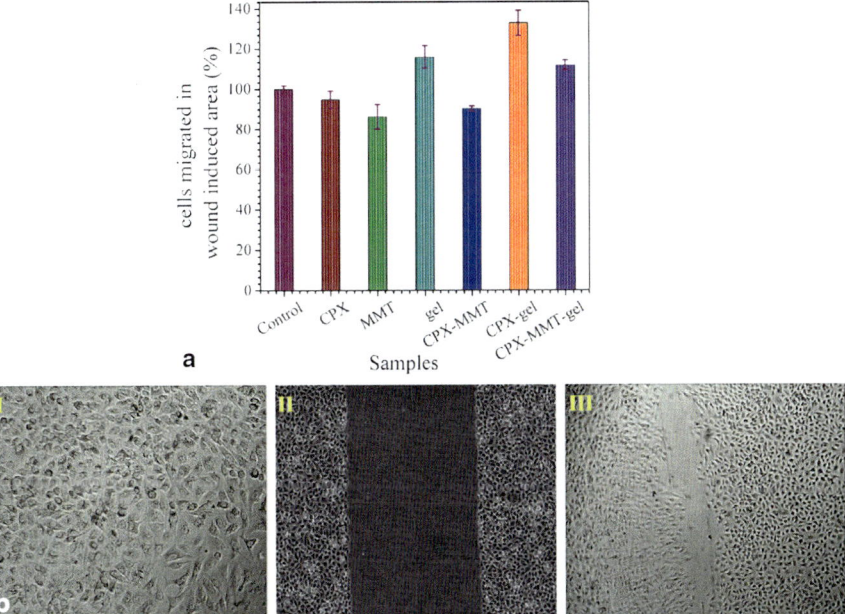

Fig. 12.4 a Quantitative proliferation. **b** Migration assay of A549 cells to carrier materials, CPX–gel, CPX–MMT, MMT and CPX–MMT–gel composite hydrogel. Confluent cell cultures of A549 were wounded with a micropipette tip. The number of cells migrated from the edge of the wound within each 250×500 μm area were counted and (**b**) positive effect on wound healing and migration of cells. (I) Confluent cells in a 12-well culture plate without wound induction, healthy cells serve as control. (II) Cells at "0" h of wound induction showing large vacant space between two cell colonies and (III) cells after 24 h of wound induction and exposure of CPX–MMT–gel composite hydrogels, cells bridge the gap significantly leaving scarce space between the colonies ($200 \times$) [33]. *CPX* ciprofloxacin, *MMT* montmorillonite

Table 12.1 Summary of recently developed silver nanoparticles incorporated wound dressings

Biocomposite	Fabrication technique	Effect of AgNP on the wound dressing and healing property	Reference
AgNP/Gelatin/PVA hydrogels	Gamma irradiation-based synthesis and crosslinking	Decreased degree of crosslinking and deteriorated mechanical properties at higher concentrations ($> 1\%$); enhanced antibacterial activity via Ag^+ release	[37]
AgNP/chitosan/ polyethylene oxide nanofibers	Electrospinning	Improved electrospinnability; enhanced tensile strength; antibacterial activity; acted as a nucleating agent	[38]
Chitosan–PVP–AgNP films	Solvent casting and air drying	Increased tensile strength; lowered water uptake capability; synergistic bioactivity along with CS	[39]
CS/silver-NPs/PVA nanofibers	Electrospinning and glutaraldehyde crosslinking	Reduced diameter of the e-spun fibers; enhanced electrospinnability; improved antibacterial ability	[40]
CS/regenerated cellulose/AgNP/gentamicin composites	Solution blending followed by drying at $30\,^{\circ}C$	Higher tensile strength values; accelerated wound healing	[41]
AgNP//bacterial cellulose (BC) gel membranes	In situ synthesis and deposition of AgNP on BC membranes	Robust hybrid nanostructure formation; reduced inflammation; promoted scald wound healing	[42, 43]

PVP poly (*N*-vinyl pyrrolidone), *NP* nanoparticle

In addition to above composite biomaterial architectures, several researchers have reported specialized silver nanoparticles (AgNP) enclathcrated biopolymeric systems for advanced wound healing applications. A list of selected recent research studies involving AgNP incorporated systems is presented in Table 12.1.

12.5   Future Outlook

The research studies involving integrated biomaterial wound dressings described in the text above mainly focused on the moisture retaining and antiinfective properties of the composite materials. Additionally, only few studies tested the in vivo efficacy of the designed polymeric systems wherein the bio-characterization was limited to extensive cytocompatibility studies. However, the potential of composite wound dressings cannot be undermined with several polymeric modifications such as pseudo-zwitterionic biomaterial dressing formation via charge balance, thermo-responsive swelling–deswelling approach, 3D-printed nanofibrils, HAM-mimicking biomaterials, SF-based composites, bioactive glass incorporated nanofibrous membranes, and polycation/polyanion LBL have the potential to translate into clinical practice requiring advanced wound care. In addition, these biomaterial composites proved that it is immensely important to employ "bioactive-biomaterials" for accelerated wound healing. Furthermore, the compatibility and incorporability of these composites towards therapeutic molecules such as antibiotics and natural compounds (e.g., curcumin and TTO) provides versatility to these archetypes which needs to be explored extensively. This confirms the nascent state of these specialized systems in current research scenario and their clinical testing, though achievable, is yet a far-fetched implication.

Acknowledgments This work was funded by the National Research Foundation (NRF) of South Africa.

References

1. Jhong JF, Venault A, Liu L, Zheng J, Chen SH, Higuchi A, Huang J, Chang Y. Introducing mixed-charge copolymers as wound dressing biomaterials. ACS Appl Mater Interfaces. 2014;6:9858–70.
2. Ngadaonye JI, Geever LM, McEvoy KE, Killion J, Brady DB, Higginbotham CL. Evaluation of novel antibiotic-eluting thermoresponsive chitosan-PDEAAm based wound dressings. Int J Polym Mater Polym Biomater. 2014;63:873–83.
3. Hsu SH, Hsieh PS. Self-assembled adult adipose-derived stem cell spheroids combined with biomaterials promote wound healing in a rat skin repair model. Wound Repair Regen. 2015;23(1):57–64.
4. Han F, Dong Y, Song A, Yin R, Li S. Alginate/chitosan based bi-layer composite membrane as potential sustained-release wound dressing containing ciprofloxacin hydrochloride. Appl Surf Sci. 2014;311:626–34.
5. Yang JM, Yang JH, Huang HT. Chitosan/polyanion surface modification of styrene–butadiene–styrene block copolymer membrane for wound dressing. Mater Sci and Eng C. 2014;34:140–48.
6. Harkins AL, Duri S, Kloth LC, Tran CD. Chitosan–cellulose composite for wound dressing material. Part 2. Antimicrobial activity, blood absorption ability, and biocompatibility. J Biomed Mater Res Part B. 2014;102B.1199–1206.
7. Bai MY, Chou TC, Tsai JC, Yu WC. The effect of active ingredient-containing chitosan/polycaprolactone nonwoven mat on wound healing: In vitro and in vivo studies. J Biomed Mater Res Part A. 2014;102A:2324–33.
8. Aramwit P, Ratanavaraporn J, Ekgasit S, Tongsakul D, Bang N. A green salt-leaching technique to produce sericin/PVA/glycerine scaffolds with distinguished characteristics for wound-dressing applications. J Biomed Mater Res B Appl Biomater. 2015;103(4):915–24. doi:10.1002/jbm.b.33264.
9. Shahverdi S, Hajimiri M, Esfandiari MA, et al. Fabrication and structure analysis of poly(lactide-co-glycolic acid)/silk fibroin hybrid scaffold for wound dressing applications. Int J Pharm. 2014;473:345–55.
10. Nayak S, Kundu SC. Sericin–carboxymethyl cellulose porous matrices as cellular wound dressing material. J Biomed Mater Res Part A. 2014;102A:1928–40.
11. Sukumar N, Ramachandran T, Lakshmikantha C. Development and characterization of cactus–dextrin–recombinant human epidermal growth factor based silk scaffold for wound dressing applications. J Indust Text. 2014;43(4):565–76.

12. Li Y, Qing S, Zhou J, Yang G. Evaluation of bacterial cellulose/hyaluronan nanocomposite-biomaterials. Carbohyd Polym. 2014;103:496–501.
13. Zhou Z, Chen J, Peng C et al. Fabrication and physical properties of gelatin/sodium alginate/hyaluronic acid composite wound dressing hydrogel. J Macromol Sci A: Pure Appl Chem. 2014;51:318–25.
14. da Cunha CB, Klumpers DD, Li WA, Koshy ST, Weaver JC, Chaudhuri O, Granja PL, Mooney DJ. Influence of the stiffness of three-dimensional alginate/collagen-I interpenetrating networks on fibroblast biology. Biomaterials 2014;35:8927–36.
15. De Cicco F, Reverchon E, Adami R, et al. In situ forming antibacterial dextran blend hydrogel for wounddressing: SAA technology vs. spray drying. Carbohyd Polym. 2014;101:1216–24.
16. Rees A, Powell LC, Carrasco GC, et al. 3D bioprinting of carboxymethylated-periodate oxidized nanocellulose constructs for wound dressing applications. BioMed Res Int. 2015:Article ID 925727 (http://dx.doi.org/10.1155/2015/925757.
17. Ye J, Shi X, Chen X, et al. Chitosan-modified, collagen-based biomimetic nanofibrous membranes as selective cell adhering wound dressings in the treatment of chemically burned corneas. J Mater Chem B. 2014;2:4226–36.
18. Fu S, Meng X, Fan J, et al. Acceleration of dermal wound healing by using electrospun curcumin-loaded poly(e-caprolactone)-poly(ethylene glycol)-poly(ecaprolactone) fibrous mats. J Biomed Mater Res B. 2014;102B:533–42.
19. Vatankhah E, Prabhakaran MP, Jin G, Mobarakeh LG, Ramakrishna S. Development of nanofibrous cellulose acetate/gelatin skin substitutes for variety wound treatment applications. J Biomater Appl. 2014;28(6):909–21.
20. Zhao R, Li X, Sun B, et al. Electrospun chitosan/sericin composite nanofibers with antibacterialproperty as potential wound dressings. Int J Biol Macromol. 2014;68:92–7.
21. Çalamak S, Erdoğdu C, Özalp M, Ulubayram K. Silk fibroin based antibacterial bionanotextiles as wound dressing materials. Mater Sci Eng C. 2014;43:11–20.
22. Naseri N, Algan C, Jacobs V, John M, Oksman K, Mathew AP. Electrospun chitosan-based nanocomposite mats reinforced with chitin nanocrystals for wound dressing. Carbohydr Polym. 2014;109:7–15.
23. Huang Y, Zhong Z, Duan B, et al. Novel fibers fabricated directly from chitin solution and their application as wound dressing. J Mater Chem B. 2014;2:3427.
24. Xu F, Weng B, Materon LA, Gilkerson R, Lozano K. Large-scale production of a ternary composite nanofiber membrane for wound dressing Applications. J Bioact Compat Polym. 2014, 29(6):646–60.
25. Xu F, Weng B, Gilkerson R, Materon LA, Lozano K. Development of tannic acid/chitosan/pullulan composite nanofibers from aqueous solution for potential applications as wound dressing. Carbohyd Polym. 2015;115:16–24.
26. Jin G, Prabhakaran MP, Ramakrishna S. Photosensitive and biomimetic core–shell nanofibrous scaffolds as wound dressing. Photochem Photobiol. 2014;90:673–81.
27. Gnanasekaran G, Sathishkumar Y, Lee YS, Kim CS, Unnithan AR. Electrospun antibacterial polyurethane–cellulose acetate–zeincomposite mats for wound dressing. Carbohyd Polym. 2014;102:884–92.
28. Chutipakdeevong J, Ruktanonchai U, Supaphol P. Hybrid biomimetic electrospun fibrous mats derived from poly(e-caprolactone) and silk fibroin protein for wound dressing application. J Appl Polym Sci. 2015;132:41653.
29. Kim MK, Kwak HW, Kim HH, et al. Surface modification of silk fibroin nanofibrous Mat with dextran for wound dressing. Fiber Polym. 2014;15:1137–45.
30. Ong CT, Zhang Y, Lim R, et al. Preclinical evaluation of tegadermTM supported nanofibrous wound matrix dressing on porcine wound healing model. Adv Wound Care. 2015;4(2):110–8. doi:10.1089/wound.2014.0527.
31. Fan Z, Liu B, Wang J, et al. A novel wound dressing based on Ag/Graphene polymer hydrogel: effectively kill bacteria and accelerate wound healing. Adv Funct Mater. 2014;24:3933–43.

32. Grumezescu AM, Holban AM, Andronescu E, et al. Anionic polymers and 10 nm Fe3O4@ UA wound dressings support human foetal stem cells normal development and exhibit great antimicrobial properties. Int J Pharm. 2014;463:146–54.
33. Kevadiya BD, Rajkumar S, Bajaj HC, et al. Biodegradable gelatin–ciprofloxacin–montmorillonite composite hydrogels for controlled drug release and wound dressing application. Colloid Surf B: Biointerface. 2014;122:175–83.
34. Ma W, Yang X, Ma L, et al. Fabrication of bioactive glass-introduced nanofibrous membranes with multifunctions for potential wound dressing. RSC Adv. 2014;4:60114–122.
35. Klinkajon W, Supaphol P. Novel copper (II) alginate hydrogels and their potential for use as anti-bacterial wound dressings. Biomed Mater. 2014;9:045008.
36. Perumal S, Ramadass SK, Madhan B. Sol–gel processed mupirocin silica microspheres loaded collagen scaffold: a synergistic bio-composite for wound healing. Eur J Pharm Sci. 2014;52:26–33.
37. Leawhiran N, Pavasant P, Soontornvipart K, Supaphol P. Gamma irradiation synthesis and characterization of AgNP/Gelatin/PVA hydrogels for antibacterial wound dressings. J Appl Polym Sci. 2014;131:41138.
38. Wang X, Cheng F, Gao J, Wang L. Antibacterial wound dressing from chitosan/polyethylene oxide nanofibers mats embedded with silver nanoparticles. J Biomater Appl. 2015;29(8):1086–95. doi:10.1177/0885328214554665.
39. Archana D, Singh BK, Dutta J, Dutta PK. Chitosan-PVP-nano silver oxide wound dressing: In vitro and in vivo evaluation. Int J Biol Macromol. 2015;73:49–57.
40. Abdelgawad AM, Hudson SM, Rojas OJ. Antimicrobial wound dressing nanofiber mats from multicomponent (chitosan/silver-NPs/polyvinyl alcohol) systems. Carbohyd Polym. 2014;100:166–78.
41. Ahamed MIN, Sankar S, Kashif PM, Basha SKH, Sastry TP. Evaluation of biomaterial containing regenerated cellulose and chitosan incorporated with silver nanoparticles. Int J Biol Macromol. 2015;72:680–86.
42. Wu J, Zheng Y, Wen X, Lin Q, Chen X, Wu Z. Silver nanoparticle/bacterial cellulose gel membranes for antibacterial wound dressing: investigation in vitro and in vivo. Biomed Mater. 2014;9:035005.
43. Wu J, Zheng Y, Song W, et al. In situ synthesis of silver-nanoparticles/bacterial cellulose composites for slow-released antimicrobial wound dressing. Carbohyd Polym. 2014;102:762–71.

Prof. Viness Pillay, a fulbright scholar and fellow of the African Academy of Sciences (AAS), is a South African National Research Foundation (NRF) Research Chair in Pharmaceutical Biomaterials and Polymer-Engineered Drug Delivery Technologies at the University of the Witwatersrand (Wits), Johannesburg, South Africa. He is also the recipient of the Olusegun Obasanjo Prize 2014 for Scientific Breakthrough and Innovation in Advanced Drug Delivery Technologies and director of the Wits Advanced Drug Delivery Platform (WADDP) Research Unit (http://www.wits.ac.za/waddp). The WADDP is the only platform in its domain in South Africa that encompasses the research, development, and commercialization of innovative prototypes in the area of neurotherapeutic delivery technologies under development by his research team. Prof. Pillay currently has several granted international patent applications, and his research has been extensively published in international peer-reviewed journals, with vast contributions to editorials, books, and book chapters as well as presentations at local and international conferences in the areas of pharmaceutical drug delivery technology, neurotherapeutics, and polymer science. He currently serves on the reviewer, editorial advisory board, and lead guest editor of several international scientific journals and books.

Chapter 13
Adverse Effects of By-products from Polymers Used for Joint Replacement

Tzu-Hua Lin, Jukka Pajarinen, Florence Loi, Taishi Sato, Changchun Fan, Zhenyu Yao and Stuart Goodman

13.1 Epidemiology of Total Joint Replacement

Total joint replacement (TJR) has become a successful surgical procedure that decreases pain and improves function for millions of patients worldwide. Indeed, total hip and knee replacements are the most cost-effective procedures in all the surgical specialties.

The reasons for TJR include osteoarthritis (OA), rheumatoid arthritis (RA), traumatic osteoarthritis (TA), tumor, and others. Kurtz et al. have reported that in the USA, total hip and knee replacement will continue to increase [1]. By the year 2010, almost 1,000,000 total hip arthroplasties (THA) and total knee arthroplasties (TKA) were performed annually in the USA [2]. By 2030, it is predicted that the number of THAs will reach 572,000, an increase of 174% compared to 2005 [3]. In the UK, 86,488 THAs were entered into the UK National Joint Registry in 2012, an increase of 7.5% from 2011 [4]. Currently, TJR is a commonly performed procedure for end-stage arthritis of the hip, knee, shoulder, ankle, and other joints.

Cram et al. reported excellent results in over 800,000 Medicare patients who underwent cemented or cementless total joint arthroplasty (TJA) in the USA between 1991 and 2008 [5]. However, TJA procedures are not without potential risks including infection, component loosening, wear and osteolysis, periprosthetic fracture, and other local and general complications. These complications are dependent on many variables including the patient (gender, age, medical comorbidities, bone quality, neuromuscular status, etc.), the surgeon (surgical approach, component position, etc.), and the characteristics of the implant [6]. Bozic et al. reported the crude complication rate for primary total hip and knee arthroplasty to be 3.6%. Pneumonia (0.86%), pulmonary embolism (0.75%), and periprosthetic joint or wound

S. Goodman (✉) · T.-H. Lin · J. Pajarinen · F. Loi · T. Sato · C. Fan · Z. Yao
Orthopaedic Research Laboratories, Stanford University, Stanford, CA, USA
e-mail: goodbone@stanford.edu

© Springer International Publishing Switzerland 2015
L. Santambrogio (ed.), *Biomaterials in Regenerative Medicine and the Immune System*,
DOI 10.1007/978-3-319-18045-8_13

infection (0.67%) were the most common complications [7]. Other authors reported complication rates as high as 7–8% [8].

13.2   Biomaterials and Biomechanics

The history of joint replacement has a close relationship to improvements in bio-materials and biomechanics. Joint replacement was first attempted with primitive materials such as ivory, celluloid, Pyrex glass, Bakelite, and acrylic prostheses, which all failed due to problems with adverse biological reactions to these materials, infection, difficulties in fixation of the prosthesis to bone, mechanical breakage of implants, and difficulties in restoring normal joint biomechanics. Sir John Charnley introduced the first highly successful joint replacement, the cemented low-friction arthroplasty (LFA) of the hip [9]. However, his first attempts at hip replacement used a cemented femoral component with a larger femoral head and thin polytetrafluoroethylene (Fluon) socket, which was associated with chronic synovitis, osteolysis, and prosthetic loosening because of the generation of excessive wear particles. Subsequently, Charnley and colleagues improved the biomechanical and tribological characteristics of the LFA and optimized the materials used in the reconstruction [10]. For total hip replacement, meticulous preoperative planning and surgical technique and the introduction of polymethylmethacrylate (PMMA) and high-density polyethylene by Sir John Charnley created the possibility for the emergence of joint replacement as a credible procedure.

13.2.1   Cemented and Cementless Joint Replacement

In conjunction with Professor Dennis Smith, a dental researcher, Charnley was the first to use a cemented femoral component that was fixed to bone with PMMA, so-called bone cement. Although a properly placed cement mantle can achieve fixation for many years, chronic inflammation and periprosthetic osteolysis (so-called cement disease) can result due to fractured cement and the generation of wear particles [11, 12]. Currently, periprosthetic osteolysis is most often due to wear of conventional polyethylene at the bearing surface [10, 13–17]. This wear process undermines the bony support of the implant and can lead to prosthetic loosening due to failure of fixation. The classic "pseudosynovial membrane" at the bone–cement interface of loose prostheses can secrete high levels of pro-inflammatory factors [17–21]. Histologically, the interface of loose cemented implants is loaded with macrophages, scattered lymphocytes, and other cells in a fibrous tissue stroma [22]. This chronic inflammatory reaction to wear debris can lead to bone destruction via activation of osteoclastogenesis, potentially compromising the long-term stability of the implant [10, 16, 21, 23, 24]. However, many cemented implants last for decades without substantial degradation of the interfaces.

In the late 1970s and 1980s, cementless joint replacement was developed to eliminate the risks of cement failure and "cement disease" [11]. Fixation of cementless implants was achieved by machining the bone to obtain a "line-to-line" fit, or subsequently a "press fit" (a slightly oversized implant literally jammed into the prepared bone bed) to obtain initial implant stability. In addition, porous coatings (small beads or wires) were sintered or plasma-sprayed onto portions of the implant surface to facilitate the ingrowth of bone and fibrous/fibro-cartilaginous tissue into the porous interstices for more long-term stabilization. Some manufacturers also implemented coatings with osteoconductive materials such as calcium phosphates on portions of the implant. Carbonated hydroxyapatite (HA) which is chemically similar to the apatite of bone and is a source of calcium and phosphate ions to the healing interface [25] was also proposed as a coating to improve osseointegration of cementless metallic prostheses [26, 27]. HA coatings provide an osteoconductive stimulus for enhancement of bone formation onto orthopedic implants.

Successful osseointegration of retrieved cementless hip implants in bone has been demonstrated in autopsy retrieval studies; however, much of the porous coating may be filled with oriented fibrous tissue instead of bone [28–31]. Numerous factors determine the robustness of osseointegration of an implant and subsequent bone remodeling including characteristics of the patient (e.g., systemic diseases, local bone quality), the implant (e.g., design, materials, coatings), surgical technique, and the biomechanical environment. In cases of distally fixed stiff implants, adverse bone remodeling occurs in relation to altered biomechanical loads to the surrounding bone, such as with the disappearance of the proximal femur after fully porous coated hip stems, so-called stress shielding [31]. Although this adverse remodeling may be marked, it does not generally lead to loosening of the implant or pain [32]. In fact, when a cementless prosthesis has undergone extensive osseointegration, it is common to see proximal stress shielding (rounding off of the proximal–medial calcar of the femur), osteopenia, and more distal oblique struts of bone connecting the cortex to the distal porous coating, so-called spot welds.

13.3 Tribology

13.3.1 Metal-on-Polyethylene Bearings

After the introduction of high-density polyethylene in hip joint replacement by Charnley, ultra-high molecular weight polyethylene has been the orthopedic bearing surface of choice since the 1960s. More recently, conventional polyethylene has undergone further alterations, including cross-linking and thermal and mechanical treatment in order to achieve improved mechanical characteristics and wear resistance. Polyethylene implants used to be manufactured by a ram extrusion method, in which the material was briefly pushed from a piston into a mold of the implant; however, the resulting polyethylene would undergo subsequent oxidative degradation in air. Compression molding of polyethylene, combined compression and heating during molding, and other techniques improved the mechanical characteristics. Cross-linking of polyethylene, obtained by irradiation at 50–100 kGy, was shown to increase wear resistance; the irradiation is followed by thermal treatment to remelt the polyethylene or cycles of annealing and heating below the melting temperature in order to help remove free radicals which cause oxidative degradation [33–35]. The wear properties of highly cross-linked polyethylene are largely dependent on the radiation dose and the thermal and mechanical treatments. Although increased amounts of radiation and remelting lead to higher wear-resistant properties due to higher cross-link densities and lower levels of free radicals, a loss of crystallinity due to heating leads to reduced elasticity and fatigue resistance, increasing the risk of cracks due to edge loading, especially with suboptimally positioned prostheses. Currently, several novel strategies such as adding vitamin E, repeated cycles of thermal and mechanical treatment, or highly hydrophilic and lubricious coatings are being explored in order to reduce the production of free radicals and oxidative degradation [36–39].

13.3.2 Hard-on-Hard Bearings

"Hard-on-hard" bearings such as metal-on-metal and ceramic-on-ceramic emerged due the presence of plastic wear and periprosthetic osteolysis of metal-on-conventional polyethylene implants, especially in active hip replacement patients [40]. These bearings have significantly less wear compared to metal-on-polyethylene articulations. Furthermore, dislocation of hip replacements with smaller femoral heads of 22 and 28 mm was still a major problem. In general, metal-on-metal bearings could employ larger femoral heads as the polyethylene layer was omitted from the bearing surface. Thus, the enlarged range of motion improved stability and decreased the dislocation rate. Wear of the articulation was supposedly dictated and limited by fluid film lubrication. However, the theoretical, extremely low friction associated with fluid film lubrication was often altered by edge loading and other issues, leading to the generation of wear particles and ions of cobalt, chromium, and other by-products. In addition, two new junctions have emerged, namely modularity between the femoral neck and head (to adjust femoral length), and, more recently, modularity between the femoral stem and neck (to adjust femoral prosthesis offset and neck ante- or retro-version). Mechanically associated crevice corrosion has led to the generation of more particles and metallic by-products, which have been shown to stimulate both the innate and adaptive immune system. In some cases, this has resulted in a chronic inflammatory response, cell death involving soft tissue and bone, and large solid or cystic pseudotumors [41, 42]. Metallic wear by-products can be cytotoxic and can lead to hypersensitivity reactions (a T cell-mediated type IV immune reaction) [42, 43].

Ceramic-on-ceramic bearing surfaces for joint replacement have been used for decades, but are less commonly used in the USA compared to Europe and Asia. These hard-on-hard bearings must be positioned and assembled precisely to avoid improper seating, impingement, chipping or fracture of the ceramic, and striped wear and squeaking, due to edge loading and stress concentration [44]. Currently used ceramics contain alumina–zirconia–chromium oxide–strontium with smaller grain sizes and are more robust than the previous alumina-on-alumina articulations. However, ceramic bearings are generally more expensive than metal-on-highly cross-linked polyethylene.

13.3.3   By-Products of Wear and Clinical Outcome

The characteristics of the materials used for TJR are crucial to the clinical outcome. The generation of by-products of wear stimulate the innate immune system; when in sufficient numbers, this process results in a chronic inflammatory and foreign body reaction that may lead to chronic synovitis, periprosthetic osteolysis, component loosening, and pathologic fracture [20, 21]. In some cases, both the innate and adaptive immune systems are activated, such as with cases displaying metallic by-products. Thus, it behooves the surgeon, biologist, and bioengineer to understand the basic tribological and immunological characteristics of TJR so that the long-term outcome of this highly successful procedure could be further optimized.

13.4   Innate Immune System

Innate immunity is an evolutionarily, ancient, non-clonal host defense mechanism found in all plants and animals, defending them against constant exposure to bacteria, viruses, fungi, and parasites present in the environment [45]. To combat microbial invasion even more effectively, vertebrates also developed the adaptive immune system, which is based on the clonal expansion of antigen-specific lymphocytes [46]. Innate immune responses are nonspecific and present at birth whereas adaptive immune responses are acquired or improved through interaction with microorganisms or other antigens [47]. Since the primary adaptive immune response is mounted about 4–7 days after exposure, the innate immune system, which responds immediately, is an important first line of defense against invasion by infectious microorganisms [48].

13.4.1   Mechanisms of the Innate Immune System

The innate immune response has two phases: immediate and induced. The first consists of physical and chemical barriers such as antimicrobial peptides and proteins (AMPs) and the second consists of the detection of invariant, conserved molecule motifs typical of microbes (called pathogen-associated molecular patterns; PAMPs) by special receptors (called pattern-recognition receptors; PRRs) and clearance of pathogens by phagocytes [49, 50].

13.4.1.1   Antimicrobial Peptides and Proteins

The innate immune system's initial barrier to invading pathogens are the skin and AMPs in the bloodstream, extracellular fluid, and epithelial secretions [49]. These AMPs are produced by leukocytes, hepatocytes, and epithelial cells once a group of PRRs called Toll-like receptors (TLRs) are activated [49]. They are typically cationic and can be antibacterial, antifungal, antiviral, and anticancer in addition to participating in inflammation, release of cytokines, homeostasis, and proliferation [51]. Examples of antimicrobial enzymes are: lysozyme, which cleaves important bonds in peptidoglycan cell walls; secretory phospholipase A_2, which penetrates bacterial cell walls to hydrolyze cell membrane phospholipids; and bactericidal permeability-increasing protein, which damages bacterial membranes, neutralizes lipopolysaccharide (LPS), and marks Gram-negative bacteria for phagocytosis by neutrophils [52–56]. Antimicrobial peptides include defensins, cathelicidin (LL-37), and histatins, and they primarily disrupt membranes [49, 52]. Defensins are predominantly expressed by neutrophils and Paneth cells (α-defensins) or epithelial cells (β-defensins) [57]. LL-37 is secreted by epithelial cells, neutrophils, monocytes, and macrophages [56, 58]. In addition to acting as an antibiotic, LL-37 is shown to be involved in immune response modulation by inducing cytokine/chemokine production, inhibiting LPS-induced pro-inflammatory cytokine production, and modulating the activity of dendritic cells (DC) and T cells [56]. Histatins are bactericidal and fungicidal, especially against fungi *Candida albicans, Cryptococcus neoformans,* and *Aspergillus fumigatus* [51].

13.4.1.2   Induced Innate Response: Detection of PAMPs with PRRs

One strategy in which the innate immune system detects infectious agents is through PAMPs. PAMPs are conserved molecular motifs produced by all microorganisms, even nonpathogenic ones, which allows the host to differentiate "non-self" from "self" [50]. This is one of the reasons why PAMPs play a large role in innate immune recognition [48]. Another reason is that PAMPs are invariant for a given class of microorganisms, thus maximizing the effect of a single PRR [50]. A third reason is that PAMPs play essential roles for the microbes so mutants or loss of PAMPs are extremely maladaptive or even fatal [48]. Examples of PAMPs include LPS and lipoteichoic acids (LTAs) of Gram-negative and Gram-positive bacteria, respectively, mannans in yeast cell walls, and formylated peptides in bacteria [59] (Fig. 13.1).

It is important to note that the immune system is somehow able to tolerate nonpathogenic microbes even though all microbes express PAMPs. Although the exact mechanisms have not yet been elucidated, it has been suggested that compartmentalization and anti-inflammatory cytokines play a major role [48].

PRRs that detect PAMPs can be found circulating in the blood stream, on cell surfaces, or in intracellular compartments [48]. Phagocytes including macrophages, neutrophils, and DCs all produce PRRs [49]. Once activated, PRRs participate in opsonization, phagocytosis itself, activation of the complement system, initiation of

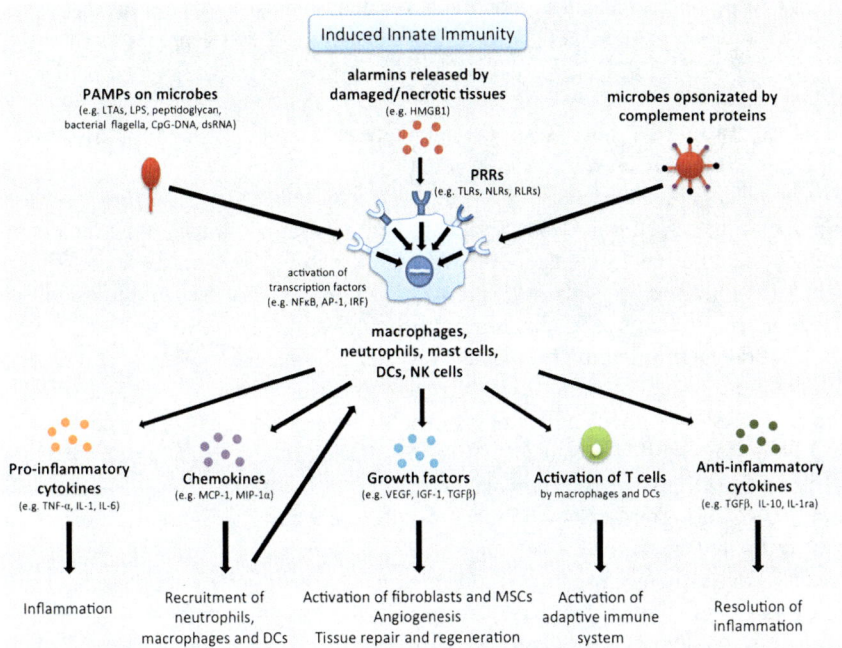

Fig. 13.1 Induced innate immunity. Pathogen-recognition receptors (*PRRs*) present on or in innate immune cells such as macrophages, neutrophils, mast cells, dendritic cells (*DCs*), and natural killer (*NK*) cells can detect the presence of potential pathogens through PAMPs, alarmins, or opsonizing complement proteins. These signals induce immune cells to activate transcription factors, release soluble factors including pro- and anti-inflammatory cytokines, chemokines, and growth factors, and activate T cells of the adaptive immune system. *PAMPs* pathogen-associated molecular patterns, *MSC* mesenchymal stromal cell

pro-inflammatory signaling pathways, activation of naïve T cells, and induction of apoptosis [48, 49, 60]. Secreted PRRs typically tag microbes and signal the complement system or phagocytic cells to eliminate them [50]. TLRs, a class of PRRs found on the surface of and within all innate immune cells and certain endothelial cells, activate pro-inflammatory signaling pathways and initiate the adaptive immune response [48]. There are ten TLRs in humans and they are capable of recognizing PAMPs of bacteria, fungi, and viruses [49]. Specifically, they can detect microbial products such as LTAs, LPS, peptidoglycan, bacterial flagella, CpG-DNA, and double-stranded RNA [48, 49]. TLRs are especially important in innate immunity because they activate transcription factors including nuclear factor kappa-light-chain enhancer of activated B cells (NF-kB), activator protein 1 (AP-1), and interferon regulatory factor (IRF) to upregulate the downstream gene expression of inflammatory and antiviral cytokines, chemokines, and antiviral type I interferons [61]. Nucleotide oligomerization domain (NOD)-like receptors (NLRs), another group of PRRs, detect pathogen fragments in the cytoplasm [49]. Like TLRs, NLRs also induce type I interferon production and activate NF-kB to induce the inflammatory response [61]. Retinoic acid inducible gene 1 (RIG-I)-like receptors (RLRs) sense viral RNAs in the cytoplasm and upregulate type I interferon production [62].

13.4.1.3   Induced Innate Response: Detection of Non-self Through Missing Self Markers

In addition to detecting PAMPs, another common strategy in which the innate immune system detects foreign agents is through the absence of self-markers. Briefly, healthy host cells constitutively express ubiquitous markers that bind to inhibitory receptors on immune cells including macrophages, natural killer (NK) cells, and DCs to prevent elimination by phagocytosis, cell lysis, and activation of the adaptive immune system [50]. For example, the expression of major histocompatibility complex class I (MHC-I) molecules on healthy nucleated cells is both constitutive and ubiquitous [50]. Activated MHC-I-specific inhibitory receptors on NK cells inhibit lytic activity by recruiting and activating src-homology phosphatases SHP-1 and SHP-2 [63]. Thus, cells with inadequately expressed MHC-I molecules, which can be due to viral infection or cellular transformation, are selectively targeted by NK cells [64]. Another example is CD47, which is constitutively expressed in many cell types, including erythrocytes. It binds to the inhibitory receptor signal regulatory protein alpha (SIRPα) on splenic macrophages to prevent phagocytosis of healthy erythrocytes while allowing senescent ones to be cleared from the bloodstream [65]. However, this strategy of targeting cells without sufficient self-markers can be evaded by pathogens that acquire self-marker genes through horizontal transfer [50].

13.4.1.4   Complement System

In the 1890s, Jules Bordet discovered a group of over plasma proteins that "complemented" the activity of antibodies by opsonizing bacteria for phagocytosis [49]. Once activated, this group of over 30 proteins can lead to direct or indirect killing of the pathogen by tagging and destroying them [66]. In addition to this, the complement is also involved in the removal of apoptotic cells, tissue regeneration, angiogenesis, synapse maturation, and lipid metabolism [66].

13.4.1.5   Inflammatory Response

In innate immunity, inflammation defends against infection by recruiting proteins and immune cells to affected tissues and increasing the flow of microbes and antigen-presenting cells (APCs) to surrounding lymphocytes in surrounding tissues, which activates the adaptive immune response [49]. First, neutrophils are recruited to infected tissues, followed by monocytes, which quickly differentiate into macrophages [67]. Neutrophils and macrophages are key cells in the inflammatory response, in part, because of the PRRs on their cell surfaces, their roles as phagocytes, and the cytokines they release such as tumor necrosis factor-α (TNF-α), interleukin-1 (IL-1), and IL-6 [46, 68]. Inflammation is beneficial because it helps remove pathogens from the system and aids as a healing process for damaged tissues [68].

13.4.2   Cells of the Innate Immune System

The cells that participate in innate immunity originate in the bone marrow; they include monocytes, macrophages, neutrophils, eosinophils, basophils, mast cells, DCs, and NK cells [69].

13.4.2.1   Monocytes/Macrophages

Macrophages are cells that are differentiated from monocytes. When local injury, infection, or other perturbing stimulus occurs, macrophages residing in the adjacent tissues become activated and subsequently orchestrate the pre-programmed innate immune response to eliminate the offender and initiate homeostatic healing. Furthermore, monocytes circulating in the bloodstream are recruited and migrate along chemokine gradients to the injured tissue, where they differentiate into macrophages, amplifying the inflammatory response [70]. Macrophages are crucial for internalizing and degrading invading microbes and senescent cells, activating T cells through antigen presentation, modulating inflammation, and signaling other inflammatory-related cells through cytokines and chemokines [59].

13.4.2.2   Granulocytes

Granulocytes are phagocytes containing densely staining granules, which store digestive enzymes and toxic proteins [49]. Of the three granulocytes, neutrophils are the most abundant and most important in innate immunity; eosinophils and basophils are the other two [71]. Often cooperating with other immune cells such as macrophages, neutrophils engulf microbes and other injured material and destroy them with the degradative substances stored in their granules [71]. Eosinophils and basophils preferentially target parasites, which are too large for macrophages and neutrophils [49]. However, eosinophils and basophils can be harmful rather than defensive when involved in allergic inflammatory reactions [49].

13.4.2.3   Mast Cells

The chief function of mast cells is to modulate allergic reactions; however, they are thought to protect the internal body surfaces from pathogens and to participate in the response against parasitic worms [49]. Recent evidence shows that mast cells can release pro-inflammatory mediators to help optimize the immune response to infection [72].

13.4.2.4   Dendritic Cells

The name for DCs is derived from their fingerlike processes, which are similar to the dendrites of nerve cells [49]. Their main function as immature DCs is to clear the bloodstream and tissues of pathogens via phagocytosis and macropinocytosis [49]. Once they interact with pathogens, they mature into APCs and can activate the adaptive immune system by presenting antigens to T cells [49].

13.4.2.5   Natural Killer Cells

NK cells are non-antigen-specific T lymphocytes that detect and kill abnormal cells such as tumor or infected cells [73]. They also release pro-inflammatory and im-munosuppressive cytokines and chemokines such as interferon-γ (IFN-γ), TNF-α, and monocyte chemoattractant protein-1 (MCP-1) [73].

13.4.3   Conclusion

Although the innate immune system is nonspecific, the mechanisms that it employs are very powerful. The skin and AMPs prevent microbes from invading the body. If microbes or other adverse stimuli evade this line of defense, the innate immune system has direct and indirect elimination methods such as recognizing PAMPs with PRRs, activating the complement cascade, and initiating and orchestrating in-flammation. Thus, the innate immune system is effective in either destroying the threat or, if an antigenic stimulus is present, initiating the adaptive immune system. In some cases, the innate immune system fails to eliminate the stimulus and chronic inflammation ensues, which could potentially jeopardize the existence of the organism. However, this scenario is uncommon; the innate immune system together with the adaptive immune system provides a robust platform for survival of the host.

13.5   Wear of Joint Replacements and the Innate Immune System

13.5.1   Introduction

Aseptic loosening refers to the sequence of events in which the bone that has previously surrounded the total joint replacement is resorbed ultimately leading to the detachment of the implant components from the surrounding bone bed. Occasionally, significant periprosthetic osteolysis may develop without loosening of the implant components.

Since the pioneering work by Willert et al., it has been widely accepted that peri-implant osteolysis and subsequent implant loosening are primarily driven by biomaterial wear debris that is released from implant-bearing surfaces, articulations between modular implant components or from the interfaces between the bone and the implant, or bone and bone cement [74–78]. Reflecting the materials commonly used in these interfaces, the wear debris is typically ultra-high molecular weight polyethylene (UHMWPE), PMMA (bone cement), or metal or ceramic particles, typically of 1 μm in size or less. Biomaterial wear debris continuously generated in these locations is carried around the implant due to the cyclic pressure waves in the pseudosynovial fluid, resulting in low-grade, chronic inflammation [79]. In addition to UHMWPE, PMMA and larger metal particles that mainly cause the activation of the innate immunity, nano-sized metal particles and metal ions may cause the activation of the adaptive immune system. The phenomenon is typically encountered with implants with metal-on-metal bearing surfaces or implants with modular components. High amounts of nano-sized metal particles and metal ions can be released from the articulating surfaces of these implants due to the combined effect of wear and corrosion. Problems with these implants display clinical, histological, and pathological features that are distinct compared to the adverse tissue reactions caused by larger metal particles or wear products from polymeric materials [80–83].

The critical role of implant wear products as the main cause of aseptic loosening is supported both by clinical observations and a wealth of experimental studies [76–78, 84]. For example, clinical studies have shown that development of aseptic loosening is directly correlated to implant wear; in cell culture, implant wear products cause biochemical changes in cells that favor bone resorption over formation; and in animal models, wear particles induce local osteolysis. [85]. In addition to wear debris, less well-understood mechanical factors contribute to the loosening process [86]. Furthermore, the concept of individual susceptibility to peri-implant osteolysis and aseptic loosening either due to hereditary factors and/or systemic and local conditions that might modulate the manner the innate immune system responds to the accumulating wear debris load, has recently evoked much interest.

Peri-implant osteolysis and subsequent implant loosening are leading causes of long-term joint replacement failures and a significant cause of morbidity. Other than revision surgery, no treatment is currently available for this condition. A detailed understanding of the underlying mechanisms is necessary for preventing peri-implant osteolysis and increasing the life-span of the joint replacements; furthermore, the studies to the mechanisms of the peri-implant osteolysis have provided detailed insights into the fundamental principles of tissue-biomaterial interactions that will prove valuable in the other areas of biomaterial science such as tissue engineering and regenerative medicine.

13.5.2   Peri-implant Histopathology

The histopathology of a well-functioning joint replacement that maintains its fixation to bone over many years of service despite moderate rates of wear is poorly understood. In contrast, the histopathology surrounding failed implants has been studied in detail [19, 23, 24, 75, 87–90]; as these tissue samples are typically collected
during implant revision surgeries, they represent only the end-point of a process that
typically takes years to develop. Despite this apparent weakness, as with any other
disease processes, examination of the histopathological samples provides a starting
point for a more profound understanding of the mechanisms of aseptic osteolysis.

The generally, the interface tissue developing between the implant and surrounding
bone due to polymer wear debris has been described as well-vascularized loose
connective tissue with a distinct tissue architecture; facing the implant is a layer of
synovial membrane-like tissue with type A macrophage-like cells and type B fibro-
blast-like cells. Underneath the pseudosynovial lining is a layer of varied thickness
containing neovascularization and ample inflammatory cell infiltrates followed by
a more fibrous layer adjacent to the bone that is covered by bone-resorbing osteoclasts.

This typical histopathological architecture is however encountered only occasionally as large areas of necrosis and fibrosis are also commonly seen; especially
the adverse tissue reaction developing due to metal ions and metal corrosion products has been described to contain large areas of necrosis. In the case of UHMWPE
wear-induced osteolysis, large amount of UHMWPE wear debris either phagocytosed by the interface tissue cells or surrounded by foreign body giant cells is typically observed when examined under polarized light. In the case of larger metal particles that appear black on routine histopathological preparations, the wear debris
load can be so extensive that it obscures the tissue architecture (commonly known
as metallosis).

The characteristic inflammatory cell populations present at the osteolytic lesions
caused by polymer wear are macrophages, foreign body giant cells, and occasional
T lymphocytes scattered among the macrophage infiltrates. These infiltrates can be
extensive and typically form sheet-like formations or well-defined granulomas. Fibroblasts are seen in the soft tissue stroma, and, in relation to bone, osteoblasts and
osteoclasts. The adverse tissue reaction to metal ions and other corrosion products
is further characterized by various types of T and B lymphocyte infiltrates as well
as plasma cells, suggesting the activation of the adaptive immune system. although
the specific terminology and detailed description of this condition are still being
defined, this specific type of tissue reaction has been termed aseptic lymphocyte-
dominated vasculitis-associated lesion (ALVAL) [83, 91].

Neutrophils are not typically encountered in relation to aseptic implant loosening
and their presence in the interface tissue has been considered as indicative of infection [92–95]. Although the specific diagnostic criteria vary, examination of intraoperative peri-implant frozen sections or neutrophil count in the preoperative pseudosynovial fluid aspirates is commonly used in the differential diagnosis between
bacterial implant infection and sterile aseptic loosening driven by biomaterial wear

debris. Recently, it has been suggested that detection of B lymphocytes and plasma cells in the periprosthetic tissues might also be a useful marker of implant-related infection but this hypothesis is yet to be validated by larger clinical studies [88].

13.5.3   Mechanisms of Wear Debris-Induced Chronic Inflammation and Osteolysis

Due to the abundance of macrophages in the histopathological samples from peri-prosthetic osteolytic lesions, the majority of research into the mechanisms of the aseptic loosening has focused on how macrophages interact with wear debris, become activated, and interact with the other cells at the bone–implant interface. Indeed, it has now been widely accepted that macrophages play the key role in the development of aseptic osteolysis [96, 97]. As macrophages are important regulators of not only inflammation but also tissue regeneration and healing, they are emerging as attractive targets for novel treatments for wear debris-induced osteolysis [98].

Macrophages phagocytose wear particles up to a diameter of about 8–10 μm, while larger foreign bodies induce macrophage fusion into foreign body giant cells [99]. Wear particle-activated macrophages assume an inflammatory, M1-like phenotype and secrete various pro-inflammatory cytokines, chemokines, and growth factors that are considered fundamental to the development of osteolytic lesions. According to the established model of aseptic osteolysis, macrophage-derived pro-inflammatory cytokines, such as TNF-α, IL-1β, and IL-6, induce osteoclast formation and activation [77, 78, 84, 86, 97, 100]. These inflammatory cytokines act both directly on osteoclast precursors and indirectly by increasing the production of receptor activator of nuclear factor kappa-B ligand (RANKL) and by suppressing the production of osteoprotegerin (OPG) from local osteoblasts and fibroblasts.

It is thought that the inflammation-induced local imbalance in the RANKL/OPG ratio along with simultaneous wear debris and pro-inflammatory cytokine-induced suppression of osteoblast function are ultimately responsible for the particle-induced osteolysis [77, 78, 84, 86, 97, 100]. At the same time, the multitude of chemokines secreted by wear debris-activated macrophages such as CCL2 (MCP1), CCL3, CCL4, and IL-8 recruit more monocytes and osteoclast precursors to the peri-implant tissues and probably also play a role in the formation of osteoclasts and foreign body giant cells [101]. The role of macrophage-derived growth factors such as macrophage colony-stimulating factor (M-CSF), granulocyte-macrophage colony-stimulating factor (GM-CSF), and vascular endothelial growth factor (VEGF) are less well understood but might further contribute to osteoclast and foreign body granuloma formation and to the development of neo-vasculature.

Among the long list of macrophage-derived inflammatory mediators that have been linked to the development of the aseptic loosening, TNF-α is among the most studied. In cell cultures, various types of wear debris ranging from UHMWPE and PMMA to titanium particles and others induce TNF-α production from human and mouse macrophages [102–104]. TNF-α directly supports osteoclast formation and function, induces upregulation of RANKL and downregulation of OPG in osteo-blasts and fibroblasts, and also suppresses osteoblast differentiation and subsequent bone formation [105]. Indeed, in animal models of particle-induced osteolysis, blockade of TNF-α signaling, either by use of a TNF-α neutralizing antibody or by deleting the TNF-α receptor, significantly reduces osteolysis [106–108]. Finally, retrieval studies have shown local production of TNF-α in the periprosthetic tissues, although contradicting studies also exist [109, 110]. Recently, elevated levels of TNF-α were detected in the circulation of patients with aseptic loosening [111]. Despite evidence of the role of TNF-α in aseptic loosening, a clinical trial investigating the usefulness of TNF-α blocking antibody in the later stages of the peri-prosthetic osteolysis did not prevent development of the lesions possibly due to small sample size or because of the multiple cytokines and the redundancy of the cytokine networks involved in the process [112].

In addition to pro-inflammatory cytokines, the periprosthetic tissue expresses high levels of various chemokines of which CCL2 is among the most studied [101, 113]. Similar to TNF-α, wear debris-stimulated macrophages produce CCL2, and CCL2 blockade reduces macrophage chemotaxis that was induced by conditioned media from wear debris-stimulated macrophages [114, 115]. In a murine model, blockade of CCL2 signaling reduced the systemic homing of reporter macrophages to the area of particle-induced inflammation and also reduced the development of osteolytic lesions [116]. In addition, CCL2 plays a role in osteoclast differentiation and macrophage cell fusion, the significance of which from the point of view of aseptic loosening is yet to be determined [117, 118].

In addition to macrophages, wear debris has effects on the various other types of cells of the interface tissue. For example, wear debris directly suppresses osteoblast differentiation and function and, to some extent, also directly upregulates RANKL production from osteoblasts, thus inhibiting bone formation and promoting bone resorption [119]. Likewise, wear particles directly activate interface tissue fibro-blasts (ITF) to produce inflammatory and chemotactic mediators as well as matrix metalloproteinases that further contribute to the local tissue destruction [120–122].

The role of the adaptive immune system and its cells, namely T and B lympho-cytes, in the inflammatory response against UHMWPE and PMMA wear is more controversial. Cells of the innate immune system can interact directly with bioma-terial debris with their various danger-sensing receptors, while T lymphocytes can recognize their corresponding antigen only if presented to them by professional APCs in a complex multistep process; it is unclear whether such presentable antigen exists in wear debris-induced inflammation. Thus, the involvement of the adaptive immune response in the host response to inorganic orthopedic materials seems un-likely. Indeed, there currently is little evidence that adaptive immunity is involved in the host response to typical orthopedic polymers [123]. For example, in animal models, the foreign body reaction and accompanied inflammation to UHMWPE or PMMA debris is not dependent on the existence of T lymphocyte [124–126]. In contrast, it has been widely speculated that adaptive immunity is involved in the adverse tissue reaction to metal wear and corrosion products, as metal ions can function as haptens forming immunogenic neo-antigens by complexing with host proteins [80–82].

The mechanisms by which various types of wear debris induce the activation of macrophages and other cells are incompletely understood. Current research suggests that innate immune PRRs, especially the family of TLRs, play a central role in wear particle recognition and subsequent cell activation. In an elegant mouse study, it was shown that the magnitude of wear debris-induced osteolysis and inflammation was significantly reduced in mouse deficient of the adaptor protein MyD88, a mediator in the downstream intracellular singling cascade of all TLRs [127]. In another study conducted using TLR knockout mice, TLR2 and TLR4 were found to mediate titanium particle-induced osteolysis, particularly if the particles were contaminated with TLR ligands [128]. Furthermore, retrieval studies have demonstrated that macrophages and other peri-implant tissue cells express various TLRs, and animal models have shown that TLR expression is upregulated in the areas of particle-induced inflammation [88, 129–132].

There is evidence that some forms of wear debris can be directly recognized by certain TLRs. Alkane polymers released from UHMWPE and oxidized by interface tissue macrophages can activate TLR2 signaling [133]. Likewise, nickel and cobalt ions have been show to induce TLR4 signaling [43, 134–136]. Despite these examples of direct TLR engagement by implant-derived wear debris, it seems likely that, in most cases, TLRs and other PRRs are activated by danger signal molecules (DAMPs) adhering to wear particle surfaces rather than direct recognition of wear debris. For example, it has been suggested that bacterial LPS and other PAMPs accumulate to the wear debris surfaces thus opsonizing them for TLR recognition [137, 138]. Potential PAMP sources may include subclinical bacterial infection and hematogenous spread of bacterial structural components from the body's surfaces [139–142]. Additionally, it has been suggested that endogenous DAMPs released from necrotic cells and damaged extracellular matrix could function in a similar manner opsonizing wear debris for TLR recognition [130, 138].

In addition to the role of TLRs in wear debris recognition, there is evidence that phagocytosed wear debris can cause endosomal damage followed by activation of intracellular danger-sensing receptors and assembly of the inflammasome protein complex [143, 144]. The signaling via these various TLRs and intracellular danger-sensing receptors culminates in the activation of NF-κB, which is one of the key inflammatory transcription factors that directly drives the expression of the pro-inflammatory cytokines and chemokines relevant to local osteolysis [145].

A summary of the cell biological mechanisms of aseptic loosening is presented in Fig. 13.2.

13.5.4 Individual Susceptibility to the Peri-implant Osteolysis

The observation that some patients develop aseptic loosening of joint replacements while others do not, even though implants produce substantial amounts of wear particles, has brought about the idea that the risk for developing periprosthetic

Wear Debris

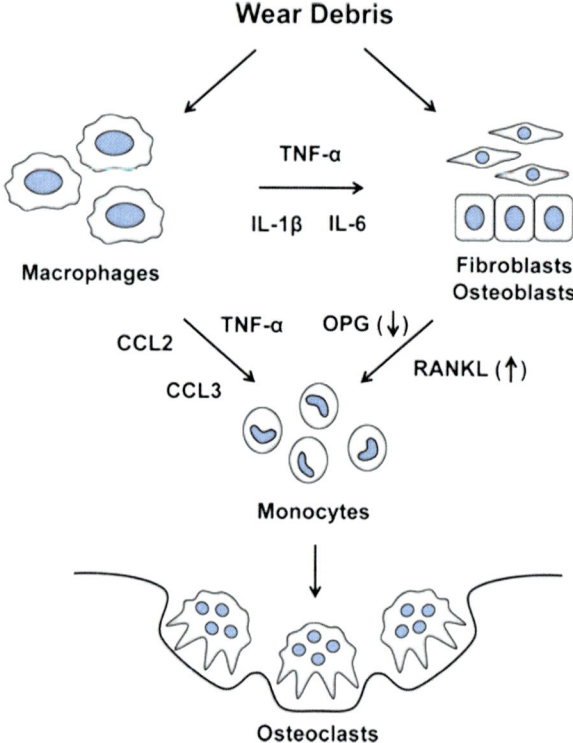

Fig. 13.2 The cell biology of the aseptic osteolysis. Macrophages activated into M1-like pheno-type by UHMWPE or PMMA wear debris produce various chemokines and pro-inflammatory mediators. Chemokines recruit additional monocytes to the interface tissue while pro-inflammatory cytokines lead to increased osteoclastogenesis and bone resorption either directly by acting on osteoclast precursors or indirectly by inducing the production of RANKL and suppression of OPG production from local fibroblasts and osteoblasts. In addition to inducing osteoclast formation, wear debris and macrophage derived pro-inflammatory cytokines suppresses osteoblast differentiation and function as well as induce MMP production from fibroblasts, further contributing to periprosthetic bone loss. The mechanisms by which wear particles are recognized by macrophages and other cells are still incompletely understood but the involvement of TLRs and other families of innate immune pattern-recognition receptors has been speculated. The signaling from these receptors culminates to the activation of the transcription factor NF-κB, which is directly responsible for the transcription of inflammatory mediators. *UHMWPE* ultrahigh molecular weight polyethylene, *PMMA* polymethylmethacrylate, *TNF-α* tumor necrosis factor alpha, *IL* interleukin, *RANKL* receptor activator of nuclear factor kappa-B ligand, *OPG* osteoprotegerin, *MMP* matrix metalloproteinase, *TLR* Toll-like receptor

osteolysis is dissimilar among individuals. For example, it has been reported that polymorphisms in the genes that encode certain pro-inflammatory cytokines or their antagonists, mediators in the RANK–RANKL–OPG signaling pathway, and others related to bone metabolism are associated with increased risk of developing aseptic osteolysis [138, 146]. Although currently not investigated, it seems likely that the risk of developing type IV hypersensitivity to metal wear and metal haptens is genetically determined. Furthermore, in vitro studies have shown that macrophage responses to wear debris stimulation depend on the activation state of the macrophages: In M1 macrophages, the inflammatory wear debris responses are enhanced, and in M2 macrophages they are suppressed [104, 147–149]. Preliminary results provide proof of principle that the biological response to wear debris is not stereotypical but context dependent; it can be speculated that the conditions that promote macrophage activation, e.g., low levels of systemic inflammation associated, for example, to atherosclerosis or obesity, might predispose individuals to more aggressive responses to wear debris and ultimately to aseptic loosening. Currently, there indeed exists some evidence that comorbid diseases such as cardiovascular diseases are associated with increased prosthesis revision rates, but the overall validity of these attractive hypotheses will be confirmed by further clinical studies [150].

13.6 Biological Approaches to Mitigate Wear Particle-Induced Osteolysis

The approaches to diminish wear particle-induced inflammation and osteolysis can be categorized by targeting the cell types, which includes macrophages, osteoclasts and osteoprogenitors/mesenchymal stromal cells (MSCs), and others. Despite the fact that multiple cell types infiltrate a generally confined region with particle exposure, macrophages are still the main immune cells that regulate the pro-inflammatory response. During inflammation, macrophages can (1) infiltrate into the inflammatory site due to enhanced chemokine production; (2) recognize the DAMP and PAMP through TLRs; (3) activate the inflammatory signaling pathways; and (4) secrete pro-inflammatory cytokines and chemokines.

The balance between osteoclastogenesis and osteoblastogenesis is critical in bone remodeling. Inhibition of osteoclast activation or induction of osteoblast differentiation can potentially mitigate wear particle-induced osteolysis. Osteoclast activation can be suppressed by blocking RANKL, TNF-α, or downstream intracellular signaling. On the other hand, osteoblast differentiation can be enhanced by increasing osteoprogenitors/MSCs infiltration or enhancing the osteogenesis signaling.

Strategies to deliver agents that regulate inflammation and osteolysis are also critical. Systemic delivery could be helpful to control the systemic response but may cause more adverse effects at higher doses. In comparison, local delivery such as coating on the implant surface may modulate the local tissue response with fewer side effects. Approaches to mitigate wear particle-induced inflammation and osteolysis are summarized in the following sections (Fig. 13.3).

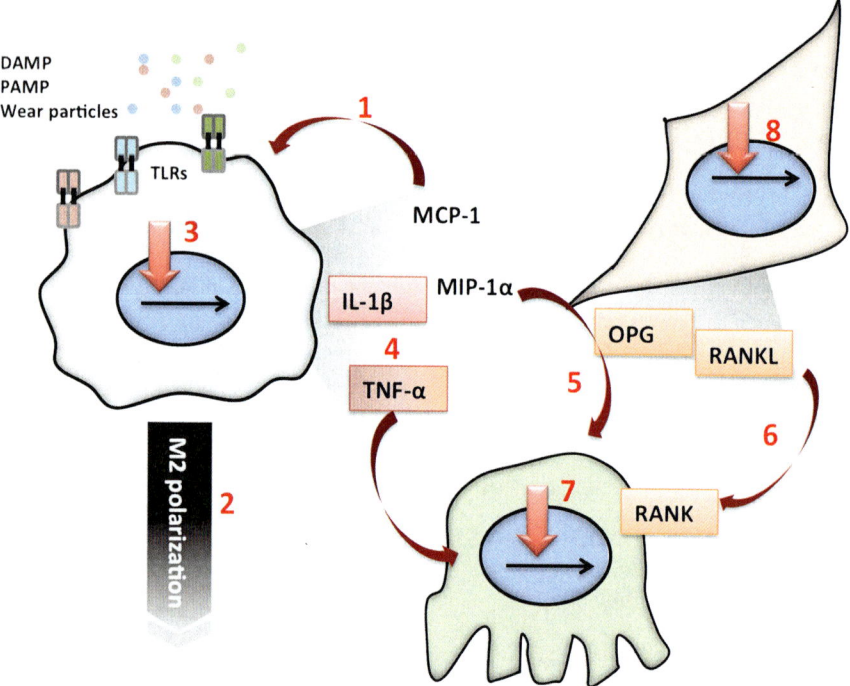

Fig. 13.3 Biological approaches to diminish particle-induced inflammation and osteolysis. Targeting particle-induced biological responses in macrophages *(gray)*, MSCs *(orange)*, and osteoclasts *(green)* may mitigate particle disease. These approaches include *1* suppressing MCP-1-mediated macrophage recruitment; *2* enhancing M2 macrophage polarization (via IL-4); *3* suppressing intracellular signaling in macrophages (NF-κB); *4* suppressing pro-inflammatory cytokines (TNF-α and IL-1β); *5* suppressing osteoclast recruitment (via MIP-1α); *6* suppressing osteoclast activation (RANKL or TNF-α); *7* suppressing active osteoclast function; and *8* enhancing MSCs osteogenesis (increase the number of MSCs or enhancement of osteogenic signaling). *TNF-α* tumor necrosis factor alpha, *IL* interleukin, *RANKL* receptor activator of nuclear factor kappa-B ligand, *OPG* osteoprotegerin, *DAMP* danger signal molecule, *PAMP* pathogen-associated molecular patterns, *TRL* Toll-like receptors, *MCP-1* monocyte chemoattractant protein-1, *MIP* macrophage inflammatory protein

13.6.1   Modulation of Particle-Induced Inflammation

Increasing evidence suggests that both the innate and adaptive immune systems are involved in particle-induced osteolysis. Macrophages are well known for their dominant role in inflammation and consequent osteolysis. Comparably, direct evidence to support the role of other immune cells in particle-induced osteolysis is limited. The biological strategies to modulate macrophage-associated inflammation are summarized below (Fig. 13.3).

13.6.1.1 Suppression of Macrophage Infiltration

MCP-1 mainly regulates infiltration of monocyte/macrophages into the site of inflammation [151]. Several in vitro studies showed that MCP-1 expression in macrophages or fibroblasts could be induced by particle exposure. Huang et al. found that MCP-1 expression was enhanced fourfold in murine macrophages (RAW 264.7) exposed to PMMA and UHMWPE particles [114]. Studies using human macrophages or fibroblasts also showed increased MCP-1 and macrophage inflammatory protein (MIP)-1α after exposure to particles [122, 152]. In addition, the conditioned medium attracted more macrophage migration, which could be blocked by MCP-1 neutralizing antibody.

In vitro and in vivo studies provide evidence that inhibition of macrophage infiltration by blocking MCP-1 signaling can mitigate particle-induced osteolysis. Mutant MCP-1 protein, which has truncated N-terminal amino acids, has been shown to inhibit macrophage migration in vitro [115, 153]. Gibon et al. demonstrated that blocking the MCP-1 receptor binding by MCP-1 receptor antagonist decreased exogenously injected RAW264.7 cell infiltration and osteolysis associated with UHMWPE infusion [116, 154]. Local therapy with MCP-1 inhibitors has been applied to treat inflammatory-associated diseases [155–160]. One potentially novel application of this technology is the coating of implants with MCP-1 inhibitor to control macrophage infiltration while limiting adverse effects [153].

13.6.1.2 Modulation of Macrophage Polarization

Macrophage polarization may also determine the outcome of the inflammatory event. Macrophages exposed to LPS and/or IFN-γ will be polarized into the M1 type (pro-inflammatory). IL-4 or IL-13 exposure will polarize macrophages into the M2 type (anti-inflammatory) [161]. Recent evidence suggested that increased M1 macrophages enhances wear particle-induced inflammation and osteolysis, and polarization of M1 into the M2 phenotype may mitigate the osteolytic response [104, 148, 149, 162]. In vitro studies suggest that IL-4 modulated PMMA and titanium particles-induced pro-inflammatory cytokine expression in human and mouse macrophages in a dose-dependent fashion [148, 149, 162]. Pajarinen et al. [104] showed titanium particle exposure induced higher chemotactic and inflammatory responses in M1 macrophages but lower responses in M2 macrophages when compared to the un-polarized M0 macrophages. In addition, the inflammatory cytokines suppression by IL-4 treatment was more prominent if IL-4 was applied to macrophages before exposure to PMMA [103]. Notably, expression of the M2 cytokine interleukin-1 receptor antagonist (IL-1Ra) was higher when M2 macrophages were polarized from M1 macrophages, compared to direct polarization from M0 macrophages.

In vivo studies also suggested that modulation of macrophage polarization with IL-4 could mitigate particle-induced osteolysis. Rao et al. reported that daily IL-4 injections could mitigate particle-induced osteolysis using an in vivo murine calvarial model [163]. The M1 to M2 ratio was increased in mice with UHMWPE particle exposure, while the ratio in mice with IL-4 injection was similar to mice with no particle exposure. Studies using the murine air-pouch model also observed that daily IL-4 or IL-13 injections reduced bone collagen loss and the number of osteo-clasts induced by polyethylene particles [164]. The effects were more pronounced when IL-4 and IL-13 were both injected, compared to injection with IL-4 or IL-13 injection alone.

13.6.1.3   Suppression of Intracellular NF-κB Signaling

In vitro studies suggest that macrophages exposed to wear particles enhance the downstream intracellular signaling of NF-κB [165, 166]. NF-κB is a key transcrip-tional factor that regulates the production of multiple pro-inflammatory cytokines and chemokines [145]. Exposure of human THP1 macrophages to titanium alloy particle or RAW264.7 cells to UHMWPE particles enhanced NF-κB activity in vi-tro [165, 166]. Suppression of NF-κB activity can be achieved by multiple strate-gies targeted at different transactivation levels [145]. Among them, suppression of DNA binding ability via decoy oligodeoxynucleotide (ODN) is one of the most potent and specific ways to suppress NF-κB transactivation. Decoy ODN is a short double-stranded DNA that mimics the transcriptional binding elements, which can suppress the DNA binding ability by competitive binding of transcription factors to genomic DNA [167]. A recent study using cytokine arrays found that NF-κB decoy ODN could suppress the expression of multiple cytokines and chemokines includ-ing TNF-α, MCP-1, MIP-1α and others, in primary mouse macrophages and human THP-1 cells exposed to UHMWPE particles [165].

13.6.1.4   Suppression of Individual Pro-inflammatory Cytokines

TNF-α is the most potent pro-inflammatory cytokine that is correlated with wear particle-induced osteolysis. Several anti-TNF-α therapies have been developed to modulate particle-induced osteolysis in murine models [108, 168–172]. Etanercept, a decoy receptor for TNF-α, could mitigate wear particle-induced inflammation and osteolysis in both in vitro and in vivo murine calvarial models [108, 168, 169]. However, no differences were observed in patients with acetabular loosening and osteolysis with etanercept treatment compared to placebo control [112]. The results may be due to the limited number of patients in the study or the compensatory roles from other pro-inflammatory cytokines.

Inhibition of IL-1β to mitigate particle-induced inflammation appears to be com-plex. Studies using the air punch model showed that expression of IL-1Ra by viral vectors reduced local inflammation and osteolysis induced by UHMWPE particles [173]. Comparably, the inflammation and osteolysis induced by intramedullary in-fusion of titanium particles showed no significant difference between wild-type and IL-1 receptor-deficient mice [171]. The debate resulting from these two studies may be due to the different models used and the multifactorial role of IL-1 in particle-induced inflammation.

13.6.1.5   Lymphocytes: A Potential Target for Particle-Disease?

Despite the fact that several different types of immune cells have been reported in tissue samples from patients with periprosthetic osteolysis [174–177], the significance of lymphocytes has not been elucidated. T lymphocytes are more likely to be involved in inflammation-mediated osteolysis because of their ability to regulate osteoclastogenesis via secretion of RANKL and IFN-γ [178]. Nevertheless, reports from particle-induced osteolysis murine models remain controversial [124–126, 179, 180]. More studies are required to determine whether lymphocytes are a viable therapeutic target in osteolysis.

13.6.2   Modulation of the Balance Between Osteoclast Activation and Bone Formation

Increased osteoclast activity and reduced bone formation lead to osteolysis. Wear particles exposure to immune cells and fibroblasts induces chemokine-directed recruitment of osteoclast precursors and further activates the infiltrated osteoclasts via RANKL secretion [181]. Recent studies suggest that wear particles reduce cell viability and osteogenic differentiation ability in human osteoblasts [182] or mouse osteoprogenitors [183–185]. Therefore, diminishing of wear particle-induced osteolysis could be approached by reestablishing the balance between osteoclast activation and bone formation.

13.6.2.1   Suppression of Osteoclast Precursor Recruitment and Activation

Yu et al. found that MIP-1α-induced osteoclast recruitment [186]. Blocking of pro-inflammatory chemokines may simultaneously reduce the infiltration of macrophages and osteoclast precursors (see Sect. 4.1.1). Infiltrated osteoclast precursors become differentiated into active osteoclasts in the presence of RANKL. RANKL is mainly secreted by fibroblasts, osteoblasts, and T lymphocytes [187]. Interaction of RANKL with RANK induces osteoclast activation via the NF-κB pathway. OPG downregulates osteoclast activation via competitive binding to RANK. Therefore, the ratio of RANKL/OPG is critical in the determination of osteoclast activity [187]. Increased RANKL [188–191] and reduced OPG expression [181] were reported in patients with osteolysis, both of these scenarios may lead to increased osteolysis.

   Studies using the murine calvarial model demonstrated that blocking RANKL signaling via administration of OPG [192, 193] or a RANKL antagonist [194] or using transgenic RANK-deficient mice [194] could efficiently prevent wear particle-induced osteolysis. A potent RANKL inhibitory antibody, Denosumab, has been used in clinical treatment of post-menopausal osteoporosis [195]. The biological effects of Denosumab in periprosthetic osteolysis patients remain to be explored.

13.6.2.2 Suppression of Activated Osteoclasts

The function of activated osteoclasts requires intracellular signaling including NF-κB and mitogen-activated protein (MAP) kinase, which could be therapeutic targets. NF-κB signaling is essential for RANKL or TNF-α-induced osteoclast activation [196, 197]. Inhibition of NF-κB activity could impair osteoclast formation, function, and survival [196]. Rakshit et al. found that exposure of osteoclasts to PMMA and titanium particle activated MAP kinase in vitro. Inhibition of p38 kinase activity, but not extracellular signal-regulated kinases (ERK) or c-Jun N-terminal kinases (JNK), suppressed the IL-6-mediated anti-osteoclastogenesis function [197].

Many inhibitors of mature osteoclast function have been developed, including inhibitors of cathepsin K [198–200], the osteoclast ATPase proton pump [201], the vitronectin receptor [202], and src tyrosine kinase. More studies are required to evaluate their potential application in wear particle-associated osteoclast activation.

13.6.2.3 Modulation of Osteogenesis by Infiltrated Osteoprogenitors/MSCs

Gibon et al. reported that infiltration of osteoprogenitors/MSCs inhibited the reduction of bone mineral density and osteoclast activation induced by UHMWPE particles in a continuous murine femoral infusion model. Inhibition of MIP-1α by its receptor antagonist diminished the protective effect of MSCs [154]. These effects might be mediated by the unique characteristic of MSCs to modulate new bone formation, attenuate osteoclast activity, and decrease inflammation [203].

Protection of endogenous MSC functions might also reduce particle-induced osteolysis [183, 184]. Lin et al. demonstrated that inhibition of NF-κB activity via decoy ODN protected MSCs viability in the presence of UHMWPE particles in vitro [204]. Recent studies suggested that NF-κB activation in MSCs or osteoprogenitors could reduce osteogenesis ability in the presence of inflammatory stimuli [204, 205]. Inhibition of NF-κB activity in osteoprogenitors/MSCs could maintain the bone-forming capacity of osteoprogenitors/osteoblasts.

13.6.3 Conclusion

Biological approaches to diminish particle-induced periprosthetic osteolysis could be achieved via modulation of the pro-inflammatory response and the balance between osteoclast and osteoblast function. Combined treatments with multiple strategies may increase the therapeutic efficiency due to the complex crosstalk among different cells in the local tissue environment. Notably, inhibition of NF-κB could mitigate particle-induced osteolysis via attenuation of macrophage-associated chronic inflammation, reduction in osteoclast activation, and protection of viability and differentiation ability of MSCs. However, systemic suppression of NF-κB activity may cause significant adverse effects because NF-κB plays a dominant role in immune functions. Local delivery such as coating on the implanted device could potentially be a safer strategy. Suppression of NF-κB activity may be complicated due to multiple feedback regulation pathways [206, 207]. Further in vivo studies are required to establish a safe and effective delivery strategy to treat periprosthetic osteolysis.

13.7   Summary

Despite the long-term success of TJR, wear of the bearing surfaces is expected. At the same time, patients with a joint replacement desire to live more active lifestyles, including performing impact-loading sporting activities. Wear of the joint surfaces continues silently for quite some time before pain and diminished function due to chronic synovitis, periprosthetic osteolysis, loosening, or fracture occurs, secondary to the biological reaction to wear debris. Although some of the newer bearing couples appear to have improved longevity (e.g., highly cross-linked polyethylene, highly polished cobalt-chrome, current ceramic femoral heads), others have lead to disastrous complications. Indeed, some of these prosthesis "innovations" have been associated with an increased early revision rate and substantial morbidity (e.g., certain metal-on-metal implants).

Polyethylene and PMMA are two polymeric materials that will continue to be used as part of a bearing couple or as a method to stabilize a prosthesis in bone, respectively. Nevertheless, these materials still degrade with time, and debris will continue to be produced. Advances in materials science, engineering, biology, and medicine will undoubtedly continue to improve joint replacement, with the hope that one day, artificial joints will last a lifetime and allow full, unrestricted activities. Until this goal is realized, researchers and clinicians should pursue a more comprehensive understanding of host-implant interactions and attempt to modulate these interactions to obtain specific, desired objectives. In particular, methods to enhance implant osseointegration and mitigate chronic inflammation are sorely needed.

Acknowledgments This work has been funded in part by the National Institutes of Health (NIH) Grants 2R01AR055650 and 1R01AR063717, the Ellenburg Chair in Surgery at Stanford University, and the Jane and Aatos Erkko Foundation.

References

1. Kurtz S, et al. Projections of primary and revision hip and knee arthroplasty in the United States from 2005 to 2030. J Bone Joint Surg Am. 2007;89(4):780–5.
2. Hussey M. US markets for large-joint reconstructive implants 2010; 2012.
3. Nho SJ, et al. The burden of hip osteoarthritis in the United States: epidemiologic and economic considerations. J Am Acad Orthop Surg. 2013;21(Suppl. 1):S1–6.
4. National Joint Registry for England, Wales and Northern Ireland. 10th Annual Report; 2013.

5. Cram P, et al. Clinical characteristics and outcomes of Medicare patients undergoing total hip arthroplasty, 1991–2008. JAMA. 2011;305(15):1560–7.
6. Prudhon JL, Ferreira A, Verdier R. Dual mobility cup: dislocation rate and survivorship at ten years of follow-up. Int Orthop. 2013;37(12):2345–50.
7. Bozic KJ, et al. Variation in hospital-level risk-standardized complication rates following elective primary total hip and knee arthroplasty. J Bone Joint Surg Am. 2014;96(8):640–7.
8. Cushner F, et al. Complications and functional outcomes after total hip arthroplasty and total knee arthroplasty: results from the global orthopaedic registry (GLORY). Am J Orthop (Belle Mead NJ). 2010;39(Suppl. 9):22–8.
9. Charnley J. Low friction arthroplasty of the hip. New York: Springer; 1979.
10. Maloney WJ, et al. Bone lysis in well-fixed cemented femoral components. J Bone Joint Surg Br. 1990;72(6):966–70.
11. Jones LC, Hungerford DS. Cement disease. Clin Orthop Relat Res. 1987;225:192–206.
12. Willert HG, Bertram H, Buchhorn GH. Osteolysis in alloarthroplasty of the hip. The role of bone cement fragmentation. Clin Orthop Relat Res. 1990;258:108–21.
13. Willert HG, Bertram H, Buchhorn GH. Osteolysis in alloarthroplasty of the hip. The role of ultra-high molecular weight polyethylene wear particles. Clin Orthop Relat Res. 1990;258:95–107.
14. Schmalzried TP, et al. The mechanism of loosening of cemented acetabular components in total hip arthroplasty. Analysis of specimens retrieved at autopsy. Clin Orthop Relat Res. 1992;274:60–78.
15. Campbell P, et al. Isolation of predominantly submicron-sized UHMWPE wear particles from periprosthetic tissues. J Biomed Mater Res. 1995;29(1):127–31.
16. Fornasier V, Wright J, Seligman J. The histomorphologic and morphometric study of asymptomatic hip arthroplasty. A postmortem study. Clin Orthop Relat Res. 1991;271:272–82.
17. Ingham E, Fisher J. Biological reactions to wear debris in total joint replacement. Proc Inst Mech Eng H. 2000;214(1):21–37.
18. Goldring SR, et al. The synovial-like membrane at the bone-cement interface in loose total hip replacements and its proposed role in bone lysis. J Bone Joint Surg Am. 1983;65(5):575–84.
19. Goodman SB, et al. A clinical-pathologic-biochemical study of the membrane surrounding loosened and nonloosened total hip arthroplasties. Clin Orthop Relat Res. 1989;244:182–7.
20. Mandelin J, et al. Pseudosynovial fluid from loosened total hip prosthesis induces osteoclast formation. J Biomed Mater Res B Appl Biomater. 2005;74(1):582–8.
21. Kadoya Y, et al. Bone formation and bone resorption in failed total joint arthroplasties: histomorphometric analysis with histochemical and immunohistochemical technique. J Orthop Res. 1996;14(3):473–82.
22. Revell PA. The combined role of wear particles, macrophages and lymphocytes in the loosening of total joint prostheses. J R Soc Interface. 2008;5(28):1263–78.
23. Santavirta S, et al. Aggressive granulomatous lesions associated with hip arthroplasty. Immunopathol Stud. J Bone Joint Surg Am. 1990;72(2):252–8.
24. Tallroth K, et al. Aggressive granulomatous lesions after hip arthroplasty. J Bone Joint Surg Br. 1989;71(4):571–5.
25. Geesink RG, de Groot K, Klein CP. Bonding of bone to apatite-coated implants. J Bone Joint Surg Br. 1988;70(1):17–22.
26. de Groot K, et al. Plasma sprayed coatings of hydroxylapatite. J Biomed Mater Res. 1987;21(12):1375–81.
27. Geesink RG, de Groot K, Klein CP. Chemical implant fixation using hydroxyl-apatite coatings. The development of a human total hip prosthesis for chemical fixation to bone using hydroxyl-apatite coatings on titanium substrates. Clin Orthop Relat Res. 1987;225:147–70.
28. Engh CA, et al. Histological and radiographic assessment of well functioning porous-coated acetabular components. A human postmortem retrieval study. J Bone Joint Surg Am. 1993;75(6):814–24.

29. Cook SD, Thomas KA, Haddad RJ, Jr. Histologic analysis of retrieved human porous-coated total joint components. Clin Orthop Relat Res. 1988;234:90–101.

30. Jacobs JJ, et al. Postmortem retrieval of total joint replacement components. J Biomed Mater Res. 1999;48(3):385–91.

31. Maloney WJ, et al. The Otto Aufranc award. Skeletal response to well fixed femoral components inserted with and without cement. Clin Orthop Relat Res. 1996;333:15–26.

32. McAuley JP, Culpepper WJ, Engh CA. Total hip arthroplasty. Concerns with extensively porous coated femoral components. Clin Orthop Relat Res. 1998;355:182–8.

33. Kurtz SM, et al. Anisotropy and oxidative resistance of highly crosslinked UHMWPE after deformation processing by solid-state ram extrusion. Biomaterials. 2006;27(1):24–34.

34. McKellop H, et al. Development of an extremely wear-resistant ultra high molecular weight polyethylene for total hip replacements. J Orthop Res. 1999;17(2):157–67.

35. Sato T, et al. Wear resistant performance of highly cross-linked and annealed ultra-high molecular weight polyethylene against ceramic heads in total hip arthroplasty. J Orthop Res. 2012;30(12):2031–7.

36. Dumbleton JH, et al. The basis for a second-generation highly cross-linked UHMWPE. Clin Orthop Relat Res. 2006;453:265–71.

37. Kyomoto M, et al. Poly(2-methacryloyloxyethyl phosphorylcholine) grafting and vitamin E blending for high wear resistance and oxidative stability of orthopedic bearings. Biomaterials. 2014;35(25):6677–86.

38. Oral E, et al. The effect of an additional phosphite stabilizer on the properties of radiation cross-linked vitamin E blends of UHMWPE. J Orthop Res. 2014;32(6):757–61.

39. Tomita N, et al. Prevention of fatigue cracks in ultrahigh molecular weight polyethylene joint components by the addition of vitamin E. J Biomed Mater Res. 1999;48(4):474–8.

40. Kamath AF, Prieto H, Lewallen DG. Alternative bearings in total hip arthroplasty in the young patient. Orthop Clin N Am. 2013;44(4):451–62.

41. Billi F, Campbell P. Nanotoxicology of metal wear particles in total joint arthroplasty: a review of current concepts. J Appl Biomater Biomech. 2010;8(1):1–6.

42. Gill HS, et al. Molecular and immune toxicity of CoCr nanoparticles in MoM hip arthroplasty. Trends Mol Med. 2012;18(3):145–55.

43. Konttinen YT, Pajarinen J. Adverse reactions to metal-on-metal implants. Nat Rev Rheumatol. 2013;9(1):5–6.

44. Jeffers JR, Walter WL. Ceramic-on-ceramic bearings in hip arthroplasty: state of the art and the future. J Bone Joint Surg Br. 2012;94(6):735–45.

45. Hoffmann J, Akira S. Innate immunity. Curr Opin Immunol. 2013;25(1):1–3.

46. Janeway CA, Jr, Medzhitov R. Innate immune recognition. Annu Rev Immunol. 2002;20:197–216.

47. Medzhitov R, Janeway CA, Jr. Innate immunity: the virtues of a nonclonal system of recognition. Cell. 1997;91(3):295–8.

48. Medzhitov R. Toll-like receptors and innate immunity. Nat Rev Immunol. 2001;1(2):135–45.

49. Murphy K, et al. Janeway's immunobiology. 8th ed., vol. xix. New York: Garland Science; 2012. p. 868.

50. Medzhitov R, Janeway CA, Jr. Decoding the patterns of self and nonself by the innate immune system. Science. 2002;296(5566):298–300.

51. De Smet K, Contreras R. Human antimicrobial peptides: defensins, cathelicidins and histatins. Biotechnol Lett. 2005;27(18):1337–47.

52. Canny G, Levy O. Bactericidal/permeability-increasing protein (BPI) and BPI homologs at mucosal sites. Trends Immunol. 2008;29(11):541–7.

53. Canny G, et al. Lipid mediator-induced expression of bactericidal/permeability-increasing protein (BPI) in human mucosal epithelia. Proc Natl Acad Sci U S A. 2002;99(6):3902–7.

54. Elsbach P, Weiss J. Role of the bactericidal/permeability-increasing protein in host defence. Curr Opin Immunol. 1998;10(1):45–9.

55. Levy O. A neutrophil-derived anti-infective molecule: bactericidal/permeability-increasing protein. Antimicrob Agents Chemother. 2000;44(11):2925–31.

56. Campbell EL, Serhan CN, Colgan SP. Antimicrobial aspects of inflammatory resolution in the mucosa: a role for proresolving mediators. J Immunol. 2011;187(7):3475–81.
57. van Wetering S, et al. Defensins: key players or bystanders in infection, injury, and repair in the lung? J Allergy Clin Immunol. 1999;104(6):1131–8.
58. Wan M, et al. Leukotriene B4/antimicrobial peptide LL-37 proinflammatory circuits are mediated by BLT1 and FPR2/ALX and are counterregulated by lipoxin A4 and resolvin E1. FASEB J. 2011;25(5):1697–705.
59. Aderem A, Underhill DM. Mechanisms of phagocytosis in macrophages. Annu Rev Immunol. 1999;17:593–623.
60. Janeway CA, Jr. Approaching the asymptote? Evolution and revolution in immunology. Cold Spring Harb Symp Quant Biol. 1989;54(Pt 1):1–13.
61. Kumagai Y, Akira S. Identification and functions of pattern-recognition receptors. J Allergy Clin Immunol. 2010;125(5):985–92.
62. Kawai T, Akira S. Innate immune recognition of viral infection. Nat Immunol. 2006;7(2):131–7.
63. Long EO. Regulation of immune responses through inhibitory receptors. Annu Rev Immunol. 1999;17:875–904.
64. Karre K, et al. Selective rejection of H-2-deficient lymphoma variants suggests alternative immune defence strategy. Nature. 1986;319(6055):675–8.
65. Oldenborg PA, et al. Role of CD47 as a marker of self on red blood cells. Science. 2000;288(5473):2051–4.
66. Ricklin D, et al. Complement: a key system for immune surveillance and homeostasis. Nat Immunol. 2010;11(9):785–97.
67. Janssen WJ, Henson PM. Cellular regulation of the inflammatory response. Toxicol Pathol. 2012;40(2):166–73.
68. Takeuchi O, Akira S. Pattern recognition receptors and inflammation. Cell. 2010;140(6):805–20.
69. Meyer KC. Immunity, inflammation, and aging. In: Rahman I, Bagchi D, editors. Inflammation, advancing age and nutrition. San Diego: Academic; 2014. p. 29–38 (Chapter 3).
70. Serbina NV, et al. Monocyte-mediated defense against microbial pathogens. Annu Rev Immunol. 2008;26:421–52.
71. Nauseef WM. How human neutrophils kill and degrade microbes: an integrated view. Immunol Rev. 2007;219:88–102.
72. Abraham SN, St John AL. Mast cell-orchestrated immunity to pathogens. Nat Rev Immunol. 2010;10(6):440–52.
73. Vivier E, et al. Innate or adaptive immunity? The example of natural killer cells. Science. 2011;331(6013):44–9.
74. Willert HG, Ludwig J, Semlitsch M. Reaction of bone to methacrylate after hip arthroplasty: a long-term gross, light microscopic, and scanning electron microscopic study. J Bone Joint Surg Am. 1974;56(7):1368–82.
75. Willert HG, Semlitsch M. Reactions of the articular capsule to wear products of artificial joint prostheses. J Biomed Mater Res. 1977;11(2):157–64.
76. Purdue PE, et al. The cellular and molecular biology of periprosthetic osteolysis. Clin Orthop Relat Res. 2007;454:251–61.
77. Gallo J, et al. Particle disease: biologic mechanisms of periprosthetic osteolysis in total hip arthroplasty. Innate Immun. 2013;19(2):213–24.
78. Goodman SB, Gibon E, Yao Z. The basic science of periprosthetic osteolysis. Instr Course Lect. 2013;62:201–6.
79. Schmalzried TP, Jasty M, Harris WH. Periprosthetic bone loss in total hip arthroplasty. Polyethylene wear debris and the concept of the effective joint space. J Bone Joint Surg Am. 1992;74(6):849–63.
80. Davies AP, et al. An unusual lymphocytic perivascular infiltration in tissues around contemporary metal-on-metal joint replacements. J Bone Joint Surg Am. 2005;87(1):18–27.
81. Willert HG, et al. Metal-on-metal bearings and hypersensitivity in patients with artificial hip joints. A clinical and histomorphological study. J Bone Joint Surg Am. 2005;87(1):28–36.

82. Korovessis P, et al. Metallosis after contemporary metal-on-metal total hip arthroplasty. Five to nine-year follow-up. J Bone Joint Surg Am. 2006;88(6):1183–91.

83. Campbell P, et al. Histological features of pseudotumor-like tissues from metal-on-metal hips. Clin Orthop Relat Res. 2010;468(9):2321–7.

84. Holt G, et al. The biology of aseptic osteolysis. Clin Orthop Relat Res. 2007;460:240–52.

85. Cordova LA, et al. Orthopaedic implant failure: aseptic implant loosening—the contribution and future challenges of mouse models in translational research. Clin Sci (Lond). 2014;127(5):277–93.

86. Sundfeldt M, et al. Aseptic loosening, not only a question of wear: a review of different theories. Acta Orthop. 2006;77(2):177–97.

87. Mirra JM, Marder RA, Amstutz HC. The pathology of failed total joint arthroplasty. Clin Orthop Relat Res. 1982 Oct;(170):175–83.

88. Pajarinen J, et al. Profile of toll-like receptor-positive cells in septic and aseptic loosening of total hip arthroplasty implants. J Biomed Mater Res A. 2010;94(1):84–92.

89. Gallo J, et al. Contributions of human tissue analysis to understanding the mechanisms of loosening and osteolysis in total hip replacement. Acta Biomater. 2014;10(6):2354–66.

90. Goodman SB, et al. Cellular profile and cytokine production at prosthetic interfaces. Study of tissues retrieved from revised hip and knee replacements. J Bone Joint Surg Br. 1998;80(3):531–9.

91. Natu S, et al. Adverse reactions to metal debris: histopathological features of periprosthetic soft tissue reactions seen in association with failed metal on metal hip arthroplasties. J Clin Pathol. 2012;65(5):409–18.

92. Zimmerli W, Trampuz A, Ochsner PE. Prosthetic-joint infections. N Engl J Med. 2004;351(16):1645–54.

93. Bori G, et al. Usefulness of histological analysis for predicting the presence of microorganisms at the time of reimplantation after hip resection arthroplasty for the treatment of infection. J Bone Joint Surg Am. 2007;89(6):1232–7.

94. Ghanem E, et al. Cell count and differential of aspirated fluid in the diagnosis of infection at the site of total knee arthroplasty. J Bone Joint Surg Am. 2008;90(8):1637–43.

95. Kanner WA, Saleh KJ, Frierson HF, Jr. Reassessment of the usefulness of frozen section analysis for hip and knee joint revisions. Am J Clin Pathol. 2008;130(3):363–8.

96. Ingham E, Fisher J. The role of macrophages in osteolysis of total joint replacement. Biomaterials. 2005;26(11):1271–86.

97. Nich C, et al. Macrophages-key cells in the response to wear debris from joint replacements. J Biomed Mater Res A. 2013;101(10):3033–45.

98. Goodman SB, et al. Novel biological strategies for treatment of wear particle-induced periprosthetic osteolysis of orthopaedic implants for joint replacement. J R Soc Interface. 2014;11(93):20130962.

99. Anderson JM, Rodriguez A, Chang DT. Foreign body reaction to biomaterials. Semin Immunol. 2008;20(2):86–100.

100. Landgraeber S, et al. The pathology of orthopedic implant failure is mediated by innate immune system cytokines. Mediat Inflamm. 2014;2014:185150.

101. Goodman SB, Ma T. Cellular chemotaxis induced by wear particles from joint replacements. Biomaterials. 2010;31(19):5045–50.

102. Nakashima Y, et al. Signaling pathways for tumor necrosis factor-alpha and interleukin-6 expression in human macrophages exposed to titanium-alloy particulate debris in vitro. J Bone Joint Surg Am. 1999;81(5):603–15.

103. Antonios JK, et al. Macrophage polarization in response to wear particles in vitro. Cell Mol Immunol. 2013;10(6):471–82.

104. Pajarinen J, et al. The response of macrophages to titanium particles is determined by macrophage polarization. Acta Biomater. 2013;9(11):9229–40.

105. Osta B, Benedetti G, Miossec P. Classical and paradoxical effects of TNF-alpha on bone homeostasis. Front Immunol. 2014;5:48.

106. Merkel KD, et al. Tumor necrosis factor-alpha mediates orthopedic implant osteolysis. Am J Pathol. 1999;154(1):203–10.
107. Schwarz EM, et al. Tumor necrosis factor-alpha/nuclear transcription factor-kappaB signaling in periprosthetic osteolysis. J Orthop Res. 2000;18(3):472–80.
108. Childs LM, et al. Efficacy of etanercept for wear debris-induced osteolysis. J Bone Miner Res. 2001;16(2):338–47.
109. Xu JW, et al. Tumor necrosis factor-alpha (TNF-alpha) in loosening of total hip replacement (THR). Clin Exp Rheumatol. 1996;14(6):643–8.
110. Koulouvaris P, et al. Expression profiling reveals alternative macrophage activation and impaired osteogenesis in periprosthetic osteolysis. J Orthop Res. 2008;26(1):106–16.
111. Chaganti RK, et al. Elevation of serum tumor necrosis factor alpha in patients with periprosthetic osteolysis: a case-control study. Clin Orthop Relat Res. 2014;472(2):584–9.
112. Schwarz EM, et al. Use of volumetric computerized tomography as a primary outcome measure to evaluate drug efficacy in the prevention of peri-prosthetic osteolysis: a 1-year clinical pilot of etanercept vs. placebo. J Orthop Res. 2003;21(6):1049–55.
113. Jamsen E, et al. Characterization of macrophage polarizing cytokines in the aseptic loosening of total hip replacements. J Orthop Res. 2014;32(9):1241–6.
114. Huang Z, et al. Effects of orthopedic polymer particles on chemotaxis of macrophages and mesenchymal stem cells. J Biomed Mater Res A. 2010;94(4):1264–9.
115. Yao Z, et al. Mutant monocyte chemoattractant protein 1 protein attenuates migration of and inflammatory cytokine release by macrophages exposed to orthopedic implant wear particles. J Biomed Mater Res A. 2014;102(9):3291–7.
116. Gibon E, et al. Selective inhibition of the MCP-1-CCR2 ligand-receptor axis decreases systemic trafficking of macrophages in the presence of UHMWPE particles. J Orthop Res. 2012;30(4):547–53.
117. Kyriakides TR, et al. The CC chemokine ligand, CCL2/MCP1, participates in macrophage fusion and foreign body giant cell formation. Am J Pathol. 2004;165(6):2157–66.
118. Kim MS, Day CJ, Morrison NA. MCP-1 is induced by receptor activator of nuclear factor-{kappa}B ligand, promotes human osteoclast fusion, and rescues granulocyte macrophage colony-stimulating factor suppression of osteoclast formation. J Biol Chem. 2005;280(16):16163–9.
119. O'Neill SC, et al. The role of osteoblasts in peri-prosthetic osteolysis. Bone Joint J. 2013;95-B(8):1022–6.
120. Yao J, et al. The potential role of fibroblasts in periprosthetic osteolysis: fibroblast response to titanium particles. J Bone Miner Res. 1995;10(9):1417–27.
121. Manlapaz M, Maloney WJ, Smith RL. In vitro activation of human fibroblasts by retrieved titanium alloy wear debris. J Orthop Res. 1996;14(3):465–72.
122. Yaszay B, et al. Fibroblast expression of C-C chemokines in response to orthopaedic biomaterial particle challenge in vitro. J Orthop Res. 2001;19(5):970–6.
123. Goodman SB. Wear particles, periprosthetic osteolysis and the immune system. Biomaterials. 2007;28(34):5044–8.
124. Goodman S, et al. T-lymphocytes are not necessary for particulate polyethylene-induced macrophage recruitment. Histologic studies of the rat tibia. Acta Orthop Scand. 1994;65(2):157–60.
125. Jiranek W, et al. Tissue response to particulate polymethylmethacrylate in mice with various immune deficiencies. J Bone Joint Surg Am. 1995;77(11):1650–61.
126. Taki N, et al. Polyethylene and titanium particles induce osteolysis by similar, lymphocyte-independent, mechanisms. J Orthop Res. 2005;23(2):376–83.
127. Pearl JI, et al. Role of the Toll-like receptor pathway in the recognition of orthopedic implant wear-debris particles. Biomaterials. 2011;32(24):5535–42.
128. Greenfield EM, et al. Bacterial pathogen-associated molecular patterns stimulate biological activity of orthopaedic wear particles by activating cognate toll-like receptors. J Biol Chem. 2010;285(42):32378–84.

129. Takagi M, et al. Toll-like receptors in the interface membrane around loosening total hip replacement implants. J Biomed Mater Res A. 2007;81(4):1017–26.
130. Lahdeoja T, et al. Toll-like receptors and aseptic loosening of hip endoprosthesis-a potential to respond against danger signals? J Orthop Res. 2010;28(2):184–90.
131. Pajarinen J. Toll-like receptors and macrophage polarization in the loosening of total hip replacements [dissertation]. Helsinki, Finland: University of Helsinki; 2012. http://urn.fi/URN:ISBN:978-952-9657-66-7. Accessed 12 Dec 2014.
132. Valladares RD, et al. Toll-like receptors-2 and 4 are overexpressed in an experimental model of particle-induced osteolysis. J Biomed Mater Res A. 2014;102(9):3004–11.
133. Maitra R, et al. Immunogenecity of modified alkane polymers is mediated through TLR1/2 activation. Plos One. 2008;3(6):e2438.
134. Schmidt M, et al. Crucial role for human Toll-like receptor 4 in the development of contact allergy to nickel. Nat Immunol. 2010;11(9):814–64.
135. Rachmawati D, et al. Transition metal sensing by toll-like receptor-4: next to nickel, cobalt and palladium are potent human dendritic cell stimulators. Contact Dermat. 2013;68(6):331–8.
136. Tyson-Capper AJ, et al. Metal-on-metal hips: cobalt can induce an endotoxin-like response. Ann Rheum Dis. 2013;72(3):460–1.
137. Greenfield EM, et al. Does endotoxin contribute to aseptic loosening of orthopedic implants? J Biomed Mater Res Part B—Appl Biomater. 2005;72B(1):179–85.
138. Greenfield EM. Do genetic susceptibility, toll-like receptors, and pathogen-associated molecular patterns modulate the effects of wear? Clin Orthop Relat Res. 2014;472(12):3709–17.
139. Xing ZQ, et al. Accumulation of LPS by polyethylene particles decreases bone attachment to implants. J Orthop Res. 2006;24(5):959–66.
140. Tatro JM, et al. The balance between endotoxin accumulation and clearance during particle-induced osteolysis in murine calvaria. J Orthop Res. 2007;25(3):361–69.
141. Kobayashi N, et al. Molecular identification of bacteria from aseptically loose implants. Clin Orthop Relat Res. 2008;466(7):1716–25.
142. Schafer P, et al. Prolonged bacterial culture to identify late periprosthetic joint infection: a promising strategy. Clin Infect Dis. 2008;47(11):1403–9.
143. Maitra R, et al. Endosomal damage and TLR2 mediated inflammasome activation by alkane particles in the generation of aseptic osteolysis. Mol Immunol. 2009;47(2–3):175–84.
144. Burton L, et al. Orthopedic wear debris mediated inflammatory osteolysis is mediated in part by NALP3 inflammasome activation. J Orthop Res. 2013;31(1):73–80.
145. Lin TH, et al. Chronic inflammation in biomaterial-induced periprosthetic osteolysis: NF-kappaB as a therapeutic target. Acta Biomater. 2014;10(1):1–10.
146. Del Buono A, Denaro V, Maffulli N. Genetic susceptibility to aseptic loosening following total hip arthroplasty: a systematic review. Br Med Bull. 2012;101:39 55.
147. Trindade MC, et al. Interferon-gamma exacerbates polymethylmethacrylate particle-induced interleukin-6 release by human monocyte/macrophages in vitro. J Biomed Mater Res. 1999;47(1):1–7.
148. Trindade MC, et al. Interleukin-4 inhibits granulocyte-macrophage colony-stimulating factor, interleukin-6, and tumor necrosis factor-alpha expression by human monocytes in response to polymethylmethacrylate particle challenge in vitro. J Orthop Res. 1999;17(6):797–802.
149. Rao AJ, et al. Revision joint replacement, wear particles, and macrophage polarization. Acta Biomater. 2012;8(7):2815–23.
150. Jamsen E, et al. Comorbid diseases as predictors of survival of primary total hip and knee replacements: a nationwide register-based study of 96 754 operations on patients with primary osteoarthritis. Ann Rheum Dis. 2013;72(12):1975–82.
151. Tuan RS, et al. What are the local and systemic biologic reactions and mediators to wear debris, and what host factors determine or modulate the biologic response to wear particles? J Am Acad Orthop Surg. 2008;16(Suppl 1):S42–8.

152. Nakashima Y, et al. Induction of macrophage C-C chemokine expression by titanium alloy and bone cement particles. J Bone Joint Surg Br. 1999;81(1):155–62.

153. Keeney M, et al. Mutant MCP-1 protein delivery from layer-by-layer coatings on orthopedic implants to modulate inflammatory response. Biomaterials. 2013;34(38):10287–95.

154. Gibon E, et al. Effect of a CCR1 receptor antagonist on systemic trafficking of MSCs and polyethylene particle-associated bone loss. Biomaterials. 2012;33(14):3632–8.

155. Ohtani K, et al. Antimonocyte chemoattractant protein-1 gene therapy reduces experimental in-stent restenosis in hypercholesterolemic rabbits and monkeys. Gene Ther. 2004;11(16):1273–82.

156. Kitamoto S, Egashira K, Takeshita A. Stress and vascular responses: anti-inflammatory therapeutic strategy against atherosclerosis and restenosis after coronary intervention. J Pharmacol Sci. 2003;91(3):192–6.

157. Egashira K, et al. Local delivery of anti-monocyte chemoattractant protein-1 by gene-eluting stents attenuates in-stent stenosis in rabbits and monkeys. Arterioscler Thromb Vasc Biol. 2007;27(12):2563–8.

158. Kitamoto S, Egashira K. Gene therapy targeting monocyte chemoattractant protein-1 for vascular disease. J Atheroscler Thromb. 2002;9(6):261–5.

159. Nakano K, et al. Catheter-based adenovirus-mediated anti-monocyte chemoattractant gene therapy attenuates in-stent neointima formation in cynomolgus monkeys. Atherosclerosis. 2007;194(2):309–16.

160. Schepers A, et al. Anti-MCP-1 gene therapy inhibits vascular smooth muscle cells proliferation and attenuates vein graft thickening both in vitro and in vivo. Arterioscler Thromb Vasc Biol. 2006;26(9):2063–9.

161. Martinez FO, et al. Macrophage activation and polarization. Front Biosci. 2008;13:453–61.

162. Im GI, Han JD. Suppressive effects of interleukin-4 and interleukin-10 on the production of proinflammatory cytokines induced by titanium-alloy particles. J Biomed Mater Res. 2001;58(5):531–6.

163. Rao AJ, et al. Local effect of IL-4 delivery on polyethylene particle induced osteolysis in the murine calvarium. J Biomed Mater Res A. 2013;101(7):1926–34.

164. Wang Y, et al. Inhibitory effects of recombinant IL-4 and recombinant IL-13 on UHMWPE-induced bone destruction in the murine air pouch model. J Surg Res. 2013;180(2):e73–81.

165. Lin TH, et al. Suppression of wear-particle-induced pro-inflammatory cytokine and chemokine production in macrophages via NF-kappaB decoy oligodeoxynucleotide: a preliminary report. Acta Biomater. 2014;10(8):3747–55.

166. Akisue T, et al. The effect of particle wear debris on NFkappaB activation and pro-inflammatory cytokine release in differentiated THP-1 cells. J Biomed Mater Res. 2002;59(3):507–15.

167. Osako MK, Nakagami H, Morishita R. Modification of decoy oligodeoxynucleotides to achieve the stability and therapeutic efficacy. Curr Top Med Chem. 2012;12(15):1603–7.

168. Schwarz EM, Looney RJ, O'Keefe RJ. Anti-TNF-alpha therapy as a clinical intervention for periprosthetic osteolysis. Arthritis Res. 2000;2(3):165–8.

169. Schwarz EM, et al. Quantitative small-animal surrogate to evaluate drug efficacy in preventing wear debris-induced osteolysis. J Orthop Res. 2000;18(6):849–55.

170. Peng X, et al. Efficient Inhibition of wear debris-induced inflammation by locally delivered siRNA. Biochem Biophys Res Commun. 2008;377(2):532–7.

171. Bragg B, et al. Histomorphometric analysis of the intramedullary bone response to titanium particles in wild-type and IL-1R1 knock-out mice: a preliminary study. J Biomed Mater Res B Appl Biomater. 2008;84(2):559–70.

172. Pollice PF, et al. Oral pentoxifylline inhibits release of tumor necrosis factor-alpha from human peripheral blood monocytes: a potential treatment for aseptic loosening of total joint components. J Bone Joint Surg Am. 2001;83-A(7):1057–61.

173. Yang S, et al. IL-1Ra and vIL-10 gene transfer using retroviral vectors ameliorates particle-associated inflammation in the murine air pouch model. Inflamm Res. 2002;51(7):342–50.

174. Lin TH, et al. Exposure of polyethylene particles induces interferon-gamma expression in a natural killer T lymphocyte and dendritic cell coculture system in vitro: a preliminary study. J Biomed Mater Res A. 2015;103(1):71–5.

175. Roato I, et al. Osteoclastogenesis in peripheral blood mononuclear cell cultures of periprosthetic osteolysis patients and the phenotype of T cells localized in periprosthetic tissues. Biomaterials. 2010;31(29):7519–25.
176. Hallab NJ, et al. In vitro reactivity to implant metals demonstrates a person-dependent association with both T-cell and B-cell activation. J Biomed Mater Res A. 2010;92(2):667–82.
177. Huss RS, et al. Synovial tissue-infiltrating natural killer cells in osteoarthritis and periprosthetic inflammation. Arthritis Rheum. 2010;62(12):3799–805.
178. Le Goff B, et al. Osteoclasts in RA: diverse origins and functions. Joint Bone Spine. 2013;80(6):586–91.
179. Sandhu J, et al. The role of T cells in polyethylene particulate induced inflammation. J Rheumatol. 1998;25(9):1794–9.
180. Childs LM, et al. Effect of anti-tumor necrosis factor-alpha gene therapy on wear debris-induced osteolysis. J Bone Joint Surg Am. 2001;83-A(12):1789–97.
181. Purdue PE, et al. The central role of wear debris in periprosthetic osteolysis. HSS J. 2006;2(2):102–13.
182. Lochner K, et al. The potential role of human osteoblasts for periprosthetic osteolysis following exposure to wear particles. Int J Mol Med. 2011;28(6):1055–63.
183. Chiu R, et al. Polymethylmethacrylate particles inhibit osteoblastic differentiation of bone marrow osteoprogenitor cells. J Biomed Mater Res A. 2006;77(4):850–6.
184. Chiu R, et al. Kinetics of polymethylmethacrylate particle-induced inhibition of osteoprogenitor differentiation and proliferation. J Orthop Res. 2007;25(4):450–7.
185. Chiu R, et al. Polymethylmethacrylate particles inhibit osteoblastic differentiation of MC3T3-E1 osteoprogenitor cells. J Orthop Res. 2008;26(7):932–6.
186. Yu X, et al. CCR1 chemokines promote the chemotactic recruitment, RANKL development, and motility of osteoclasts and are induced by inflammatory cytokines in osteoblasts. J Bone Miner Res. 2004;19(12):2065–77.
187. Khosla S. Minireview: the OPG/RANKL/RANK system. Endocrinology. 2001;142-(12):5050–5.
188. Horiki M, et al. Localization of RANKL in osteolytic tissue around a loosened joint prosthesis. J Bone Miner Metab. 2004;22(4):346–51.
189. Haynes DR, et al. The osteoclastogenic molecules RANKL and RANK are associated with periprosthetic osteolysis. J Bone Joint Surg Br. 2001;83(6):902–11.
190. Gehrke T, et al. Receptor activator of nuclear factor kappaB ligand is expressed in resident and inflammatory cells in aseptic and septic prosthesis loosening. Scand J Rheumatol. 2003;32(5):287–94.
191. Mandelin J, et al. Imbalance of RANKL/RANK/OPG system in interface tissue in loosening of total hip replacement. J Bone Joint Surg Br. 2003;85(8):1196–201.
192. Goater JJ, et al. Efficacy of ex vivo OPG gene therapy in preventing wear debris induced osteolysis. J Orthop Res. 2002;20(2):169–73.
193. Ulrich-Vinther M, et al. Recombinant adeno-associated virus-mediated osteoprotegerin gene therapy inhibits wear debris-induced osteolysis. J Bone Joint Surg Am. 2002;84-A(8):1405–12.
194. Childs LM, et al. In vivo RANK signaling blockade using the receptor activator of NF-kappaB: Fc effectively prevents and ameliorates wear debris-induced osteolysis via osteoclast depletion without inhibiting osteogenesis. J Bone Miner Res. 2002;17(2):192–9.
195. Tella SH, Gallagher JC. Biological agents in management of osteoporosis. Eur J Clin Pharmacol. 2014;70(11):1291–301.
196. Soysa NS, Alles N. NF-kappaB functions in osteoclasts. Biochem Biophys Res Commun. 2009;378(1):1–5.
197. Rakshit DS, et al. Wear debris inhibition of anti-osteoclastogenic signaling by interleukin-6 and interferon-gamma. Mechanistic insights and implications for periprosthetic osteolysis. J Bone Joint Surg Am. 2006;88(4):788–99.

198. Bossard MJ, et al. Mechanism of inhibition of cathepsin K by potent, selective 1,5-diacyl-carbohydrazides: a new class of mechanism-based inhibitors of thiol proteases. Biochemistry. 1999;38(48):15893–902.

199. Lark MW, et al. A potent small molecule, nonpeptide inhibitor of cathepsin K (SB 331750) prevents bone matrix resorption in the ovariectomized rat. Bone. 2002;30(5):746–53.

200. Shakespeare WC, et al. Novel bone-targeted Src tyrosine kinase inhibitor drug discovery. Curr Opin Drug Discov Dev. 2003;6(5):729–41.

201. Visentin L, et al. A selective inhibitor of the osteoclastic V-H(+)-ATPase prevents bone loss in both thyroparathyroidectomized and ovariectomized rats. J Clin Invest. 2000;106(2):309–18.

202. Lark MW, et al. Antagonism of the osteoclast vitronectin receptor with an orally active nonpeptide inhibitor prevents cancellous bone loss in the ovariectomized rat. J Bone Miner Res. 2001;16(2):319–27.

203. Le Blanc K, Mougiakakos D. Multipotent mesenchymal stromal cells and the innate immune system. Nat Rev Immunol. 2012;12(5):383–96.

204. Lin TH, Sato T, Barcay KR, Waters H, Loi F, Zhang R, Pajarinen J, Egashira K, Yao Z, Goodman SB. NF-kappaB Decoy Oligodeoxynucleotide Enhanced Osteogenesis in Mesenchymal Stem Cells Exposed to Polyethylene Particle. Tissue engineering Part A 2014.

205. Chang J, et al. NF-kappaB inhibits osteogenic differentiation of mesenchymal stem cells by promoting beta-catenin degradation. Proc Natl Acad Sci U S A. 2013;110(23):9469–74.

206. Stein B, et al. Cross-coupling of the NF-kappa B p65 and Fos/Jun transcription factors produces potentiated biological function. EMBO J. 1993;12(10):3879–91.

207. Lee EG, et al. Failure to regulate TNF-induced NF-kappaB and cell death responses in A20-deficient mice. Science. 2000;289(5488):2350–4.

Stuart Goodman MD, PhD is the Robert L. and Mary Ellenburg professor of surgery and professor with tenure in the Department of Orthopaedic Surgery at Stanford University; he has a courtesy appointment with the Department of Bioengineering. Dr. Goodman's clinical practice and research concentrates on adult reconstructive surgery; his basic science interests center on biocompatibility of orthopedic implants and musculoskeletal tissue regeneration and repair.

Chapter 14
Artificial Antigen-Presenting Cells: Biomimetic Strategies for Directing the Immune Response

Randall A. Meyer and Jordan J. Green

14.1 Introduction

Immunotherapy can be broadly defined as modulation of the immune system to achieve a medicinal benefit. Several examples of modern immunotherapies include the use of cytokines for general immune activation in disease such as chronic granulomatous disorder [1] or the use of immunosuppressive drugs such as steroids in the treatment of autoimmune disorders [2]. As our understanding of the immune system continues to develop, so does our capability to intervene therapeutically. Over the past few decades, there has been a significant increase in new immunotherapies developed to treat a variety of diseases. Among these is the manipulation of autologous antigen-presenting cells (APCs) to achieve a therapeutic effect. Although partly successful in some limited cases such a prostate cancer [3], autologous antigen-presenting cell-based therapy suffers from several drawbacks such as limited potency due to susceptibility to immune regulation and the high costs and labor involved with cellular therapy [4]. Therefore, there is an impetus to develop an alternative strategy to achieve the same desired goal of adaptive immune system modulation.

One promising technology is the artificial APC (aAPC) to serve as a biomimetic platform to modulate the adaptive immune system. aAPCs are synthetic constructs that are designed to recapitulate the natural process of T cell activation by biological APCs. Through the presentation of crucial APC signal proteins on a surface, aAPCs

J. J. Green (✉)
Departments of Biomedical Engineering, Materials Science and Engineering, Ophthalmology, and Neurosurgery, Translational Tissue Engineering Center, and Institute for Nanobiotechnology, Johns Hopkins School of Medicine, Baltimore, USA
e-mail: green@jhu.edu

R. A. Meyer
Department of Biomedical Engineering, Translational Tissue Engineering Center, and Institute for Nanobiotechnology, Johns Hopkins School of Medicine, 400 N Broadway, Smith 5017, Baltimore, MD 21231, USA

© Springer International Publishing Switzerland 2015

257

L. Santambrogio (ed.), *Biomaterials in Regenerative Medicine and the Immune System*, DOI 10.1007/978-3-319-18045-8_14

are capable of directing a T cell response very similar to the natural APC. aAPCs possess certain advantages over autologous biological APCs. Generally, an aAPC is based on a plastic particle that is much easier to produce and maintain, thus making the aAPC technology capable of being an "off-the-shelf" therapy [4]. In addition, aAPCs are not derived from an immune cell progenitor and thus are resistant to the immune suppression that can inhibit the activity of natural APCs. aAPCs have also been proven to be equally or more effective than the autologous APC at immune system stimulation in certain scenarios. Taken together, these traits of the aAPC make it an attractive platform for immunotherapy with applications to treat various diseases such as cancer, infectious disease, and autoimmune disorders.

In this chapter, we summarize the relevant immunobiology as well as several criteria that should be considered in the design of aAPCs, including fully synthetic aAPCs as well as engineered biological cells that function as cellular aAPCs. In addition, we review the approaches that have been taken to develop the aAPC as well as the broad application of these constructs in various biomedical scenarios. Continued development of aAPCs will greatly benefit medicine as an engineered synthetic strategy to control the power of the immune response.

14.2 The Natural APC

A strong understanding of the natural APC/T cell interaction is required in order to generate the most effective aAPC for antigen-specific T cell modulation. Here, we summarize the basic APC physiology as well as the important aspects of the APC/T cell interaction that should be considered in the design of this technology.

14.2.1 The Role of APCs in Innate and Adaptive Immunity

Adaptive immune modulation interactions occur primarily through the professional APC. Nearly, every cell in the body acts as an "APC" due to their ability to present major histocompatibility complex (MHC) class I restricted antigens on the surface. In addition, B cells and cytotoxic T cells can present antigens through the MHC class II surface protein in order to be activated further by helper T cells. Despite these capabilities to present antigens on the surface, generally, these cell types are subject to effector functions of the immune system and do not heavily impact the course of an immune response. Professional aAPCs have the capabilities to strongly influence the immune system through the presentation of antigens restricted to MHC class I and class II as well as companion signals to either activate or suppress the T cell response. The two types of cells that have been identified as "professional APC" include dendritic cells (DCs) and macrophages. Generally, DCs are considered the primary APC of the body with specific physiological adaptation for this function [5]. The DC will be used in subsequent explanation of APC physiology.

APCs have a special function in immunity as the bridge between innate and adaptive immune systems. Immature DCs generally reside in the periphery and continuously screen environmental antigens through rapid macropinocytosis. Although extracellular antigens are generally exclusively presented by MHC class II molecules (to engage helper CD4+ T cells), APCs can partake in cross-presentation of these antigens to present them on MHC class I molecules (to interact with cytotoxic CD8+ T cells). DCs are activated into their mature, immunogenic form through encountering a danger signal. This can include pathogen-associated molecular patterns (PAMPs), inflammatory cytokines, or previously activated T cells. Upon activation, the DC will halt rapid cellular uptake, upregulate antigen-presenting MHC molecules, produce more costimulatory surface signals, and synthesize chemokine homing receptors for lymph-node-targeted chemotaxis. Upon entering the lymph nodes, the DC will interact with target T cells to activate them to carry out their effector functions [6]. Activation will also result in a positive feedback loop with macrophages and natural killer cells in the innate immune system through the production of cytokines interleukin (IL)-12 and interferon gamma (IFNγ) [7]. In the absence of the correct danger signal, DCs have also been shown to promote tolerance through regulatory T cell induction [8].

14.2.2   Signal 1, Signal 2, and Signal 3: APC Communication with a T cell

The most important aspect in the design of an aAPC is recapitulation of the signals that a natural APC relays to a T cell upon cell contact. These signals are in part mediated by cell surface receptor interactions as well as soluble cytokines. Together, three signals enable tight control of the nature of the immune response (Fig. 14.1).

Signal 1 is considered the "recognition" signal. At the protein level, this signal occurs as the result of the interaction of the T cell receptor (TCR) and the antigen loaded in the MHC. This signal is central to the specificity of the immune system as it establishes the identity of target antigen. The T cell will seek in its subsequent effector function. In addition to the TCR (CD3) signal, the T cell differentiation marker (CD-4 for helper T cells and CD-8 for cytotoxic T cells) participates in the signal 1 protein machinery. It has been proposed that these co-receptors help to concentrate the MHC protein to permit for continued signaling despite relatively low affinity of the MHC/TCR interaction [9].

Signal 2 can be described as the "outcome" signal. While signal 1 establishes the identity of the antigen in question, signal 2 directs the nature of the T cell response to the APC/T cell binding event. For T cell activation, the classic stimulatory pathway is through CD28 on the T cell surface [9]. In addition, 41BB and CD40 on the surface of T cells have been shown to have an activation effect in the context of CD4+ T cells engaging a CD8+ T cell to further activate it [10]. Other important types of signal 2 are the ones that induce anergy in target T cells. This is important for the establishment of peripheral tolerance for autologous antigens. Some of the proteins on the T cell that have been shown to be involved in this suppressive form

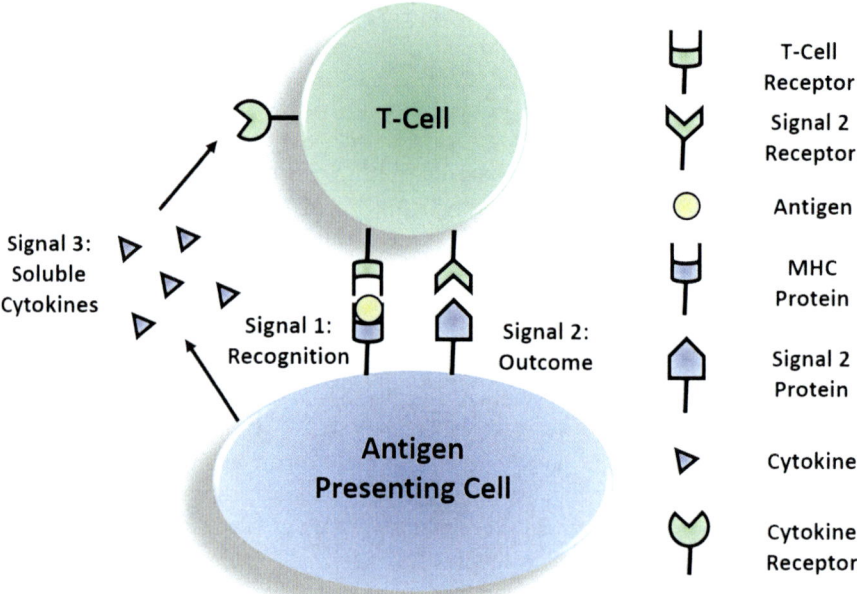

Fig. 14.1 The natural APC relays several signals to T cells to decide the fate of the interaction. Signal 1 is what is used in APC/T cell recognition and ensures antigen specificity in the APC's effect on the immune system. Typically, it is mediated by the surface bound TCR on the T cell and the antigen bound to the MHC on the surface of the APC. Signal 2 determines the overall outcome of the interaction. This can be used to activate the T cell to proliferate and target the antigen in question, become anergic or apathetic to the antigen presented, or differentiate into a regulatory T cell to actively suppress an immune response to that antigen. Signal 2 is mediated by a variety of surface bound proteins. Signal 3 consists of soluble cytokines released by the T cell during the interaction and provides further direction for the nature of the T cell response. *MHC* major histocompatibility complex, *TCR* T cell receptor

of signal 2 include the Fas receptor for apoptosis, CTLA-4, and PD-1 for inhibition of immune activation, as well as the absence of signal 2 for induction of anergy [11].

Signal 3 provides additional direction for the T cell response and typically consists of soluble cytokines secreted by the APC during T cell engagement. Signal 3 has been shown to direct the character of the type of T cell response that is elicited upon activation. Generally, this signal has been determined to be a product of polarization to one of two types of DC for the Th1 response or the Th2 response. One of the most well-characterized cytokines involved in this process is IL-12 secreted by APCs upon engagement. This has been linked to the development of a Th1-type response, directed toward the elimination of intracellular pathogens. Recently, a fourth potential signal has also been characterized in DCs [12]. DCs have been shown to induce upregulation of chemokine receptors in T cells by the use of Vitamin A and Vitamin D. IL-12 has also been shown to have an effect on the homing capabilities of T cells [13].

14.2.3 The APC/T cell Interface

Given the majority of the intercellular signaling that occurs between the APC and the T cell occurs at the APC/T cell interface, this is a critical factor that must be taken into account in the mimicry of the APC by the aAPC. Typically, when a natural APC engages a T cell, it has been shown that there is a dynamic rearrangement of surface proteins involved in immune signaling. The mature complex that forms has been termed the immunological synapse (IS). This synapse has been shown to have a distinct molecular composition with the TCR and the CD28 costimulatory molecule in an inner central supra-molecular activation cluster (cSMAC), integrin-based adhesion molecules such as lymphocyte function-associated antigen (LFA)-3 concentrated outside the cSMAC in the peripheral supra-molecular activation cluster (pSMAC), and actin-rich exterior zones termed distal supra-molecular activation clusters (dSMAC) (Fig. 14.2). The role of the immune synapse in TCR signaling and subsequent acquisition of effector function is not completely understood, but it has been confirmed to play an important role. Although a concentric ring formation is the most well studied, it is not the only formation that the IS can assume [14].

T cell/APC IS formation is a highly specific process that occurs only in the event that the TCR and MHC bound antigen match with appropriate specificity. T cells are

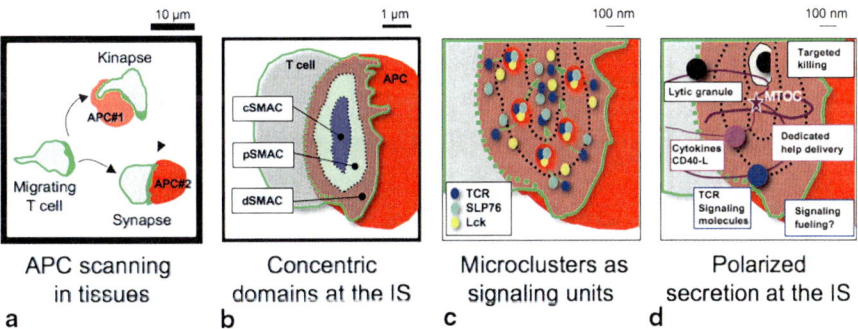

Fig. 14.2 The natural antigen-presenting cells (APC)/T cell interaction is a dynamic process at the cellular and molecular level. **a** Natural T cells will rapidly scan through multiple APCs through the formation of immune kinapses. If there is sufficient compatibility between the T cell and the APC, the two cells will form a much tighter interaction termed the immunological synapse. **b** The immunological synapse consists of three concentric domains of surface receptor termed the cSMAC, dSMAC, and pSMAC. **c** Productive T cell signaling begins as microclusters of TCRs and various signaling proteins form in the dSMAC, bind to the APC, and are transported to the cSMAC where signaling ends. **d** After interaction with an APC, the T cell will polarize its secretory vesicles to the site of the IS formation to allow for either targeted killing or targeted delivery of helper cytokines. Reproduced with permission from Ref. [15]. *cSMAC* central supra-molecular activation cluster, *dSMAC* distal supra-molecular activation clusters, *pSMAC* peripheral supra-molecular activation cluster, *TCR* T cell receptor, *IS* immunological synapse

specialized for high-throughput scanning of multiple APCs to find a correct match. Such temporary interactions between the T cell and the APC are termed immune kinapses. In the event that a T cell finds its specific APC partner (as dictated by the time scale of the TCR and MHC–antigen interaction), an IS forms. Despite the fact that the TCR and costimulatory molecule CD28 are concentrated in the center of the IS, the signaling events are generally not localized to this region. Instead, the TCR and associated costimulatory signals form as microclusters (on the order of 100 nm in size) which quickly travel to the center of the IS propelled by actin polymerization. TCR signaling at this time is characterized by low affinity and rapid signaling events that are sustained as the cluster moves. The TCR signaling is strongest in the dSMAC and weakens as the cluster reaches the cSMAC. The ultimate fate of the TCR once it reaches the cSMAC is unclear but depending on the antigen, it has been shown to continue signaling or be eliminated from the surface membrane [15].

Another important consequence of the IS formation is the polarization of the T cell membrane for appropriate effector function. Upon IS formation, the T cell will translocate its secretory granules to the portion of the membrane that is engaged with the APC in a region of the cSMAC. For helper T cell effector functions, this allows for the localized delivery of stimulatory cytokines and signals to target cells such as B cells and other APCs. In the context of killer T cells, this relocation to the cSMAC can be very important as it allows for localized release of cytotoxic compounds that are prevented from escaping into the peripheral cell space by the "sealed" regions of the pSMAC. As such, a CD8+ cell can be highly selective in the target cell it destroys [15].

14.3   Design Considerations for the aAPC

Given the complex nature of the APC interaction with the T cell, there are many parameters to consider in the fabrication of aAPCs. Many aAPC characteristics have been engineered and have led to insight into effective biomimicry conditions. These parameters include size and shape of the aAPC, surface proteins for relaying the critical signal 1 and signal 2, protein release to mimic signal 3, and recreation of the IS through the use of a fluidic membrane (Fig. 14.3).

14.3.1   Size and Shape of the APC

One of the most important parameters described in the literature for the design of an aAPC is size. Generally, it has been considered that the optimal size for an aAPC is at the micron/cellular level. For aAPCs, it has been shown that size can have an important effect on aAPC efficacy [16]. For in vitro efficacy using a poly(lactic-co-glycolic acid) (PLGA)-based aAPC, it has been shown that micron scale aAPCs could induce a threefold increase in IL-2 production of CD8+ T cells for the same protein dose as nanoscale aAPCs [17]. However, due to biodistribution

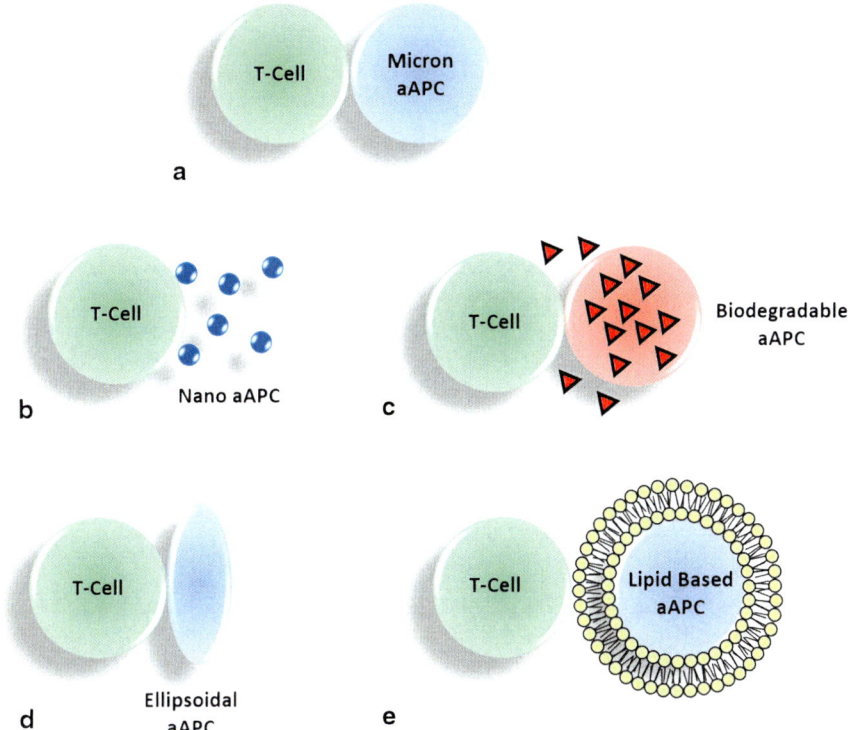

Fig. 14.3 Various parameters have been considered in the design of an optimal aAPC. **a** The classic aAPC consists of a micron scale artificial bead bound to surface proteins to serve as signal 1 and signal 2. **b** Newer aAPC designs have focused on the use of nanoparticles as a scaffold for immune protein immobilization. Although not as efficient as microparticles at stimulating T cells, the nano-aAPC has the added benefit of more favorable in vivo properties such as biodistribution. **c** Classic aAPC platforms utilized spherical microparticles, whereas new data suggest that a nonspherical shape such as an ellipsoid is preferable due to the higher radius of curvature for aAPC interaction. **d** Biodegradable aAPCs can be used to enable local delivery of costimulatory cytokines and signal 3. **e** New aAPC designs can enable the dynamic rearrangement of surface proteins through the use of lipid-based aAPCs either with liposomes or supported lipid bilayers. *aAPCs* artificial antigen-presenting cells

considerations, micron sized aAPCs may not be the optimal choice for in vivo use. Recently, it has been shown that nanodimensional aAPCs can induce nearly a tenfold increase in IFNγ production upon in vivo immunization compared to micron scale aAPCs [18]. This was linked to a favorable biodistribution of the nanoscale aAPC, in that it demonstrated superior draining to the lymph nodes, the natural site of T cell/APC interaction. Continued investigation into the impact of size will be important in the design of the optimal aAPC.

Another important biophysical parameter that has recently come to light in the design of the aAPC is shape. As was described previously, the surface area of contact between the T cell and the APC is critical for proper stimulation. Rigid, spherical aAPCs allow for minimal surface area contact with the target T cell. However, rod-shaped aAPCs can increase the available area of contact between the aAPC and the T cell through the increased radius of curvature of the long axis. This was shown to have a very powerful effect on the activation of CD8+ lymphocytes. Spherical and ellipsoidal aAPCs, with the same volume and the same protein on their surfaces, were compared for their capability to induce antigen-specific proliferation, and the ellipsoidal aAPCs were shown to increase the expansion of CD8+ T cells by up to 20-fold [19]. In addition, live-cell confocal imaging revealed a higher prevalence of aAPC conjugates as well as a preference for the long axis of the aAPC in conjugation. An in vivo effect was also seen with the ellipsoidal aAPC-mediating stronger tumor immunity compared to the spherical aAPC [19]. Although traditionally neglected in terms of aAPC design, shape is another important design consideration in the fabrication of aAPCs.

14.3.2 Proteins for Antigen Presentation and Costimulation: Recreation of Signal 1 and Signal 2

The choice of surface protein and the density at which it is presented have also been shown to be important for aAPC function. For signal 1, there have been two predominantly used strategies to trigger the TCR. The first is the use of an anti-CD3 antibody as a direct trigger for the TCR (CD3). Although this is a simple way to deliver signal 1 and is used ubiquitously throughout the research community for the study of T cell biology, it lacks the antigen specificity that a natural APC possesses. Therefore, this strategy can be used to trigger T cells in vitro, but in vivo there would be the complication of antigen specificity. The second, and more popular for disease-based applications of aAPCs, is the direct use of a MHC class I (for CD8+ T cells) or MHC class II (for CD4+ cells) loaded with the cognate antigen. Although slightly more complicated than the use of a monoclonal antibody, the added benefit of antigen specificity is absolutely necessary for the use of aAPCs in therapeutic applications [4].

Recreation of the second signal delivered upon APC engagement is generally accomplished by the use of a monoclonal antibody to directly trigger a costimulatory molecule on the surface of the T cell. The most popular targets for signal 2 are CD28 and 41BB. Although classic aAPCs have focused on the use of anti CD28, 41BB has been shown to effectively stimulate lymphocytes [20]. More recent advances in aAPC technology have shown that a blend of these two signal protein may be optimal in the design of an aAPC. A 25:75 blend of CD28 and 41BB induced a—three- to fivefold greater expansion of CD8+ cells than anti-CD28 or anti-41BB alone [21]. In addition, it was demonstrated that these cells had a strong effector and memory function in vitro.

Surface density of the aAPC proteins is another parameter to be considered in the design of an effective aAPC. Recent advances in nanofabrication have allowed for nanometer scale control over surface density and orientation of stimulatory proteins. Utilizing anti-CD3 as a signal 1, Matic et al. demonstrated that 1000 proteins/μm^2 spacing of anti-CD3 proteins was optimal at triggering T cell activation as evidenced by increased production of IL-2 and enhanced rates of proliferation (upon the addition of anti-CD28 as a costimulatory signal) compared to 120 and 60 proteins/μm^2 [22].

14.3.3 Soluble Cytokines and Recreation of Signal 3

Although not critical for aAPC function, the capability to deliver a third signal through soluble cytokines has been of interest in the design of APCs. The strategy to achieve this in biological APCs has been to transfect the cell with a gene to produce a soluble cytokine. Although this is an interesting possibility in the design of engineered biological APCs, it can be difficult to control the amount of the soluble cytokine that is released. Currently, the technology has been limited to use in studying the biology of certain cytokines released during the APC/T cell engagement including IL-21 which has been shown to have opposing effects of activation and suppression [23]. The engineering of a third signal into aAPCs can be accomplished through the use of a controlled release system as the core particle. In this manner, the core of the particle can be designed to optimize signal 3, whereas the particle surface is designed to optimize signal 1 and signal 2. Controlled release from the core can be accomplished through the use of a biodegradable polymer such as PLGA. The advantage of the use of signal 3 in a biodegradable acellular system is that the rate of delivery can be controlled based on the type of biomaterial. Currently, there has been limited use of a third signal in the design of an acellular aAPC. Steenblock et al. demonstrated the feasibility of this approach by encapsulation and controlled release of IL-2 in a PLGA-based aAPC [24]. It was shown that there was a three- to fourfold increase in the expansion of T cells upon stimulation of encapsulated IL-2 as opposed to exogenous IL-2 due to local concentration release effects. This approach has the potential to lead to much higher local concentrations of soluble cytokines at a T cell surface than the concentrations that can be achieved through systemic in vivo administration [24].

14.3.4 Fluidic Membranes and the Recreation of the Immunological Synapse

As the IS has been identified as a key component affecting the signaling and activation of the T cell, enabling dynamic rearrangement of proteins on the surface of an aAPC should be a design consideration. Cellular-based aAPCs offer a naturally fluid plasma membrane to achieve this goal. The majority of acellular aAPCs to date have been predominantly based on the use of immobilized proteins on a particle surface. Liposome-based aAPCs have been utilized as a tool to investigate aspects of the IS, but not in a therapeutic fashion [25].

Supported lipid bilayers (SLBs) appear to be one viable strategy to achieve a fluidic in a particulate aAPC. Planar SLBs were utilized in a study that illustrated the effects of receptor patterning on TCR signaling [26]. In addition, a wide variety of fabrication methods exist for the synthesis of SLBs on a particle surface and have been utilized various biomedical applications [27]. Continued investigation into the potential of an aAPC with a fluidic membrane will be of interest in the next generation of aAPC technology.

14.4   Cellular aAPCs

While the focus of this chapter is on acellular APCs, such as those fabricated from beads or particles, engineered biological cells, or what we refer to as cellular aAPCs, have also been constructed. Cellular aAPCs derived from the genetic modification of existing cell lines remain a popular choice for the modulation of T cell activity for therapeutic purposes (Table 14.1). Although cell-based therapies are immensely expensive compared to acellular therapies, many of these cell-based strategies have been entered into clinical trials for various diseases. The following is a summary of the key advances in the field of cellular-based aAPCs.

14.4.1   K562 Cell Line

The most popular cell line utilized in the design of a cellular aAPC is the K562 human leukemia line [28]. This cell type is frequently used due to the absence of any endogenous MHC molecules on its surface, thus allowing for the expression of the only chosen MHC for antigen-specific cell therapy. In addition, this cell line does not express any of the costimulatory molecules characteristic of a natural aAPC. This allows for greater control over the type of stimulatory protein that is delivered upon aAPC/T cell engagement. They also lack expression of inhibitory molecules that can be used by APCs to suppress the activity of T cells. In addition, these cells can easily be genetically modified and have a history of safe use in humans. Taken together, these advantages make the K562 cell line an attractive platform for cellular aAPC development [28].

Several different categories of the K562 cell-line-based APC have been developed based on the application and immune cell to be targeted. One type of K562 aAPC that has been developed was a cell line to activate CD4+ T cells. Through transduction of genes encoding the human leukocyte antigen (HLA)-DR MHC class II protein as well as the costimulatory signals for CD28, the authors were able to generate a stable aAPC cell line with expression of all desired proteins on the surface [29]. The authors were able to induce exogenous antigen presentation through the added transduction of a gene encoding the Fcγ receptor, thus enabling receptor-mediated endocytosis. Fold expansion of antigen-specific T cells was noted to be

Table 14.1 Examples of the various existing aAPCs and their efficacies

Type	Material	Size	Signal 1	Signal 2	Efficacy	Reference
Cellular	K562 cells	10–20 μm	HLA MHC class II	B7-1	1000-fold expansion of CD4+ T cells	[29]
	K562 cells	10–20 μm	HLA MHC class I	B7-1	1,000,000-fold expansion of CD8+ T cells over 1.5 years	[46]
	K562 cells	10–20 μm	Anti-CD3	Anti-CD28/41BB-L	1000–10,000-fold expansion of CD8+ T cells	[30]
	NIH/3T3 murine fibroblasts	18 μm	HLA MHC class I	B7-1	Twofold greater CD8+ expansion compared to autologous APC	[34]
	NIH/3T3 murine fibroblasts	18 μm	HLA MHC class I	B7-1	100-fold expansion of CMV CD8+ T cells over 14 days	[48]
Acellular	PS	5 μm	K^b MHC class I	B7-1/B7-2	Threefold secretion of IL-2 compared to soluble protein	[37]
	PS	6 μm	HLA MHC class I	Blend of Anti-CD28/anti 41BB	Fivefold increase of CD8+ T cells of blended signal 2 compared to individual signal 2	[21]
	PGA	7 μm	Anti-CD3	Anti-CD28	Twofold greater CD8+ T cell expansion compared to plate-bound ligand	[38]
	PLGA encapsulating IL-2	6 μm	Anti-CD3	Anti-CD28	Threefold greater secretion of IL-2 by T cells compared to soluble signal	[24]
	Nonspherical PLGA	4–5 μm	D^b MHC class I IgG dimer	Anti-CD28	20-fold greater of expansion of CD8+ T cells of ellipsoidal compared to spherical particles	[19]
	Liposomes	60 nm	Anti-CD3	Anti-CD28	175-fold expansion of CD8+ T cells over 14 days	[14]
	Magnetic Dynal® Beads	4.5 μm	HLA MHC class I	Anti-CD28	1000-fold expansion of antigen-specific CD8+ T cells	[41]
	Magnetic Dynal® Beads	4.5 μm	HLA MHC class I	Anti-Fas	Fivefold reduction in antigen-specific CD8+ T cells	[52]
	SWNT bundles	5–10-nm tubes/100-μm bundles	Anti-CD3	None	Tenfold increase in secretion of IL-2 compared to soluble antibody	[43]
	QDs	30 nm	D^b MHC class I	Anti-CD28	15-fold expansion of CD8+ T cells over 7 days	[44]

PS polystyrene. *PGA* poly glycolic acid. *PLGA* poly(lactic-*co*-glycolic acid). *SWNT* single-walled carbon nanotubes. *QDs* quantum dots. *HLA* human leukocyte antigen. *MHC* major histocompatibility complex. *APC* antigen-presenting cell. *IL* interleukin. *CMV* cytomegalovirus

greater the 1000-fold over the course of 150 days with repeated stimulation by the cellular aAPCs. In addition, the authors did not note the generation of regulatory T cells [29].

The K562 has also been applied to the stimulation of CD8+ cytotoxic T cells as well as CD4+ cells, in numerous applications. One interesting example is the use of a costimulation strategy with both CD28 and 41BB. The authors utilized a K562 cell line genetically modified to express the Fcγ receptor for immobilization of an anti-CD3 and an anti-CD28 antibody [30]. The authors then transduced these cells with a gene encoding the 41BB ligand for costimulation. With this strategy, the authors were able to generate up to 1000–10,000-fold expansion of CD8+ T cells over the course of 25 days. The ultimate fold expansion was nearly 100 times greater than the cellular APCs bearing anti-CD3 and anti-CD28 alone [30].

Other examples of the use of the K562 cell line have also been reported for different, specific applications. CD19-transduced K562 cells were utilized to stimulate T cells with chimeric antigen receptors designed to react specifically to CD19. Through zinc-finger nuclease-based elimination of the endogenous TCR, the authors were able to generate an APC and companion T cell line that could specifically target B cell lymphoma [31]. A similar strategy was used to generate natural killer cells that exerted a cytotoxic effect on CD19+ cells [32]. K562-based aAPCs were also genetically modified further to secrete the soluble cytokine IL-21 as a signal 3 [23]. Although not used for a biomedical application, the authors were able to use this technology to study the biological effects of this cytokine.

14.4.2   Murine- and Drosophilla-Based aAPCs

In addition to human-based cell systems, other xenogenic systems have been developed using similar transduction strategies. One of the earliest cell-based aAPC systems to be utilized for T cell expansion was the *Drosophilla*-based cell line [33]. Using a similar strategy to that of the K562 cells of genetic transduction with a murine MHC molecule class 1 protein, a costimulatory molecule for CD28, and intercellular adhesion molecule 1 (ICAM-1) for integrin-based binding to LFA-3 on T cells, these *Drosophilla* cells were able to mediate expansion of murine CD8+ T cells. Due to the foreign nature of the cells, the authors also noted a stimulatory effect on B cells, which could then be relayed to the CD8+ T cells [33].

Another alternative to the K562 is the mouse fibroblast line NIH/3T3. These cells were transduced with the genes for an HLA MHC protein and a costimulatory molecule for CD28 [34]. In addition, these cells were genetically modified to express the ICAM-1 protein for adhesion to the T cell through LFA-3. The aAPC was shown to efficiently mediate antigen-targeted activation of influenza primary CD8+ T cells. Compared to the autologous APC, stimulation with this aAPC resulted in a twofold increase in the number of CD8+ T cells derived from peripheral blood. In addition, these cells were able to mediate an antigen-specific cytotoxicity against melanoma cells [34].

14.5   Acellular aAPCs

Despite the broad utility that cellular aAPCs have demonstrated, there are still a number of concerns over their use. Genetic modification of cells has the potential to result in cancerous growth due to the unpredictable nature of the site of insertion into the target genome. In addition, cell-based therapies are time-consuming and costly to produce. While possessing the advantage of supreme cell mimicry, cellular aAPCs also lack the antigen flexibility that an "off-the-shelf" particle-based system can provide. Switching the type and nature of the protein signal is much simpler and can be engineered more precisely, using an acellular aAPC system with a defined conjugation chemistry. To that end, numerous acellular aAPCs have been developed and characterized for the activation and suppression of target T cell populations [35] (Table 14.1).

14.5.1   Polymeric Particle aAPCs

Some of the earliest work in the development of aAPCs was with nonbiodegradable polystyrene particles. Immobilization of protein ligands for aAPC function involved a simple surface adsorption. Polystyrene-based aAPCs have been used in antigen-specific and antigen-nonspecific stimulation of T cells [36]. One example of the use of polystyrene is in the expansion of ovalbumin-specific T cells. Tham et al. utilized a polystyrene bead displaying an MHC I molecule loaded with an ovalbumin antigen as well as recombinantly produced B7-1 and B7-2, the natural ligand for CD28 costimulation. In vitro stimulation results demonstrated a threefold increase of proliferation and IL-2 production of CD8+ T cells compared to soluble stimulation proteins [37].

Over recent years, emphasis has shifted from nonbiodegradable polymers to biodegradable polymers for aAPC construction. These materials have gained popularity in recent years due to their capabilities of controlled degradation and thus predictable elimination from the body. In addition, biodegradable polymeric aAPCs offer the advantage of sustained release of a third soluble signal. Taken together, these advantages have resulted in effective aAPCs for a wide variety of applications.

One of the earliest examples of the biodegradable aAPC was the use of polyglycolic acid microparticles as a platform for APC protein immobilization [38]. For the generation of aAPCs, the authors demonstrated a conjugation-free surface adsorption of anti-CD3 and anti-CD28 antibodies. The aAPC was evaluated against a plate coated with these antibodies and was determined to twofold more efficient at eliciting T cell proliferation. Combining this aAPC with another degradable particle containing granulocyte macrophage colony-stimulating factor (GM-CSF), the authors were able to demonstrate in vivo efficacy by reduced tumor burden of a fibrosarcoma cancer implantation [38].

More recent biodegradable systems have focused on the use of PLGA-based particles for aAPC synthesis. Steenblock et al. developed a PLGA aAPC using both

MHC dimer and anti-CD3 as a signal 1 and anti-CD28 as a signal 2 [17]. The authors were able to demonstrate a threefold production of IL-2 by CD8+ T cells compared to soluble signal protein. In addition, the study demonstrated the utility of a biodegradable polymer to have sustained release of a third signal, IL-2. Comparing IL-2 released locally by the particle to bulk delivery of IL-2, there was a twofold increase in the production of IFNγ by target T cells [17].

A major advantage of these biodegradable polymeric systems is the ability to modify particle shape by one of many established protocols. As was described earlier, the impact of aAPC shape on function and T cell stimulatory capacity has recently come to light. Combining this approach with the biodegradable polymer-based approach of sustained cytokine or chemokine release may result in an even more powerful aAPC which may be useful for clinical applications.

14.5.2   Liposome aAPCs

Liposomes have also been of interest as a platform for aAPC generation. Owing in part to their fluid membranes, a liposome-based aAPC has the potential to allow for recreation of the IS during T cell signaling. One of the first aAPC platforms involved the use of liposomes reconstituted from cellular membranes [39]. While the use of liposomes as aAPCs has not found widespread application for therapeutic application or for the expansion of clinically relevant T cells, there are examples of efficacy. One was the use of GM-1-enriched liposomes for clustering of anti-CD3 and anti-CD28 stimulation ligands mediated by neutravidin bound cholera toxin. The authors compared this liposome-based aAPC to a rigid, bead-based aAPC and found close to a 1.5-fold increase in the expansion of Mart-1 antigen-specific T cells, with an overall expansion of about 175-fold over the course of 14 days [40].

14.5.3   Magnetic aAPCs

Another popular choice of materials for construction of aAPCs is paramagnetic materials. Due to the ease of purification and processing, these magnetic particles have found widespread application in the generation of aAPCs. One study published by Oelke et al. focused on the antigen-specific expansion of CD8+ T cells [41]. The strategy was to couple a HLA MHC class I loaded with antigen as a signal 1 and an anti-CD28 molecule as a signal 2. The authors demonstrated 1000-fold expansion of specific cells responding to the target antigen. The T cells also demonstrated excellent targeted effector activity in a cell lysis assay [41].

One interesting and recent application of magnetic materials is in the induction of TCR clustering by magnetic-field-driven aggregation of nano-aAPC bound to the receptors. Through antibody staining of TCR before and after an applied magnetic field, the authors demonstrated a doubling of cluster size and overall reduction in the number of clusters per T cell [42]. At a maximal dose, these magnetically clustered nano-aAPCs were able to mediate a 15-fold antigen-specific expansion of target T cells compared to the non-clustered nano-aAPCs which only mediated a fivefold expansion of cells over 7 days. Ex vivo stimulation of these T cells under the influence of the magnetic field also led to a reduced tumor burden and enhanced survival of mice in an adoptive transfer immunotherapy model [42].

14.5.4 Other Materials

Another material used in the application of aAPCs is single-walled carbon nanotubes (SWNTs). This material has a large surface area that can be used for activation of T cells. The authors demonstrated a more than tenfold production of IL-2 of T cells by anti-CD3 loaded onto the surface of these carbon nanotubes versus free antibody in solution [43]. There was also a demonstrated effect of surface area for activation. Through expansion of the available SWNT surface area for activation, the authors were able to demonstrate enhanced activation of T cells for a similar dose of protein to fabricate the aAPCs [43].

Quantum dots have also been utilized recently for the stimulation of T cells. Through the use of MHC class I protein and anti-CD28 immobilized on the surface of 30 -nm quantum dots, the authors of this study were able to demonstrate an antigen-specific expansion of T cells [44]. Over the course of 7 days, there was a 15-fold expansion of target T cells and no expansion of non-cognate T cells. In addition, these quantum dot aAPCs mediated an excellent in vivo tumor reduction in an adoptive transfer model [44].

14.6 Biomedical Applications of the aAPC

14.6.1 Cancer Immunotherapy

One of the primary applications of aAPCs over the past years has been in the induction and expansion of cytotoxic T cells for cancer immunotherapy. Both cellular and acellular aAPCs have been utilized for this purpose. One study that highlights the use of aAPC for this purpose was the application of micron scale latex particles bonded to MHC dimer to serve as a signal 1 and a blend of anti-CD28, 41BBL, and CD83 as signal 2 [45]. The bead-based system was able to mediate a 250-fold expansion of melanoma CTL from naïve splenocytes. In addition, a strong in vivo effect was demonstrated in both a murine lung metastasis model and a subcutaneous melanoma adoptive transfer model [45].

Efficacy of aAPCs in the context of cancer immunotherapy has also been shown in the context of human cytotoxic lymphocytes. Butler et al. demonstrated impressive long-term CD8+ expansion and functionality of human CD8+ T cells [46]. The platform utilized was the K562 cells line transduced to express a human MHC

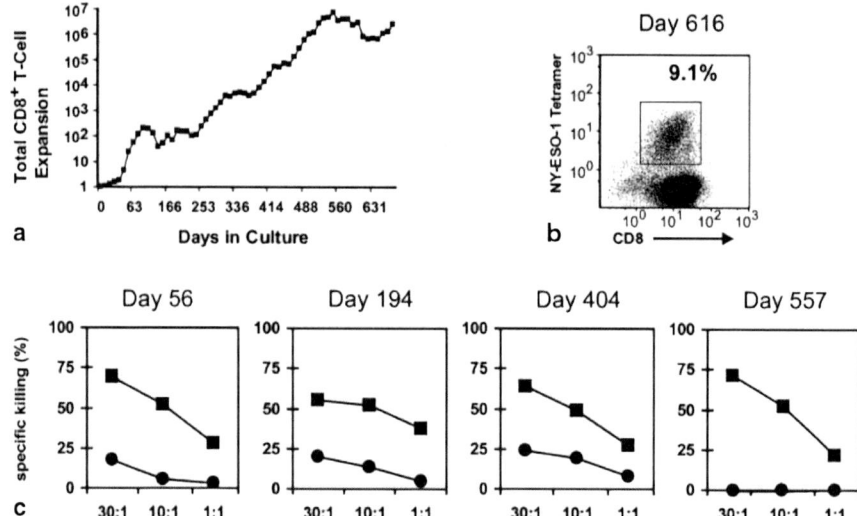

Fig. 14.4 Cellular aAPCs have demonstrated the potential to generate long-lasting, antitumor T cell responses. a Human CD8+ T cells were stimulated with the K532 aAPC cell line over the course of nearly 2 years, and proliferation was noted to be more than 1,000,000-fold over this time period. b MHC tetramer staining demonstrates an antigen-specific enrichment of T cells directed toward human cancer antigen NY-ESO-1. c T cell effector function was evaluated over the 2-year stimulation period, and expanded T cells were able to maintain their antigen-targeted killing effector function during the entire time of stimulation. Reproduced with permission from Ref. [46]

molecule as well as costimulatory molecules for CD28. Similar to previous studies with this cell-based aAPC, there was a rapid short-term expansion of antigen-specific T cells (~100,000-fold over 5 weeks) [46]. However, there was impressive long-term efficacy noted as well. Over the course of 1.5 years, there was more than a 1,000,000-fold expansion of antigen-specific T cells which were verified to have effector function in a cancer cell lysis assay [46] (Fig. 14.4). Following the successful preclinical studies, this cellular aAPC has been transferred to clinical trials for ex vivo expansion of T cells for immunotherapy of advanced melanoma [28].

14.6.2 Infectious Disease

Beyond cancer therapy, the aAPC platform has also been investigated in the context of infectious disease. One pathogen which has been investigated extensively for potential treatment by aAPC-based immunotherapy is cytomegalovirus (CMV). Adoptive immunotherapy in the context of CMV has been previously established using autologous APCs [47]. Despite this clinical efficacy, the need to establish an autologous APC for each patient would render the therapy very expensive. As a result, an aAPC was developed for the expansion of CMV-specific CD8+ T cells. The platform utilized was the murine fibroblast cell line expressing a human-based

MHC protein. The aAPC was shown to induce a near 100-fold expansion over 10–14 days of culture with CD8+ cells from CMV+ human donors [48]. These T cells were also shown to have specific effector function against target cells. In addition, the authors of this study demonstrated the capability to generate memory T cells from this population [48].

aAPC platforms have also been investigated in the context of other viral infections. One of the published examples is the use of aAPCs to treat Epstein–Barr virus (EBV). Lu et al. utilized a latex bead coated with a human MHC dimer loaded with cognate EBV peptide, as well as anti-CD28 and ICAM-1 as costimulatory molecules and adhesion molecules, respectively. The authors demonstrated a selective cytotoxicity against cells pulsed with the target EBV-associated antigen [49]. Another platform which could benefit from the use of aAPCs is human immunodeficiency virus (HIV-1). Clinical benefit has already been established utilizing an autologous APC platform [50]. Additional research into the use of expanded CD8+ T cells to treat the active or latent reservoir of HIV infection will likely be of benefit in the coming years.

14.6.3   Autoimmunity and Killer aAPCs

In addition to the stimulation of T cells to attack a target disease such as cancer or infection, there has been research devoted to researching how aAPCs can mimic the regulatory function of normal aAPCs through deletion of T cells in an antigen-specific fashion (Fig. 14.5). The term "Killer aAPCs" has been designed from a paramagnetic bead conjugated to MHC dimer to serve as a signal 1 and anti-Fas as a signal 2 to mimic the engagement of Fas by the FasL on the APC to induce activated T cell apoptosis [51]. This technology has been used in a demonstrative capacity using Mart-1, the melanoma antigen, and CMVpp65, the CMV antigen. Culturing antigen-specific T cells with the cognate killer aAPC resulted in more than a fivefold reduction of the T cell population [52]. This was not observed in non-cognate populations of T cells. In addition, this elimination was shown to occur within 30 min of the beginning of coculture [52]. This technology is promising for autoimmune disorders as several different target antigens have been implicated in diseases such as type I diabetes, multiple sclerosis, and primary biliary cirrhosis [51].

14.7   Conclusion

aAPC technology enables the capability for modern science to harness and control the immense power of the mammalian adaptive immune system. Many platforms have been developed for this technology. Cell-based aAPCs, including the K562 leukemia cell line and murine fibroblast lines, have shown direct and powerful ac-

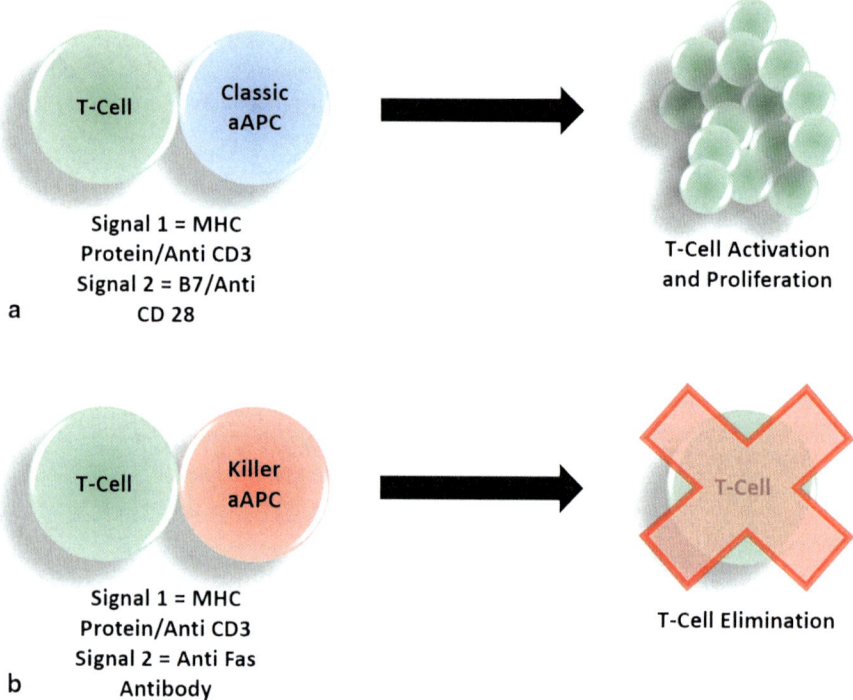

Fig. 14.5 aAPCs can be applied to modulate many aspects of the immune system similar to natural APCs. **a** Traditional aAPCs incorporate recognition signal 1 proteins as well as costimulatory signal 2 molecules such as recombinant B7 or anti-CD28. The result is T cell activation and proliferation. **b** aAPCs can also be used to downregulate the immune response. By maintaining the same signal 1 proteins, and changing the signal 2 protein to be an apoptosis inducing ligand, aAPCs can be used to deplete T cells in an antigen-specific/nonspecific manner. *MHC* major histocompatibility complex, *aAPC* artificial antigen-presenting cell

tivation of T cells in an antigen-specific manner. There are several ongoing clinical trials with different cell-based aAPCs for various diseases [28]. Acellular, biomaterials-based aAPCs have proven nearly as effective as cell-based aAPCs and have enabled this powerful arm of immunotherapy to be translated to an "off-the-shelf" technology. These aAPCs also enable precise tuning of their physical, chemical, and biological properties. As a result, therapeutic aAPC platforms are becoming more amenable to clinical use. Continued development of both of these platforms will be interesting to follow in the coming years.

One of the key advantages of the aAPC design is its ability to have an impact in any disease that has an immune component. As a result, the aAPC technology has found application in many illnesses for which immune-targeted antigens have been identified. Cancer immunotherapy has benefitted greatly from the aAPC due to the capability but inadequacy of the natural immune response to a cancerous growth. In addition, aAPCs have been developed for several viral infections including EBV and CMV. Despite a focus on aAPCs for T cell activation, there has also been work in the development of aAPCs to deplete T cells in an antigen-specific fashion. This could have an impact for autoimmune diseases and organ transplantation where the current option is systemic immunosuppression.

aAPCs represent a successful biomimicry of the APC in its regular function of immune cell engagement. With a thorough understanding of the function and physiology of APCs, we can engineer more effective aAPCs for a wide variety of applications. Ongoing development of aAPCs for biomedical and therapeutic purposes will greatly benefit human medicine through engineered synthetic control of the natural immune response.

References

1. Delsing CE, Gresnigt MS, Leentjens J, et al. Interferon-gamma as adjunctive immunotherapy for invasive fungal infections: a case series. BMC Infect Dis. 2014;14(1):166.
2. Lodhi S, Lamb K, Meier- Kriesche H. Solid organ allograft survival improvement in the United States: the long- term does not mirror the dramatic short- term success. Am J Transpl. 2011;11(6):1226–35.
3. Cheever MA, Higano CS. PROVENGE (Sipuleucel-T) in prostate cancer: the first FDA-approved therapeutic cancer vaccine. Clin Cancer Res. 2011;17(11):3520–6.
4. Oelke M, Krueger C, Giuntoli RL 2nd, Schneck JP. Artificial antigen-presenting cells: artificial solutions for real diseases. Trends Mol Med. 2005;11(9):412–20.
5. Knight SC, Stagg AJ. Antigen-presenting cell types. Curr Opin Immunol. 1993;5(3):374–82.
6. Rossi M, Young JW. Human dendritic cells: potent antigen-presenting cells at the crossroads of innate and adaptive immunity. J Immunol. 2005;175(3):1373–81.
7. Ohteki T, Koyasu S. Role of antigen-presenting cells in innate immune system. Arch Immunol Ther Exp—Engl Ed. 2001;49:S47–52.
8. Adalid-Peralta L, Fragoso G, Fleury A, Sciutto E. Mechanisms underlying the induction of regulatory T cells and its relevance in the adaptive immune response in parasitic infections. Int J Biol Sci. 2011;7(9):1412.
9. Gascoigne NRJ, Zal T. Molecular interactions at the T cell–antigen-presenting cell interface. Curr Opin Immunol. 2004;16(1):114–9.
10. Capece D, Verzella D, Fischietti M, Zazzeroni F, Alesse E. Targeting costimulatory molecules to improve antitumor immunity. BioMed Res Int. 2012;2012:926321.
11. Von Bubnoff D, De La Salle H, Wessendorf J, Koch S, Hanau D, Bieber T. Antigen- presenting cells and tolerance induction. Allergy. 2002;57(1):2–8.
12. Kaliński P, Hilkens CM, Wierenga EA, Kapsenberg ML. T-cell priming by type-1and type-2 polarized dendritic cells: the concept of a third signal. Immunol Today. 1999;20(12):561–7.
13. Kalinski P. Dendritic cells in immunotherapy of established cancer: roles of signals 1, 2, 3 and 4. Curr Opin Investig Drugs. 2009;10(6):526.
14. Alarcon B, Mestre D, Martinez-Martin N. The immunological synapse: a cause or consequence of T-cell receptor triggering? Immunology. 2011;133(4):420–5.
15. Valitutti S, Coombs D, Dupre L. The space and time frames of T cell activation at the immunological synapse. FEBS Lett. 2010;584(24):4851–7.
16. Mescher M. Surface contact requirements for activation of cytotoxic T lymphocytes. J Immunol. 1992;149(7):2402–5.
17. Steenblock ER, Fahmy TM. A comprehensive platform for ex vivo T-cell expansion based on biodegradable polymeric artificial antigen-presenting cells. Mol Ther. 2008;16(4):765–72.

18. Fifis T, Gamvrellis A, Crimeen-Irwin B, et al. Size-dependent immunogenicity: therapeutic and protective properties of nano-vaccines against tumors. J Immunol. 2004;173(5):3148–54

19. Sunshine JC, Perica K, Schneck JP, Green JJ. Particle shape dependence of CD8+ T cell activation by artificial antigen presenting cells. Biomaterials. 2014;35(1):269–77.

20. Chacon JA, Wu RC, Sukhumalchandra P, et al. Co-stimulation through 4-1BB/CD137 improves the expansion and function of CD8(+) melanoma tumor-infiltrating lymphocytes for adoptive T-cell therapy. PloS One. 2013;8(4):e60031.

21. Rudolf D, Silberzahn T, Walter S, et al. Potent costimulation of human CD8 T cells by anti-4-1BB and anti-CD28 on synthetic artificial antigen presenting cells. Cancer Immunol, Immunother. 2008;57(2):175–83.

22. Matic J, Deeg J, Scheffold A, Goldstein I, Spatz JP. Fine tuning and efficient T cell activation with stimulatory aCD3 nanoarrays. Nano Lett. 2013;13(11):5090–7.

23. Ansen S, Butler MO, Berezovskaya A, et al. Dissociation of its opposing immunologic effects is critical for the optimization of antitumor CD8+ T-cell responses induced by interleukin 21. Clin Cancer Res. 2008;14(19):6125–36.

24. Steenblock ER, Fadel T, Labowsky M, Pober JS, Fahmy TM. An artificial antigen-presenting cell with paracrine delivery of IL-2 impacts the magnitude and direction of the T cell response. J Biol Chem. 2011;286(40):34883–92.

25. Prakken B, Wauben M, Genini D, et al. Artificial antigen-presenting cells as a tool to exploit the immunesynapse'. Nat Med. 2000;6(12):1406–10.

26. Mossman KD, Campi G, Groves JT, Dustin ML. Altered TCR signaling from geometrically repatterned immunological synapses. Science. 2005;310(5751):1191–3.

27. Mandal B, Bhattacharjee H, Mittal N, et al. Core–shell-type lipid–polymer hybrid nanoparticles as a drug delivery platform. Nanomed: Nanotechnol, Biol Med. 2013;9(4):474–91.

28. Butler MO, Hirano N. Human cell- based artificial antigen- presenting cells for cancer immunotherapy. Immunol Rev. 2014;257(1):191–209.

29. Butler MO, Ansen S, Tanaka M, et al. A panel of human cell-based artificial APC enables the expansion of long-lived antigen-specific CD4+ T cells restricted by prevalent HLA-DR alleles. Int Immunol. 2010;22(11):863–73.

30. Maus MV, Thomas AK, Leonard DG, et al. Ex vivo expansion of polyclonal and antigen-specific cytotoxic T lymphocytes by artificial APCs expressing ligands for the T-cell receptor, CD28 and 4-1BB. Nat Biotechnol. 2002;20(2):143–8.

31. Torikai H, Reik A, Liu PQ, et al. A foundation for universal T-cell based immunotherapy: T cells engineered to express a CD19-specific chimeric-antigen receptor and eliminate expression of endogenous TCR. Blood. 2012;119(24):5697–705.

32. Imai C, Iwamoto S, Campana D. Genetic modification of primary natural killer cells overcomes inhibitory signals and induces specific killing of leukemic cells. Blood. 2005;106(1):376–83.

33. Sun S, Cai Z, Langlade-Demoyen P, et al. Dual function of Drosophila cells as APCs for naive CD8+ T cells: implications for tumor immunotherapy. Immunity. 1996;4(6):555–64.

34. Latouche J-B, Sadelain M. Induction of human cytotoxic T lymphocytes by artificial antigen-presenting cells. Nat Biotechnol. 2000;18(4):405–9.

35. Steenblock ER, Wrzesinski SH, Flavell RA, Fahmy TM. Antigen presentation on artificial acellular substrates: modular systems for flexible, adaptable immunotherapy. Expert Opin Biol Ther. 2009;9(4):451–64.

36. Turtle CJ, Riddell SR. Artificial antigen-presenting cells for use in adoptive immunotherapy. Cancer J. 2010;16(4):374–81.

37. Tham EL, Jensen PL, Mescher MF. Activation of antigen-specific T cells by artificial cell constructs having immobilized multimeric peptide—class I complexes and recombinant B7–Fc proteins. J Immunol Methods. 2001;249(1):111–9.

38. Shalaby WS, Yeh H, Woo E, et al. Absorbable microparticulate cation exchanger for immunotherapeutic delivery. J Biomed Mater Res Part B: Appl Biomater. 2004;69(2):173–82.

39. Engelhard VH, Strominger JL, Mescher M, Burakoff S. Induction of secondary cytotoxic T lymphocytes by purified HLA-A and HLA-B antigens reconstituted into phospholipid vesicles. Proc Natl Acad Sci U S A. 1978;75(11):5688–91.

40. Zappasodi R, Di Nicola M, Carlo-Stella C, et al. The effect of artificial antigen-presenting cells with preclustered anti-CD28/-CD3/-LFA-1 monoclonal antibodies on the induction of ex vivo expansion of functional human antitumor T cells. Haematologica. 2008;93(10):1523–34.

41. Oelke M, Maus MV, Didiano D, June CH, Mackensen A, Schneck JP. Ex vivo induction and expansion of antigen-specific cytotoxic T cells by HLA-Ig-coated artificial antigen-presenting cells. Nat Med. 2003;9(5):619–24.

42. Perica K, Tu A, Richter A, Bieler JG, Edidin M, Schneck JP. Magnetic field-induced T cell receptor clustering by nanoparticles enhances T cell activation and stimulates antitumor activity. ACS Nano. 2014;8(3):2252–60.

43. Fadel TR, Steenblock ER, Stern E, et al. Enhanced cellular activation with single walled carbon nanotube bundles presenting antibody stimuli. Nano Lett. 2008;8(7):2070–6.

44. Perica K, De Leon Medero A, Durai M, et al. Nanoscale artificial antigen presenting cells for T cell immunotherapy. Nanomed: Nanotechnol Biol Med. 2014;10(1):119–29.

45. Lu X, Jiang X, Liu R, Zhao H, Liang Z. Adoptive transfer of pTRP2-specific CTLs expanding by bead-based artificial antigen-presenting cells mediates anti-melanoma response. Cancer Lett. 2008;271(1):129–39.

46. Butler MO, Lee J-S, Ansén S, et al. Long-lived antitumor CD8+ lymphocytes for adoptive therapy generated using an artificial antigen-presenting cell. Clin Cancer Res. 2007;13(6):1857–67.

47. Walter EA, Greenberg PD, Gilbert MJ, et al. Reconstitution of cellular immunity against cytomegalovirus in recipients of allogeneic bone marrow by transfer of T-cell clones from the donor. N Engl J Med. 1995;333(16):1038–44.

48. Papanicolaou GA, Latouche JB, Tan C, et al. Rapid expansion of cytomegalovirus-specific cytotoxic T lymphocytes by artificial antigen-presenting cells expressing a single HLA allele. Blood. 2003;102(7):2498–505.

49. Lu X-L, Liang Z-H, Zhang C-E, Lu S-J, Weng X-F, Wu X-W. Induction of the Epstein-Barr Virus latent membrane protein 2 antigen-specific cytotoxic T lymphocytes using human leukocyte antigen tetramer-based artificial antigen-presenting cells. Acta Biochim Biophys Sin. 2006;38(3):157–63.

50. Brodie SJ, Lewinsohn DA, Patterson BK, et al. In vivo migration and function of transferred HIV-1-specific cytotoxic T cells. Nat Med. 1999;5(1):34–41.

51. Schütz C, Oelke M, Schneck JP, Mackensen A, Fleck M. Killer artificial antigen-presenting cells: the synthetic embodiment of a 'guided missile'. Immunotherapy. 2010;2(4):539–50.

52. Schütz C, Fleck M, Mackensen A, et al. Killer artificial antigen-presenting cells: a novel strategy to delete specific T cells. Blood. 2008;111(7):3546–52.

Randall A. Meyer is a graduate student in the Biomedical Engineering Department at Johns Hopkins University and is working to develop artificial antigen-presenting cells for cancer immunotherapy.

Dr. Jordan J. Green is an associate professor of biomedical engineering, ophthalmology, neurosurgery, and materials science and engineering at Johns Hopkins University and is the director of the Biomaterials and Drug Delivery Laboratory at JHU.

Index

© Springer International Publishing Switzerland 2015
L. Santambrogio (ed.), *Biomaterials in Regenerative Medicine and the Immune System,*
DOI 10.1007/978-3-319-18045-8